Patient
ZERO

A CURIOUS HISTORY OF THE WORLD'S WORST DISEASES

LYDIA KANG, MD · NATE PEDERSEN

WORKMAN PUBLISHING * NEW YORK

*To the countless frontline and essential workers
who have put their lives at risk for all of our sakes.
We are eternally grateful.*

Library of Congress Cataloging-in-Publication Data
Names: Kang, Lydia, author. | Pedersen, Nate, author.
Title: Patient zero : a curious history of the world's worst diseases / Lydia Kang, MD, Nate Pedersen.
Identifiers: LCCN 2021027209 | ISBN 9781523513291 (hardcover) | ISBN 9781523513291 (ebook)
Subjects: LCSH: Epidemics—History. | Emerging infectious diseases —History.
Classification: LCC RA648.5 .K36 2021 | DDC 614.4/9—dc23
LC record available at https://lccn.loc.gov/2021027209

ISBN 978-1-5235-1329-1

Design by Janet Vicario
Photo research by Sophia Rieth

COVER CREDITS: (Center petri dish) Helga_foto/Shutterstock. *Starting in upper left, clockwise:*
(Paper texture background) Design Cuts; (Palm Tree) Creative Market; (Heart Illustration) CSA Images/
Vetta/Getty; (Tick and Mosquito engravings) Bodor Tivadar/Shutterstock; (Magnified Cells)
Eduard Muzhevskyi/Shutterstock; (Microscope) Morphart Creation/Shutterstock; (Human Body)
Grafissimo/DigitalVision Vectors/Getty; (Palm Tree) Creative Market; (Monkey) duncan1890/
DigitalVision Vectors/Getty; (Scientist) H. Armstrong Roberts/Retrofile RF/Getty; (Syringe)
A Sourcebook of French Advertising Art.; (Red Cells) Vlue/Shutterstock; (Airplane) IM_photo/
Shutterstock; (Globe) CSA Images/Vetta/Getty; (Botanical Engraving) Creative Market;
(Dead Body) mariusFM77/E+/Getty; (Human Evolution) Man_Half-tube/DigitalVision Vectors/Getty;
(Superspreader) Lester V. Bergman/Corbis Documentary/Getty.

Workman books are available at special discounts when purchased in bulk for premiums and sales
promotions as well as for fundraising or educational use. Special editions or book excerpts can also be
created to specification. For details, contact the Special Sales Director at specialmarkets@workman.com.

Workman Publishing Co., Inc.
225 Varick Street
New York, NY 10014-4381
workman.com

WORKMAN is a registered trademark of Workman Publishing Co., Inc.

Printed in the United States of America
First printing September 2021

10 9 8 7 6 5 4 3 2 1

CONTENTS

INTRODUCTION

Humans have endeavored to understand the vast mysteries of life, both around us and within, since our first moments as a species. After all, our very lives depended on identifying what attacked us, and why. Once we largely tamed the world outside of us, we began relentlessly exploring the intricacies of our own bodies: How do our circulatory systems work? Why must we age? What makes disease appear in some but not others? As we've learned more about infectious diseases through the years, it's no surprise that our goal has always been to understand them, to control them, and to kill them—before they kill us.

In this regard, we've come a long way from Galen, the second-century Roman physician-philosopher who thought blood arose from food within the liver and gave off "sooty vapors" in the lungs, though we're not entirely removed from wondering if the planets and stars affect our fecundity or if it's healthy to bathe during a full moon. The history that preceded germ theory (see page 38) was rife with questionable rationale, such as celestial events, wrathful gods, or diseased "mists" blamed for plague pandemics. Nowadays, we look not to the stars when a new outbreak lands on our shores; we track down cases, investigate where diseases took advantage of time, space, the human immune system, and the complex intricacies of society to break in—and break out.

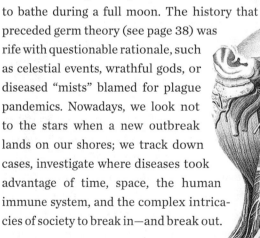

Dissection engraving based on a drawing in Antonio Scarpa's Tabulae neurologicae, *published in 1794.*

The discovery of new pathogens began hundreds of years ago, with world-altering breakthroughs around the turn of the twentieth century by the likes of Robert Koch (see page 49), Louis Pasteur (see page 47), and countless others. Through the advent of petri dishes and use of laboratory animals, we learned to grow pathogens under controlled conditions. Microscopes made the unseen suddenly visible, revealing parasites that wriggled, tiny rounded cocci (round bacterium) in baby-pink stained hues, and curly bacilli, all showing us we were not nearly as alone in our skins as we once thought. Later the unseen viruses, slipping through filters that trapped other microbes, would be revealed with electron microscopes.

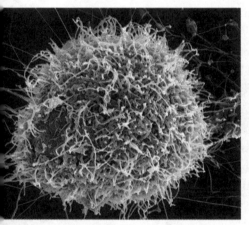

Ebola virus particles in a colored image from a scanning electron microscope.

As technology progressed, the secrets of our micro-invaders began to be revealed. We discovered nuclear genetic material, or the DNA and RNA that proved to be the blueprints by which these pathogens lived out their fate of latching on to our cells, invading them, and hijacking our base components to make more of themselves. If that weren't enough, we figured out that our often invisible tormentors change the rules. Constantly. We tend to think of our species's origin story as a unique miracle of evolution. But it turns out that bacteria and viruses have been evolving right along with us, often *within* us, and at a far faster clip than we do.

Part of our investigations of pathogens involve not just the identification of diseases they spawn, many of which have been with us for millennia, but novel invaders we've never seen before. (Or, in the case of diseases like Legionnaires', bacteria that were anonymously picking us off for years until we finally figured out what they were.) Like so many other diseases that came before, we learned the structure of new ones, how their surface proteins attach to ours to gain purchase into our bodies, and how they spread, down to the precise details. It used to take decades to identify a new disease; now it takes days. But a genetic sequence doesn't hold all the answers, and a flurry of questions always remain, like: Is six feet apart enough to prevent a loud talker from spreading COVID-19? How many layers of cotton fabric will prevent passage from a cough? How often must healthcare providers use hand sanitizer during the doffing procedure of personal protective equipment in the hospital?

The pathogens still find a way to get us, of course, but we fight back by figuring out how critically ill patients can be saved and by quickly separating effective medicines from those that are pure quackery (see page 201). We develop and produce vaccines, sometimes at a dizzying pace, all while dealing with the fallout of information that can be wrong, correct, changeable, frightening, and on occasion, laughable.

But regardless of the social reactions, no epidemic or pandemic arrives without the inevitable questions: How did it start? Why did it spread? And how do we stop it?

Even with the most mundane infection, we *must* know the nature of its origin. We may have caught something as simple as a cold and, as we hack up phlegm and call in sick to work, the question that nags is: Where did we catch it? Was it the coworker who refused to take a sick day, coughing and snorting in the cubicle next to us? Or when our doctor tells us we have hepatitis C and we later learn it was from a contaminated syringe full of pain medicine we got during an ER visit? Or even when a new disease like COVID-19 arises and upends the lives of billions of people, we still calmly need to know: Who is to blame? From where did it come? Is there a patient zero?

In many cases, we can answer at least the first two components of the disease equation, as more than 60 percent of human pathogens are zoonotic diseases, or those that have jumped from animals to humans. And, tellingly, the vast majority of novel infectious diseases that have emerged in the last seventy years resulted from zoonoses. There is no shortage of theories and explanations for why: Our voracious human appetites have brought us closer to fowl, bushmeat, and wild animals from whence these microbes lurk; animals that remain after their habitats are disrupted by agriculture and urbanization also tend to be those that host zoonotic diseases; and climate change has expanded the range of vector-borne diseases, like the Zika virus and Lyme disease. In short, pathogens are constantly on the lookout for greener pastures, and sometimes those pastures are created by us, bringing us closer to bats or the catlike civet. Then the pathogens themselves change, adapt, and sometimes truly thrive in the new landscape of our blood and lungs, far better real estate in which to reproduce and, yes, survive.

Many of the chapters in this book are about the origins of diseases, as these stories are the ones that ultimately teach us how to survive a novel, or not so novel, pathogen. In some cases, there is a clear so-called patient zero to a fresh outbreak, perhaps the most well-known of whom is "Typhoid Mary" Mallon (see page 212), an Irish immigrant to the United States who spread disease in New York City to numerous families for whom she worked as a cook. And then there is patient zero's namesake, a French Canadian flight attendant named Gaétan Dugas (see page 135) who was wrongly singled out as the originator of the HIV outbreak in the United States.

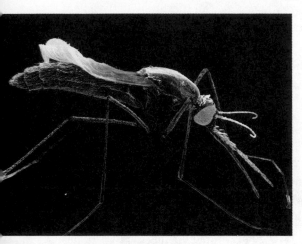

A mosquito (Anophelese stephensi) *captured in a colored image from a scanning electron microscope.*

But the truth is we don't know the great majority of patient zeroes for a given disease, because the disease has been with us since time immemorial, or it erupted and spread too fast for us to pin down its origins, or other factors like politics or wars got in the way. Where possible, we present the person who can claim the dubious honor of being the first host to spread a disease, be it brand-new or very old but breaking out in a new population. In mining these beginnings, we reveal what investigators have always sought when researching disease: Why this person? How did they get it? How did that particular microbe evolve to cause such devastation? How can we stop it from happening again?

There is no question that the concept of a patient zero can devolve into a blame-based narrative around who seeded a given epidemic or pandemic, reducing a complex infectious disease chain and ecological evolutionary process to a single person. But that is not remotely the point of this book. We subscribe to the idea of a patient zero in efforts to understand how a single person or people play their part in a complicated dance of host and victim that rarely, though devastatingly, can lead to hundreds of thousands, if not millions, of lives lost. For this book, the idea of a patient zero is not to point a definitive historical finger, but to present the stories of diseases and their origins as examples of the ways in which outbreaks and pandemics are infinitely more complex, unexpected, and often more unexplainable than we realize.

To fully understand a given infectious disease and prevent human suffering, examining its origins may reveal an entangled cause-and-effect story, as well as the human impact in its creation. The globalization of trade and travel, poverty, limited access to healthcare, misguided rules and laws by governments, racism, sexism, bigotry, cultural mores, poor education, and lack of basic public health measures are among the multiple fault lines from which devastation ensues.

◆ ◆ ◆

For better or worse, plagues and pandemics are written into our existence as a species. Epidemics are the points where ecosystems, pathogen evolution, and human evolution collide. One might argue that there is no silver lining when it comes to disease, but without them we would not have made crucial discoveries that benefited humanity. Some were happy accidents, such as the discovery of penicillin in 1928 by Alexander Fleming (see page 351). In the case of ergot (see page 12), a fungal disease of rye grain that caused a hallucinatory outbreak in France in 1951, it led to the discovery of a more famous hallucinogen—LSD. Some of the origin stories found in the pages that follow take us back to the Bible's Old Testament, where the plagues of Egypt are examined with modern eyes (see page 96), including how ecological factors might have caused many if not all of them that then led to subsequent plagues. Myths and legends can also be linked to pandemics, such as the nineteenth-century panic over "vampire slayings" that were more likely the methodical work of tuberculosis (see page 327).

Diseases can sometimes feel like the director in a show we've been marathon-watching (or doom scrolling) for a long time, with the rich and famous sometimes playing the stars. Syphilis has been with us for centuries (see page 198), afflicting the likes of Al Capone and Ivan the Terrible. Alexander Hamilton was involved in a political kerfuffle over the yellow fever outbreak in Philadelphia in 1793 before getting the disease himself. (It would not be the first or the last time an outbreak was politicized, or a prominent politician would fall ill in a pandemic.) And sometimes, reality feels more far-fetched than the movies, as when the last smallpox victim on the planet fell ill after the virus escaped from a lab (see page 329). Or when the influenza virus responsible for the 1918 flu pandemic was dug up from permafrost and re-created in a highly guarded lab just to understand exactly why it was so deadly (see page 235).

Just as we search for patient zeroes in new outbreaks, we cannot stop looking back at the beginnings of old ones. Feeding our fascination with plagues,

The angel of death depicted during the Antonine Plague of Rome (165–180 CE).

some of those ancient diseases (like the *actual* plague, the bubonic one) have managed to stay with us. And the so-called "romantic disease," tuberculosis, which probably evolved with us millions of years ago, has infected the likes of Tina Turner, Cat Stevens, and Ringo Starr, and currently infects one-fourth of the world's population today.

But while these stories can be incredibly satisfying, especially when all the pieces of a puzzle fit together seamlessly, in reality it's not always the case. And part of what makes them interesting is our distance from them. When inside an actual pandemic, it's not nearly so entertaining. But if there is one thing that living through a global outbreak teaches us, it's that in science and history every story has missing elements, cracks, imperfect pieces. And if we're honest with ourselves, we realize that *we* are one of those unstable factors in a pandemic. Does it matter, for example, that the pork we ate came from a huge industrial operation that uses antibiotics? What about how much we travel globally, or if we choose to vaccinate, or how often we take antibiotics for a sinus infection that probably doesn't require it?

No matter how gruesome or entertaining it is, history can't help but reveal to us that our choices as a species have direct consequences in how we get attacked by disease, in what ways, and how often. How we interact with each other and our environment comes back to us via mosquito bites, sex, flawed healthcare systems, or chance. And sometimes, the consequences appear in the form of a novel virus that upends everything in our lives, leaving masses of dead in its wake.

In the meantime, hold on to those masks and precious bottles of hand sanitizer. Over and over throughout history, we've largely survived pandemics, and we'll survive this one too. But it surely won't be our last.

INFECTION

Colored scanning electron micrograph image of Ergot (Claviceps purpurea) growing on wheat.

ERGOTISM
Patient ZERO:
Unknown

Cause: *Claviceps purpurea* (fungus)

Symptoms: Double vision, hallucinations, burning limb pains, convulsions, gangrene

Where: Often in places where rye was farmed

When: Primarily before the 1800s, but occasionally thereafter

Transmission: Bread made with contaminated flour

Little-Known Fact: Experimentation with the ergot fungus led to the creation of LSD.

One of life's simpler pleasures is perfectly baked French bread. The French long ago elevated bread baking to an art form. A crispy-chewy baguette from the local boulangerie could be the closest thing to bliss in a typical French repast. It is against this culinary backdrop that a small village named Pont-Saint-Esprit suffered a deadly nightmare from the innocent (and culturally ubiquitous) daily purchase of a freshly baked baguette.

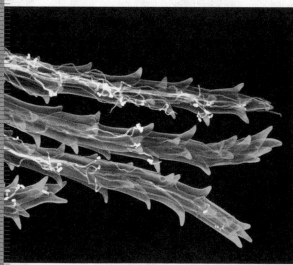

On August 15, 1951, dozens of people in Pont-Saint-Esprit suddenly became violently sick, complaining of stomach pain, nausea, and insomnia. Over the next few days, hundreds more joined them. If that weren't odd enough, the stricken reeked of a strange, disagreeable odor reminiscent of dead mice.

A colored micrograph image of wheat infected with ergot fungus.

Nighttime in the village was a strange event. Numerous insomniacs walked the streets, happily greeting each other in the midst of vivid, euphoric hallucinations. Some reported hearing heavenly choruses and seeing brilliant colors.

But things quickly took a turn for the horrific, as some hallucinations became nightmares. A terrified young girl believed she was being devoured by tigers. A garage owner became convinced he was a circus performer and began walking across the cable of a suspension bridge over the Rhône River.

The stories kept coming, and they were increasingly outlandish. A woman grieved over her children, who she thought had been ground into sausages. A man couldn't understand why he was suddenly shrinking. The head of the farmers cooperative was seized by the compulsion to fill page after page with poetry—if he stopped, he was sure he would have to jump out a window.

Some of the locals actually did take the leap. A man, convinced that his brain was being devoured by red snakes, jumped to his death. Another man also threw himself from a window, breaking both legs in the fall but surviving, then trying in excruciating desperation to keep running from whatever vision plagued him. Another would-be suicide attempt involved someone about to leap into the Rhône River, screaming, "I am dead and my head is made of copper and I have snakes in my stomach and they are burning me!" Thankfully, his friends stopped him before he ended up in the river.

Even the local animals were behaving erratically. A dog chewed on stones until it wore away all of its teeth. Local ducks started marching around the village in straight lines.

To say it was a strange time was to do disservice to strange times. So what was happening? The world's worst collective bad trip? In a sense, well, yes. The Kafkaesque scenes of hallucinations were the result of a visit to the local bakery gone horribly wrong. Remember that perfectly baked French baguette? It turns out, under specific circumstances, it can spawn mass psychosis.

Modern-day Pont-Saint-Esprit, France

The summer of 1951 in southern France had been particularly wet—the wettest in recent memory. A very cold and damp winter was followed by a long, wet spring. In other words, it was the ideal environment for ergot, a parasitic fungus, to thrive. Capable of growing on grains such as wheat and barley, ergot is particularly prevalent on rye because the growth cycle of rye corresponds with the spread of *Claviceps purpurea* spores, and both do well when cold winters are followed by wet summers. Fortunately, ergot is now easy to recognize when you know what to look for—but this wasn't the case for our medieval ancestors, who regularly suffered through epidemics of "St. Anthony's Fire," their name for the outbreaks of convulsions and hallucinations we know now was caused by ergotism. Unaware that those ergots growing in their grains were parasitic infestations with dire consequences if ingested, plenty of medieval millers unknowingly ground ergot-infected rye into flour. Worse, ergot thrived after harsh winters, which meant it did particularly well during times when a populace was already suffering from famine.

Infected grains produce oversized violet or black grains—"ergots"—that stand out from the rest of the crop. The name "ergot" comes from the Old French word *argot*, meaning "cockspur," as the shape of the sclerotium, or pod, of the fungus resembles the spurs on rooster legs. The fungus consumes the grain as it grows. In the winter, the infected grains form tiny mushroom-like structures to disperse their spores. Farmers typically recognize and dispose of ergot-infected grain. But that summer in 1951, a mistake was clearly made.

The village physician in Pont-Saint-Esprit, Dr. Hadar Gabbai, immediately suspected ergotism as the culprit of the mass hallucinations and identified a village bakery owned by M. Briand as the potential source of the outbreak, as everyone afflicted had eaten bread from there. The French government launched an investigation into the bakery's

Claviceps purpurea growing on barley with the classic "cockspur" shape

chain of supply. As the bakery's bread was made with only yeast, water, salt, and flour, it was easy to rule out everything but the flour as a potential disease vector.

The flour had originally come from a distribution center in Bagnols-sur-Cèze, which in turn acquired it from two regional mills, one in Châtillon-sur-Indre and one owned by Maurice Maillet in Saint-Martin-la-Rivière. Investigators learned that the flour from Maillet's mill had already received numerous complaints. And sure enough, Maillet had accepted some dodgy end-of-season grain from a local baker, M. Bruere, who wanted to trade his grain for a lesser quantity of milled flour. Bruere's bags of grain contained late-season rye, known to be at risk for ergot contamination, as well as dust, moths, and weevils. However, Maillet figured since the questionable grains would be such a small portion of all the grain that went into his mill that it wouldn't decrease the overall quality of his flour. On that score, he was very, very wrong.

The flour from Maillet's mill then made it to the Roch Briand bakery in Pont-Saint-Esprit, where it was baked into hundreds of loaves of ergot-tainted bread, which were then purchased and eaten by hundreds of Pont-Saint-Esprit villagers (and, it turned out, their pets and local wildlife, too).

The result was a terrifying outbreak of ergot poisoning, the likes of which hadn't been seen in France for 140 years. Within two days, 230 people came down with ergotism. Some got off with comparatively mild symptoms reminiscent of food poisoning. But others experienced terrifying hallucinations. By the time those seriously afflicted finally processed the poison a few days later (without the help of medical interventions), almost 300 people had consumed the fungus. Four had died.

While highly unusual in the modern era, mass outbreaks of ergot poisoning were disturbingly common in the Middle Ages. In ninth-century Germany, the commentator in the *Annales Xantenses* described an outbreak of gangrenous ergotism as: "a great plague of swollen blisters consumed the people by loathsome rot, so that their limbs were loosened and fell off before death."

For centuries, no one connected the outbreaks of mass hallucinations in medieval villages with the consumption of fungus-infected rye. (The connection wasn't made until the seventeenth century.) Those strange oversized black rye grains were not terribly unusual, just the occasional outlier in a rye harvest—nothing *that* out of the ordinary. They must have looked strange to farmers—the infected rye would sometimes be three times as long as the regular grains—but in the thirteenth century, food wasn't exactly plentiful. Slightly weird-looking grain was likely to be harvested and processed.

The likely result a few days after baking the tainted rye into bread was a group of villagers showing symptoms of ergot poisoning: burning pains in their limbs, convulsions, even gangrene.

Then came the hallucinations.

Ergot's toxic impacts generally fall into two categories: gangrenous ergotism and convulsive ergotism; both are terrifyingly self-explanatory. Convulsive ergotism, with its trademark convulsions and hallucinations, is what struck Pont-Saint-Esprit, but there are plenty of historic accounts that describe equally horrific gangrenous ergotism where dry rot set into limbs so severely they eventually fell off.

Geography seemed to play a role in whether gangrenous or convulsive ergotism was a more likely outcome from an outbreak of the disease. With the unusual exception of Pont-Saint-Esprit, most outbreaks in France and those anywhere west of the Rhine River tended to be gangrenous, whereas outbreaks to the east of the Rhine tended to be convulsive. Both Great Britain and the United States

Gangrenous ergotism can cause dry gangrene, where constricted arteries slow blood flow to such a degree that tissues begin to die from the lack of oxygen and limbs can even fall off.

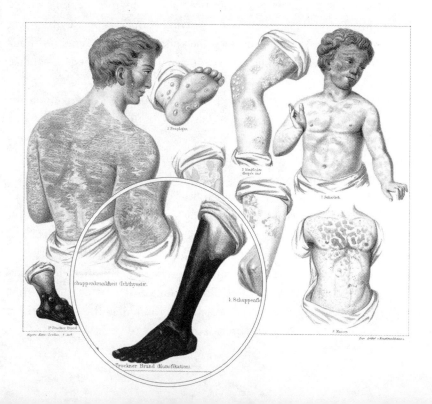

primarily had gangrenous ergotism outbreaks. Only six times in documented history have both the gangrenous and convulsive forms of ergotism occurred in the same outbreak.

◆ ◆ ◆

How the disease operates is a complex interplay between the microbiology of the ergot fungus and the human body. In gangrenous ergotism, once the ergot alkaloid is activated in the body, it creates a contraction of the smooth muscles that line the internal organs and arteries. The arteries then become constricted and blood flow slows considerably. If this carries on long enough (by continuing to eat infected rye grains), tissues begin to die from a lack of blood and oxygen, developing a "dry" form of gangrene that begins as a tingling in the extremities and can advance to the point where limbs literally start falling off.

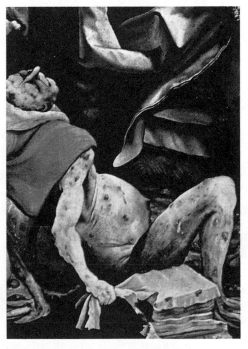

If, by virtue of geography and your biochemistry, your body activates the ergot alkaloid into the convulsive form of the disease, a different set of symptoms appears. These include muscular convulsions, double vision, and eventually the trademark hallucinations of ergotism.

Tragically, regardless of the type of outbreak, ergot poisoning also caused pregnant women to go into premature labor, as the fungus is a potent abortifacient, or abortion-inducer.

In the Middle Ages, the disease was given a variety of names connected to

A victim of advanced ergotism, from the painting Temptation of Saint Anthony, *by Matthias Grunewald (ca. 1512–15)*

fire, as the first symptom experienced by most people stricken with it was a burning sensation in their extremities. Ignis Sacer, or "holy fire," was one name for it; another was St. Anthony's Fire.

The story of St. Anthony—a religious hermit in the fourth century who eschewed personal possessions to live a solitary, religious existence in the desert—inspired and comforted those suffering from ergot-induced hallucinations. Tormented by visions from the devil, who repeatedly tried to tempt him with pleasures of food and flesh, St. Anthony rejected all these enticements, inspiring sufferers of ergotism wrestling with disturbing visions. Over time, St. Anthony's saintly name was attached to the infection as well.

Outbreaks of the "fire" were common in the medieval era. Between 990 and 1130, an estimated 50,000 people are suspected to have died from the disease in southern France alone. It was so common that a brotherhood of monastic healers, the Antonites (who took their name from St. Anthony), sprung up all over Europe to help care for those stricken with ergotism. Founded in 1095, the brotherhood would eventually expand their efforts to an international network of 370 hospitals, a testament to the sheer scale and impact of ergot outbreaks in the Middle Ages. In so doing, the brotherhood became managers of a large, complex, and highly specialized medical and social welfare system.

Research recently conducted by C. N. Nemes, a German medical historian, offered rare insight into how the disease was handled. Ergot patients would often arrive at one of these monastic Antonite hospitals where the brothers would interview and observe them. The brothers had to first decide if the patient was actually suffering from "holy fire" or some other disease against which they couldn't help. Based on their own observations and by comparison with previously admitted patients, if the monks confirmed that the person was indeed suffering from ergotism, they were admitted to the hospital. If the victim had the gangrenous form of the disease, they were allowed to stay for life with guaranteed access to food and medical care, a substantial benefit to medieval sufferers of ergotism. If the afflicted had hallucinations but not gangrene, they were allowed to stay for nine days before they had to give up their bed for the next patient. (Nine days was sufficient for the body to process the poison and for symptoms to subside on their own.) While in residence at the hospital, patients were required to participate in daily religious life, including confessing their sins, taking vows of chastity, and keeping up with daily prayers. There were perks, however—one of which was drinking the renowned wine made by the monks.

That Antonite wine was also the supposed cure developed by the monks for sufferers of St. Anthony's Fire. The recipe, which would have been a carefully guarded secret at the time, has been lost to history. Scholars have surmised that the wine would have contained vasodilating and analgesic herbs. Ergot contains two potent alkaloid chemicals that cause tightening of the blood vessels (vasoconstriction). In the extreme form, this can cut off circulation to body parts, causing pain and, eventually, gangrene. Vasodilators in the wine would combat the vasoconstriction caused by ergot alkaloids. Meanwhile, any analgesics in the wine would help with the burning pain of St. Anthony's Fire.

Another theory is that the wine contained crushed poppy seeds, which would have basically injected the wine with opium as well (the seeds themselves don't contain much opium, but they have opiate-containing residues left over from the harvesting process). If that was the case, it's no wonder the wine was so famous. With a lack of anything more effective at their disposal, easing someone into a deep and lasting sleep with opium was perhaps the best they could do in the medieval era.

The real benefits to staying with the monks, however, were probably just the healing effects of passing time combined with a separation from your home village's contaminated food supply. The monks would have had their own carefully tended gardens, and any patients would have been eating from the same gardens as the monks did. Also, and this is critical, the Antonite monasteries didn't grow rye for their bread. Coincidence? Perhaps. While ergot can grow on several types of grain, it prefers rye as a host. Or did the monks actually figure out that those weirdly shaped rye grains were contributing to the disease?

◆ ◆ ◆

Ergot has also been utilized for beneficial medical uses, and for . . . mental exploration. In 1582, Adam Lonicer, a German botanist, wrote about successfully using three ergots to stimulate uterine contractions of labor in pregnant women. He was writing about something that had already been discovered at that point by midwives, who had observed that pregnant sows, when fed grain infected with ergot, went into labor earlier than normal. In small doses, ergot actually works surprisingly well as a muscle and blood vessel constrictor. Midwives began using small doses of ergot to help speed along childbirth and also to stanch bleeding after delivery at some point in the Middle Ages, a practice copied by physicians assisting in childbirth after Lonicer published his findings in 1582. Strangely, a definitive connection wasn't made between consuming ergot-infected grains and all its

ZOONOSES
Making the Leap

Impact: In the last seventy years, the vast majority of new infectious diseases have been caused by a virus, parasite, bacterium, or prion that jumped from animal to human.

When: Since forever

What: Microbes that originate in nonhumans, including Ebola, COVID-19, SARS, rabies, bird flu, swine flu, anthrax, and pork tapeworm

What Happened Next: Understanding how "spillover" events occur, and preparing for the next pandemic caused by a novel zoonotic infection

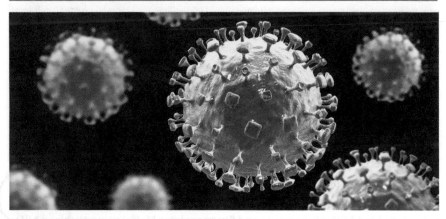

A computer-generated illustration of a Nipah virus particle.

I n September 1998, a handful of people in the Malaysian Peninsula came down with brain inflammation, or encephalitis. Unlike an infected toe we can point at and say, "Hey, that hurts there," encephalitis can rob us of all objectivity. We become confused; we can't keep our eyes open because of overwhelming drowsiness. An electrical storm can wreak havoc on our synapses, leading to convulsions. Fevers cook our bodies in an effort to battle any invading pathogen, making us even more delirious. More than 30 percent of those infected by this mysterious Malaysian infection died (for reference, a normal flu season typically kills 0.1 percent of those infected).

A health ministry worker fogs a pig farm in Malaysia to combat the Nipah virus in 1998.

The immediate and obvious connection among the cluster of cases was one thing: pigs.

The sick were pig farmers, pig sellers, butchers, or people who delivered pigs. At first, it was thought to be Japanese encephalitis (JE), which is spread via mosquitoes and usually causes a benign infection in pigs and some water birds (these are the reservoirs, or natural animal hosts that maintain the microbe's ongoing survival). But this was different. The pigs in Malaysia were getting really sick; with JE, they don't get nearly as ill. More people were dying than in typical JE cases. There was also a clear relationship between pigs and infections, whereas mosquito-borne JE is capable of spreading anywhere (and to anyone), regardless of contact with pigs. And the pigs themselves appeared to be spreading the disease to each other via sneezes and coughs that announced that a pig farm was now infected.

Investigators were initially stumped. They took a sample from a man in the town of Sungai Nipah stricken with the infection and who later died. After growing the sample in a lab, it was taken to the Centers for Disease Control (CDC) in the United States for examination. Under an electron microscope, investigators saw a virus particle filled with spaghetti-like genetic material. It was unlike anything they'd ever seen.

Blame Not the Bats!

Bats, never the most popular animal in the best of times, are having a particularly bad historical moment. This is on account of the growing revelation by the general public that many zoonoses are linked to bats, such as SARS and MERS. But some other truly awful diseases, including Hendra, Marburg, rabies, and Nipah, are also bat-related, leading to calls for bat cullings and roost burnings.

Sadly, these measures deeply misunderstand the magnificence of bats.

First, bats have been around for tens of millions of years, which speaks to their sheer ability to evolve and survive. They live on every continent (except Antarctica), they are the only mammal that flies, and they voraciously eat all sorts of insects we loathe, like mosquitoes. A single brown bat can eat twenty mosquitoes *in a minute*. Do you like mangoes, bananas, peaches, cashews, almonds, or dates? Thank bats for their help pollinating them (as well as their assist with seed dispersal). Vampire bats also adopt orphans, believe it or not, and most bat species groom themselves and each other, as meticulously as cats and primates might. Bats also appear to be cancer resistant and live long lives (decades, in some cases). Other mammals their size, like mice, live a fraction of their life span.

It might feel surprising that bats don't seem any worse for wear carrying so many lethal human diseases, but part of their tolerance may be on account of their metabolism. Put simply, with all its flying around, a bat's inner candle burns hotter than ours. That high metabolism results in more cellular DNA damage, but bats have somehow adapted to repair the damage. And that sophisticated repair mechanism may protect them against infections and the ravages of inflammation.

We should think about bats as allies. As conservationist Kristin Lear at the University of Georgia recommends, "Don't kill bats. They might actually be the key to learning how to fight these viruses in the future." If bats can live with these viruses without succumbing to them, perhaps they can teach us how, too.

Scientists named it Nipah virus.

To stop the outbreak, Malaysia ordered the culling of more than a million pigs, no easy task and one with devastating economic consequences. Nearly 300 people were already infected, but the pig culling seemed to stop the spread—until it popped up in neighboring Bangladesh. Only there, with the country's majority Muslim population, pig farming wasn't a large industry. So where were those infections coming from?

In Bangladesh, infected people shared certain risk factors, including the handling of dead animals, like chickens, as well as tree climbing and coming into contact with a person infected with the Nipah virus. There was one other risk factor that raised eyebrows among investigators: drinking raw date palm sap.

The custom of sap drinking was particularly confounding, as it's not a typical concern related to the spread of infectious diseases. Also, what on Earth does date palm sap have to do with pigs?

The answer, it turns out, is bats.

Capable of harboring a cornucopia of viruses, bats are plentiful in both Bangladesh and Malaysia. In fact, a full quarter of Bangladesh's 113 mammal species are bats; Malaysia is home to more than seventy species of them. Both countries host a species of so-called giant "flying foxes" (*Pteropus* species), which are some of the largest bats in the world. With a sooty-black wingspan of 5 feet (1.5 m) and a face that's an amalgam of an opossum's and a baby fox's, it's either amazing or the stuff of nightmares.

Scientific fieldwork in Malaysia revealed that these large fruit-eating bats had antibodies to the Nipah virus. The antibodies were found in their urine as well as fruit partially eaten by the bats. In Bangladesh, there is another *Pteropus* species of flying fox that also carries the virus. This bat in particular loves lapping up sugar-laden date palm sap as it drips into containers of trees that are tapped for their sap. That contaminated sap would often be fed to domesticated farm animals, which in turn spread the virus to humans.

Bats frequently roost under the fruit trees found on pig farms in Malaysia. Guess where the pigs like to hang out? In a shady pigpen beneath a fruit tree. From start to finish, here's how the infection happens: Flying foxes eat fruit (messily), then pee, poop, and drop bits of masticated fruit laden with viruses that they carry without much harm to themselves. The half-eaten fruit and bat waste drops into the pigsty, where the pigs get infected and create a large amount of virus, a process called amplification. This particular virus happens to infect humans as well, resulting in infected and sickened pig handlers, pig farmers, butchers, and transporters. And boom. The human population suddenly finds itself with a new, deadly virus on its hands.

◆ ◆ ◆

The story behind the Nipah virus is just one of many stories that illustrates how zoonoses work. In addition to Nipah, zoonoses is how Ebola, SARS, rabies, bird flu, and pork tapeworm are thought to have been transmitted from animals to humans.

Another disease that has wreaked much havoc on American soil for some time is Lyme disease. Now well-known as an infection caused by a bacterium of the *Borrelia* species, Lyme disease annually infects nearly 300,000 people in the US and 65,000 in Europe. The bacterium, a wavy spirochete, is transmitted via the bite of a tick from the *Ixodes* genus, and in America, *Ixodes scapularis*, or the deer tick. The nymphal ticks are the size of a poppy seed—easy to miss for a few days if you've been bitten. Which is good for the tick and the bacteria, because the bacteria need at least thirty-six hours to transmit from tick to human. When it does, most victims go on to get the classic "bulls-eye" rash that heralds Lyme disease.

The rash is the least problematic of the symptoms. People infected with Lyme disease can have headaches, fevers, fatigue, joint and neck pain, and cardiac problems, among many other issues. Even after treatment with antibiotics, patients have reported ongoing symptoms for years.

Lyme disease as we know it is relatively new to the medical field. It was only just discovered as a full disease entity in 1976, when a cluster of cases in Lyme and Old Lyme, Connecticut, were initially blamed on juvenile rheumatoid arthritis. By 1981, Willy Burgdorfer discovered the spirochete responsible for those cases (hence the name of one of the *Borrelia* species, *Burgdorferi*).

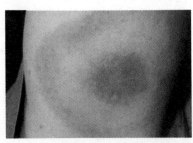

Classic erythema migrans "bulls-eye" rash in Lyme disease

But Lyme has been around for a lot longer. Ötzi the Iceman, the 5,400-year-old mummy found frozen in the Alps in 1991, tested positive for Lyme disease. And researchers have mapped out its evolutionary tree, tracing the disease back to a staggering 60,000 years ago.

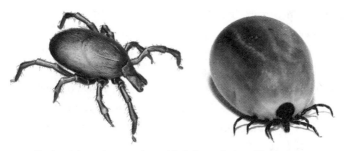

Ixodes ricinus, *the castor bean tick, before and after a blood meal*

But things have changed in 60,000 years. Cases of Lyme disease have gone up since its discovery in 1976 and have tripled since 1990. Why? How did the disease make itself known in the 1970s, but not before? The answer lies in how we've changed our environment. After North America was colonized by Europeans and vast swaths of land were deforested for farming and homes, the deer population decreased drastically, and with it the deer tick's range and ability to bite people. But in the last century, deer were reintroduced and thrived. Trees were replanted and many forests protected. The deer enjoyed nibbling our backyard hosta; in return, we planted more hosta and happily eliminated large predators like wolves from many areas. The ticks and *Borrelia* came with the deer. More humans meant more contact between us and the animals that deer ticks like to bite.

And the deer are doing a fantastic job of spreading Lyme disease. It's estimated that the areas where we can be infected stretch another 18 miles (29 km) per year. Once only found in Connecticut, Lyme disease has been detected in all states except Hawai'i.

There are other factors at play as well. *Borrelia* likes to infect white-footed mice, chipmunks, shrews, and small birds. There are theories that habitat changes and a population decrease in wolves, coyotes, and foxes meant more small mammals to spread the infection. There's also climate change to consider. A warmer spring means larval ticks emerge weeks earlier, increasing the time frame when people can get bitten. It can also mean that Lyme disease may spread ever northward each year.

Recent genomic mapping of *Borrelia burgdorferi* gathered from around the country shows that not only is the spirochete 60,000 years old, but also the most recent epidemic of Lyme spreading all over the country didn't originate from a single source. The diversity of the genome points to the possibility that Lyme popped up again in the Midwest not from the outbreak that began in Connecticut, but because it was always there—just waiting for the right conditions to spread. Conditions that we, as humans, created.

Many people are aware of the deer tick–Lyme connection, a zoonotic threat that is nearby, in some cases waiting for us in our actual backyards. And laypeople are getting much more adept at labeling infectious diseases as zoonoses thanks to the COVID-19 pandemic. Swine and bird flu (see 1918 Influenza, page 224) are fairly well-known diseases that originated in animals. There are others not as commonly known, including leprosy, now known as Hansen's disease (see page 178), and HIV (see page 130). Some may recall the terror attacks in the US using anthrax (see page 294) in the early 2000s. Found in animals like livestock, anthrax is another zoonotic disease. Pregnant women are warned not to clean litter boxes because of a parasite called toxoplasmosis, which can cause miscarriage and is carried by our beloved, if aloof, felines. Other pets, such as those peach-faced lovebirds in your living room, could carry and transmit psittacosis, or *Chlamydia psittaci*. (Yes, chlamydia. Same genus, different species and disease.) Your adorable pet bunny is capable of giving you tularemia, or "rabbit fever." A beloved dog licking your face can leave you with *Capnocytophaga*, which

Psittacosis *can be common in cockatiels, a type of parrot.*

has resulted in some recent newsworthy amputations among dog owners.

It's a stretch to say that everything furry and adorable can kill you, but the sheer variety of zoonoses lurking about can give us pause. Consider the vast numbers of animals that harbor pathogens capable of jumping to humans: ferrets, mice, racoons, wallabies, opossums, camels, sheep, pangolins, cats, dogs, birds, leopard geckos, sea lions, muskrats, monkeys, chimpanzees, gorillas, fruit bats, porcupines, hedgehogs, pigs, buffalo, beavers, donkeys, minks, poultry, parrots, civets, jackals, wolves, baboons, walruses, crocodiles.

And the list could go on, and on. In fact, more than 60 percent of infectious diseases are zoonotic. Not just viruses and bacteria, but parasites of all manners and sizes. Many of these diseases are firmly established in the animal world but can occasionally pop into ours, rather inconveniently and sometimes to devastating global effect.

Civets are nocturnal animals native to tropical Asia and Africa.

So-called enzootic diseases include rabies. With enzootic transmission, people get infected by a bite or scratch from a rabid animal, but the infected person won't transmit it to other people. Diseases that spread via enzootic transmission haven't been eradicated because they just simmer here and there in the animal population, showing up with enough frequency that we've developed vaccines where possible to protect our pets and ourselves when we get infected.

Some zoonotic diseases that jump to people become highly infectious among humans, leading to large-scale outbreaks such as Ebola in West Africa from 2014 to 2016. And sometimes an outbreak explodes, spreading worldwide, at which point we call it a pandemic, like COVID-19. In these cases, we don't need a rabid dog to bring an infection to us every time—we do just fine spreading it on our own after that first transmission, as was the case with HIV.

What's both fascinating and frightening is how, in the last seventy years, the vast majority of new infectious diseases are zoonotic. Meaning these pathogens were living in other creatures and somehow found a new, happy home in us, wreaking havoc on our way of life along the way. Though it's hardly done with malicious intent. After all, pathogens are programmed to find the very best way to survive.

As humans we tend to think that we're at the top of the food chain, but that's not entirely accurate; we are consumed—or things are attempting to consume us—constantly. Salmonella, hantavirus, huge parasitic worms called helminths, or *Ascaris lumbricoides* (warning: online searches of this will result in images you can't unsee). The truth is, we are prey to many organisms. They may not fall into tidy predatory groups that include lions and sharks that are capable of consuming

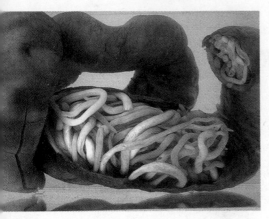

Parasitic worms in a section of intestine removed from a young boy.

us from the outside. Rather we are under siege from things that seek to feed on us from the inside, in a microscopic way. These attackers can be just as devastating than the larger predators, if not more so, precisely because we *can't* see them.

With zoonotic diseases, we share the burden with animals. That's what makes zoonoses particularly fascinating yet unpredictable. Sometimes diseases need a partner, called a vector, to find their end host. Vector-borne diseases include malaria and Lyme, which use mosquitos and ticks, respectively, to deliver either a microscopic parasite (in the case of malaria) or bacteria (in the case of Lyme) to an unsuspecting human. Which, of course, takes zoonoses to another level of complicated. This triangulation of an infectious organism traveling between animal, temporary host, and end host is, at its very simplest, a balance of multiple factors that drive how and why the disease becomes a problem in humans. This forces us to tackle other concerns. What makes that tick or mosquito thrive? In which climates? How does climate change factor in? Should we be worried about genetically modified mosquitoes? How do we stop deforestation, which allows mosquitoes to flourish? What other factors make us so delicious for the biting?

◆ ◆ ◆

The existence of zoonoses shouldn't be surprising. We are part of a complex ecological system. As our existence has evolved from hunter-gatherers to modern-day hominids that number in the billions and live in structures, we have disturbed things. We farm meat and grain in massive quantities, mine deeply for mineral and petroleum resources, and raze natural habitats to build and farm. Specifically, we force animals and their pathogens into our realm—in beef and chicken slaughterhouses, on the edges of burning rain forests, from wild game being hunted, and on farms whose pigsties sit beneath tree branches full of guano-producing fruit bats. We also travel far and wide as a species. In less than twenty-four hours, we can travel from one side of the planet to the other—and pathogens are happy to hitch a ride.

Alpha-Gal

The Meat-Lover's Worst Nightmare

Imagine you're sitting down to a delicious, dry-aged, medium-rare prime rib at your favorite steakhouse. Being the omnivore you are, you relish every bite. Sometime around two o'clock in the morning, you awake from sleep. Your heart is racing, you feel like you're about to have diarrhea, and your body is covered in itchy hives. If you're extremely unlucky, you could get a life-threatening anaphylactic allergic reaction. You might consider a food allergy, but you've eaten steak countless times with no problems. Could you actually be allergic to the steak?

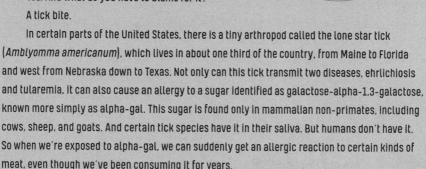

Yes. And what do you have to blame for it?

A tick bite.

In certain parts of the United States, there is a tiny arthropod called the lone star tick (*Amblyomma americanum*), which lives in about one third of the country, from Maine to Florida and west from Nebraska down to Texas. Not only can this tick transmit two diseases, ehrlichiosis and tularemia, it can also cause an allergy to a sugar identified as galactose-alpha-1,3-galactose, known more simply as alpha-gal. This sugar is found only in mammalian non-primates, including cows, sheep, and goats. And certain tick species have it in their saliva. But humans don't have it. So when we're exposed to alpha-gal, we can suddenly get an allergic reaction to certain kinds of meat, even though we've been consuming it for years.

The theory follows that when a lone star tick bites us, it's likely already bitten a mouse, rabbit, or deer with alpha-gal, or its saliva has enough alpha-gal by itself. During its feeding, the tick regurgitates its spit into our skin with a little bit of alpha-gal. For some unlucky people, the immune system of our skin causes an allergy response, and voilà: a new and relatively rare allergy to meat.

Luckily, chicken and fish are still on the menu.

Add to this that mammals evolve slowly compared to pathogens. It takes animals anywhere from months to decades to reproduce, and longer still for those young to grow and become sexually mature. Whereas the life cycle of a pathogen can happen in a radically shorter period of time, providing them with the ability to quickly bend to selection pressure. Consider how the seasonal flu changes every year, to the point where we need a new vaccine annually. West Nile virus changed from genotype NY99 to WN02—three nucleotide changes on the RNA— allowing it to transmit more effectively via mosquitoes in the span of three years. Chikungunya virus, which causes fever and joint pain among other symptoms, also altered one nucleotide between 2005 and 2006, making it more infectious in one species of mosquitoes that had not previously been identified as a vector for this virus. Perhaps more chilling, this mutation happened independently on Réunion Island in the Indian Ocean, in West Africa, and in Italy.

Changes in weather, in hosts, in their microbiomes, and in landscape, all affect zoonotic ability to jump to humans. In the early and late 1990s, for example, the El Niño phenomenon resulted in larger rainfalls, which led to increased vegetation, a food source for rodents. More rodents led to increased contact with humans and thus to hantavirus outbreaks in the southwestern US. Temperate changes have also allowed the dengue virus to show up in Texas (jumping north and west from the Caribbean), and Lyme disease to spread to Canada.

In Africa, after the Italian invasion of Eritrea in 1889, a paramyxovirus (an RNA virus, whose family includes mumps and measles) infected livestock populations. The disease was called rinderpest, or "cattle plague." Being naïve, or unexposed, to the virus, cattle herds were decimated across the continent, leading to a famine. Local flora repopulated grazing areas, along with wildlife, both of which favored the tsetse fly, a parasitic insect that feeds on the blood of vertebrates. The tsetse fly also carries a parasite that causes a deadly disease called sleeping sickness, or trypanosomiasis. Former cattle herders who waded into the brush to hunt became infected and, because the tsetse fly

A tsetse fly, known to feed on human blood and can pass parasites that can cause disease.

also flourishes near water, river villages were wiped out. In some areas, 20 percent of the population was infected at a time when there was no cure, and the disease was invariably fatal.

◆ ◆ ◆

With imagery like flying fox bats and tick-laden chipmunks, it can seem like pathogens are opportunistic invaders, waiting for us to disturb our world in small ways so they can find a way inside of us. Calling them *invaders* sounds highly predatory, and in many ways they are. But in other ways, they aren't. For every zoonotic disease that jumps to humans, there are untold others (possibly many millions) that don't because the fit just isn't right on numerous levels. There are myriad complex factors involved before a new pathogen is passed from a vertebrate animal to a human, a process called "spillover," which is rare. In truth, the barriers to spillover are significant. First, there has to be enough of the animals who are the natural host or reservoir, in the right population density, located in the right place. In the case of Nipah virus, that means lots of bats roosting together close to a pigsty.

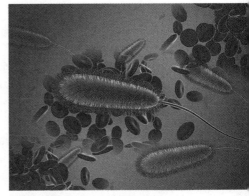

Second, there have to be enough of the animals carrying the infections with a high enough pathogen load, and they must shed that pathogen in large enough quantities for it to jump to another species. With rabies, for example, where the virus accumulates in the salivary glands, bites are a great way to transmit an abundance of virus. If the host is a food product like beef, trying to control how much pathogenic *E. coli* lives in the rectum of the cattle at the time of slaughter is important. If there is a vector, say a flea, how frequently that flea bites both hosts, as well

Pathogenic E. coli *can be naturally found in the intestines of cattle, sheep, and goats.*

as the flea's chances of survival and its patterns of movement, must be just right for pathogen release.

Third, the pathogen has to be able to survive outside the animal host. Q fever can survive in the air for a long while, leading to outbreaks miles and miles away from the livestock hosts, whereas influenza A doesn't survive long, making close contact necessary. Finally, there are variables involving the new human host.

Did they get a high enough dose of the virus? Did the usual barriers to fighting infection—healthy skin, the many layers of protection in the respiratory tract, low stomach pH—somehow not work? Are their immune systems weakened by old age or chronic diseases?

And the aforementioned factors don't even comprise all the issues that must align perfectly before a new infection can establish itself in a human. Don't forget that a pathogen also wants to replicate inside its host and spread to other people. So if it kills its host quickly, it won't be able to effectively find another host. That balance—infecting and killing—is tantamount to survival. This explains why some zoonoses do better than others. For example, with Lyme disease, the bacteria live comfortably in deer and small rodents, which can then infect humans, but humans can't spread it to each other. We are considered a "dead-end host." It doesn't absolve the disease of its awfulness, but it does mean it can't sweep through an airport and infect an entire plane of passengers.

And it's important to understand that pathogens can remain inside humans for a while, spread and mutate, go back to animal species, and mutate again. Then jump right back to humans. This is how influenza works sometimes.

When trying to put the idea of zoonosis in perspective for our complicated, highly populated, tech-powered age, it's helpful to view it as more dynamic than simply predator versus prey. Humans are instrumental in allowing emerging infections to harm us. We have altered the ecology of many habitats and, by doing so, have brought animals right into our path. We are the evolutionary masters of our own suffering in a sense, creating the conditions that allow these pathogens to take advantage of new niches to live in, new bodies upon which to prey.

Pathogens are simply making their way in the world free of emotion, taking advantage of the rapidly changing ecology around them, from the tiny universe of gut microbiomes to hurricanes sinking coastline cities. Our world is theirs, too. They change when necessary, they jump, consume ravenously, kill what's in their way, and adapt, all in order to multiply. It sounds a lot like another species on the verge of hitting an 8 billion population milestone.

Hint: us.

EBOLA
Patient ZERO
Mabalo Lokela

Cause: Ebola hemorrhagic fever (EHF)

Symptoms: Fever, vomiting, diarrhea, bleeding from eyes, ears, mouth

Where: Yambuku, Democratic Republic of Congo (DRC)

When: August 1976

Transmission: Blood and body fluids, including vomit, diarrhea, sweat, tears, and semen

Little-Known Fact: Ten or fewer Ebola virus particles could cause infection.

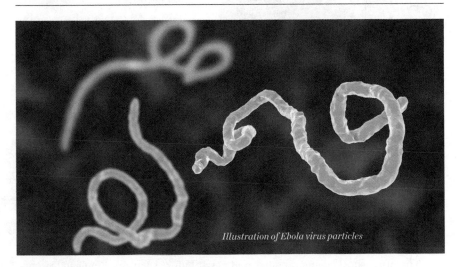

Illustration of Ebola virus particles

n August 1976, Mabalo Lokela, the head schoolteacher in the remote village of Yambuku, Zaire (now the Democratic Republic of Congo), checked into the local mission clinic complaining of a fever. He had just returned from vacation, one of the few he had ever taken. During his time off, he'd gone hunting in the rain forest and visited some family and friends in nearby villages. After he returned he started to feel unwell, so he visited the village clinic to see if they had any antibiotics (often they had none in stock) that might help.

Yambuku was far from the nearest town of any size. To get to a bigger town you had to travel over poorly maintained, sometimes dangerous dirt roads through the surrounding jungle. Yambuku's clinic, a charity hospital run by Catholic nuns from Belgium, offered the best medical care nearby. But there was no doctor on staff and the nuns, while experienced with a variety of local ailments, had no formal medical training.

The nuns examined Lokela but weren't sure what was wrong with him. A diagnosis of malaria seemed a safe bet. They had seen plenty of malaria, and fevers were a common symptom. After scrounging through their scant medical supplies, they found some doses of quinine and gave Lokela an injection. The schoolteacher headed for home, feeling slightly better.

Yambuku villagers, Zaire, 1976

For a few days, the quinine seemed to work and Lokela enjoyed a brief reprieve from his symptoms. But they soon returned with a vengeance. Within a week Lokela was too weak to get out of bed, his body racked by repeated waves of vomiting and diarrhea. Desperate for help, Lokela's wife went to the clinic and begged the nuns to visit their hut. When they arrived, they were horrified by what they saw.

In the dark confines of the small hut, Lokela was lying on a bed just barely above the floor, covered in sweat, blood oozing from his eyes, ears, and mouth. Just as the nuns entered, Lokela vomited a mixture of blood and bile.

The nuns, shocked and repelled, had never before seen anything like this, but it definitely wasn't malaria. They had no idea what to do or how to even begin helping Lokela. The nuns began to suspect that Lokela had contracted something new, some sort of devastating, rapacious fever. They tried to make Lokela as comfortable as they could but he died, in agony, just a few days later on September 7, 1976.

What the nuns didn't know—and what Lokela's family and friends who visited him during his final days couldn't possibly know—was that his hut had become a hugely dangerous biohazard zone. Every bit of blood or vomit that exited poor Lokela's body was a viral bomb. Those microorganisms contained a biosafety level 4 pathogen (see sidebar, page 29) that, if acquired by another human, could cause massive internal and external bleeding, likely resulting in death.

Mabalo Lokela had just become the world's first victim of EHF, Ebola hemorrhagic fever, or Ebola for short.

◆ ◆ ◆

After his gruesome death, Lokela's body was prepared for burial in the usual way for Yambuku villagers. His blood-soaked corpse was bathed and handled by family and friends, who then sat with him for twenty-four hours before burying him at a public funeral in a grave near the family hut.

Afterward, the village breathed a brief sigh of relief. Whatever strange new illness had taken Lokela, it had passed now. But the relief was temporary. Not long after the burial, Lokela's wife, mother, mother-in-law, sister, and daughter all visited the mission clinic with similar symptoms. Soon, twenty-one friends and family members who had attended the funeral were infected with the same disease. Eighteen of them died.

Fear quickly turned to panic when the nuns at the mission hospital also started getting sick. Young, old, healthy, frail—it didn't matter—the disease found everyone. And catching it was essentially a death sentence. People in Yambuku began dying in droves, with whole families wiped out by the violent, terrifying disease.

A Flemish nun visits the graves of her colleagues who died from Ebola during the 1976 Yambuku Ebola outbreak in Zaire.

The mission hospital was closed by the end of September—no one was left to run it. Eleven of the original seventeen staff members were dead of Ebola. The Mother Superior, one of the few remaining survivors, sent radio signal after radio signal out into the world pleading for help. With the mission hospital closed, family members of the ill got desperate and began taking them to nearby villages seeking medical help, unwittingly contributing to the spread of the disease.

In the first week of October, help finally arrived, albeit briefly. Word of the frightening new epidemic finally made its way to the Zairean government, who promptly sent in the army to seal off the village from the outside world. By that time, more than one hundred people had died. Zairean doctors were flown in by helicopter, took blood samples of some infected patients, and quickly fled back to the capital of Kinshasa, outside the initial outbreak zone. Global health officials were notified of a lethal new disease that was beginning to spread.

The blood samples collected by the Zairean doctors were quickly sent around the world, where, with a mixture of excitement and horror, international scientists realized they were looking at a brand-new virus. An international team of health workers were deployed by the World Health Organization (WHO) to the village. They quickly confirmed that the virus was indeed new, viral, and highly infectious. It was named Ebola hemorrhagic fever, after the nearby Ebola River. Those visiting doctors witnessed firsthand how shockingly devastating the disease was, and how monstrously lethal. Of those infected with Ebola, 50 to 90 percent died from it. As the 1976 outbreak spread from Yambuku to more than fifty villages in Zaire, as well as the capital city Kinshasa, 318 people came down with the disease.

A staggering 280 of them died, giving the new disease a terrifying 88 percent mortality rate.

◆ ◆ ◆

Ebola virus disease (EVD), as it's now called, is caused by an infection of any of the viruses within the Ebolavirus genus. Of the six Ebola viruses known today, four of them can infect people. Scientists still aren't completely certain where the virus came from, but the general consensus is that the virus is animal-borne, infecting people in so-called "spillover" events when humans have contact with an infected animal. The most common culprits appear to be nonhuman primates and fruit bats.

The virus spreads between humans through close and direct physical contact with a person infected with Ebola, or their body fluids. To enter the bloodstream of a new potential victim, the virus needs to find a way in through broken skin, or through mucous membranes of the eyes, nose, or mouth. The semen of a previously infected person who has recovered also remains a potential source of spread. Given those opportunities for transmission, and the fact that it takes very few virus particles to actually cause disease, it is not surprising that the virus

Biosafety Levels

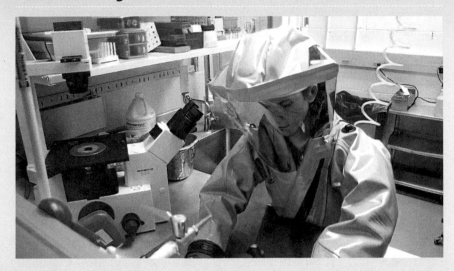

Ebola is rated as a biosafety level 4. If that sounds scary, good, because it is. Level 4 is the highest biosafety level in the United States. Each level comes with its own specific controls for the containment of microbes and biological agents when studied in a laboratory. What determines each level of biosafety precautions? The transmissibility of the microbe and the severity of the disease caused by the microbe.

According to CDC guidelines, here are the safety measures needed for a laboratory to be rated at biosafety level 4, able to study frequently fatal viruses, such as Ebola, with high risks of transmission:

- Airtight
- Self-closing, double-door access
- Controlled access
- Sharp hazards warning policy
- Handwashing sink
- Sealed penetrations
- Physical containment device
- Positive pressure protective suit
- Laboratory bench

- Autoclave (a pressure chamber allowing precise control over pressure and temperature; used for sterilizing materials)
- Chemical shower out
- Personal shower out
- Supply and exhaust HEPA filter
- Effluent decontamination system (a system that uses heat and pressure to destroy any microbes in laboratory wastewater)

is highly infectious between people. In mouse models, a single virus particle can cause disease. In many cases, Ebola has spread to families or healthcare workers taking care of Ebola victims, or to household members, who, according to custom, would touch the newly deceased with bare hands, as well as afterwards during funeral preparations. And that's why you see images of healthcare workers at Ebola outbreak sites in head-to-toe PPE (personal protective equipment), the medical equivalent of a hazmat suit. It's so dangerous that there are intensely detailed methods of "donning and doffing" the PPE in the presence of trained observers to ensure it's done correctly.

Once a person contracts Ebola, the first symptoms are typically the sudden onset of fever and chills. Fatigue, headaches, vomiting, diarrhea, and loss of appetite, as well as a general malaise and weakness follow next. Within a week of infection, a rash may develop over the face, neck, chest, and arms. The vomiting and diarrhea, which are usually severe, continues, leading to fluid loss and related complications, such as dehydration and even shock. The steady outpouring of bodily fluids, loaded with infectious particles, also provides the virus ample opportunity for spreading to other people who are helping to care for the patient or are in the same household.

The virus can sometimes lead to hemorrhage, which gives Ebola its other name: hemorrhagic fever virus. Patients may have mild bleeding (or none at all) during the course of the illness. Or they may start hemorrhaging, with blood

An Ebola "burial team" in Disco Hill, Liberia, 2015

showing up in their feces, as well as oozing from IV and needle puncture sites. Bruising can take place as well. The whites of patients' eyes may become reddened from hemorrhage. Once the disease turns terminal, significant hemorrhaging can occur, including blood pouring out from the eyes, nose, ears, and rectum. Unless fully protected with proper PPE, it's extremely difficult to avoid getting infected yourself if you're in close proximity to an Ebola victim whose body is teeming with virus. Which is why the first outbreak in Zaire had such a staggeringly high death toll.

◆ ◆ ◆

As reports of the 1976 epidemic in Zaire circulated, the World Health Organization noted a startlingly similar outbreak just a couple of months earlier in South Sudan. There, an outbreak of a mysterious new disease had led to the deaths of 151 people out of 284 cases. While the South Sudanese outbreak had not received the international attention focused on Zaire at the time, it was clear to WHO scientists who examined blood samples that they were looking at two outbreaks of the same disease. A major investigation was soon launched to examine the link between the two outbreaks and determine what preventative measures could be put in place.

Ebola, meanwhile, became a global headline and cultural touchstone of the late 1970s and early 1980s. Images of scientists in hazmat-like suits descending onto disease-stricken African villages were all over newspapers and evening news broadcasts. The horrors of the disease itself hardly required any embellishment as people literally bled out while their internal organs disintegrated.

Many of the significant discoveries about Ebola came from those first two outbreaks in 1976. Scientists determined that the South Sudanese and Zairean epidemics of Ebola were actually unrelated—coincidental, but technically two different subtypes of the disease, Ebola-Zaire and Ebola-Sudan. The Zaire subtype was more lethal, with a 90 percent mortality rate. Ebola-Sudan killed only 50 percent of its victims, which is not as bad, but still horrific.

Both 1976 African epidemics were tragically made worse by unsanitary conditions in local hospitals, as well as traditional funerary practices in the larger region. In the early days of the outbreak, for example, the village missionary clinic in Yambuku continued its practice of using the same five hypodermic needles repeatedly without sterilization each day. Between 300 and 600 patients cycled through the clinic each day after the outbreak began, so the reuse of these contaminated needles contributed significantly to the early spread of the virus. Local cultural funerary practices compounded the spread until global health officials were able to implement containment and quarantine measures.

◆ ◆ ◆

Ebola's origins remain mysterious to this day. After the initial 1976 outbreaks, virus hunters collected and tested a veritable Noah's ark of animals in an attempt to locate the source of the disease. These included bedbugs, mosquitoes, pigs, cows, bats, monkeys, squirrels, mice, and rats. But none of them showed any signs of the Ebola virus in either Zaire or South Sudan.

Eventually, scientists were able to demonstrate that nonhuman primates—monkeys, chimpanzees, and gorillas—as well as antelopes and porcupines were susceptible to Ebola and seemed capable of transmitting it to humans. That discovery was underscored by the revelation that Mabalo Lokela, patient zero in the Zairean outbreak, had eaten "bushmeat" (the meat of wild animals, antelope in his case; see sidebar) while on vacation right before he was ravaged by the disease.

In South Sudan, the first person to contract Ebola in 1976 worked in a cotton factory infested with bats. Although the factory bats ultimately tested negative for Ebola following the outbreak, scientists still suspect bats as potential carriers of the disease. Lab-based experiments have shown that bats infected with Ebola do not die or grow ill from the pathogen. This means they may play a role in harboring the virus in a dormant state, contributing to its occasional outbreaks in human populations (see Zoonoses, page 12). Four more strains of Ebola have been identified since the two 1976 outbreaks: Taï Forest, Bundibugyo, Bombali, and Ebola-Reston. Only Taï Forest and Bundibugyo, however, have yet been able to infect humans. The Taï Forest strain emerged in 1994, when a scientist contracted Ebola while conducting a necropsy on the carcass of a wild chimpanzee in Côte d'Ivoire. Luckily, the thirty-four-year-old Swiss scientist was quickly transported back to Switzerland, where containment measures prevented an outbreak and rigorous treatment saved her life. The Bundibugyo strain first appeared in 2007 in the Democratic Republic of Congo and Uganda, then again in the Democratic Republic of Congo in 2012.

Ebola-Reston was the source of a brief but terrifying global scare in 1989. Crab-eating macaques were captured in the Philippines and shipped to Reston, Virginia, for medical testing. When the monkeys arrived, they were found to be harboring a new strain of Ebola. It killed nearly all of them.

The residents of Reston, located near the nation's capital in Washington, DC, were petrified. A biosafety level 4 virus had just arrived on American shores, and near to a major metropolitan area to boot. The world breathed a collective sigh of relief, however, when the strain was identified as unique to nonhuman primates. This means humans can't get Ebola-Reston. We develop antibodies to it if we are exposed, but we don't actually generate symptoms.

Macaque monkeys shipped to Virginia in 1989 were the source of a new Ebolavirus variant, named Ebola-Reston.

If all this sounds a little familiar to you, here's why: the Reston outbreak was dramatized for Richard Preston's best-selling 1994 novel *The Hot Zone* and the 2019 miniseries of the same title from *National Geographic.*

The worst outbreak of Ebola, however, was yet to come. In December 2013, a two-year-old child named Emile Ouamouno in a small village in Guinea suddenly came down with symptoms of a bad disease: high fever, vomiting, and diarrhea. The boy's mother, pregnant at the time, took him along with his older sister to seek care at their grandmother's house.

Before long, the disease crossed over the nearby borders into Sierra Leone and Liberia, though it wasn't immediately recognized as Ebola. The area frequently experienced outbreaks of diseases with similar symptoms, and initial guesses centered on cholera. Local Guinean doctors told world health officials (who had been alerted to the burgeoning epidemic) they believed a cholera epidemic might be underway. But not all of the symptoms synced up with cholera. The nosebleeds that accompanied the fever, combined with the vomiting and diarrhea, hinted at the disease being something else.

Hiccups, of all things, finally gave it away.

When the organization Doctors Without Borders began examining reports of the Guinean outbreak, they noticed hiccups listed as one of the reported symptoms. (A curious side effect of hemorrhagic fever is the hiccup.) Even then Ebola seemed unlikely, as there had never before been an Ebola outbreak in Guinea. Because Guinean health officials didn't have materials available in their country

Bushmeat

Bushmeat is thought to be how Ebola "jumps" to humans.

Bushmeat is a term used to describe most meat from wild animals in Africa. In Africa, wild places such as forests and savannah are often referred to as "the bush," which is where the name for bushmeat comes from. While undeniably important as a food source in impoverished regions, bushmeat is also considered highly problematic. Harvesting game contributes significantly to the population decline and even extinction of wild animals in Africa. Over 1 million tons of bushmeat are thought to be consumed in Africa each year. Impacted animals include gorillas and chimpanzees (as well as other primates), elephants, antelopes, crocodiles, fruit bats, porcupines, cane rats, snakes, and pangolin. And eating bushmeat is risky for humans as a potential source of zoonotic disease transfer. For that matter so is hunting bushmeat, where hunters come into direct contact with the bodily fluids of the killed animals. A thirty-five-year study in the Cameroon-Congo Basin concluded that the following zoonotic pathogens were transferred to humans through the hunting or consumption of bushmeat: arboviruses (dengue, yellow fever), monkeypox, HIV, anthrax, salmonella, simian foamy viruses, and Ebola.

While not definitely proven, Ebola has likely transferred from bats (eaten as bushmeat during hunting-related activities or by these animals consuming partially eaten fruits and pulp) to humans some thirty times between 1976 and 2014.

It's a complex problem with no quick solutions, and not just in Africa alone. Wild game is consumed all over the world, with similar consequences. It's a crucial food source in some regions and is often rooted in cultural traditions. Until a better solution is adopted, many people all over the globe will continue to run the risk of zoonotic disease transfer. And who knows what strange new disease, just as scary as Ebola, might be part of the next spillover event?

to test for the disease, blood samples of infected patients were sent to laboratories in Senegal and France. The results came back decisive and sobering: It was Zaire ebolavirus.

Doctors Without Borders sent in teams immediately, along with emergency committees activated by local governments. The World Health Organization announced the outbreak. Field hospitals with isolation chambers were quickly assembled, and foreign doctors wearing full-body PPE suits swept into the region.

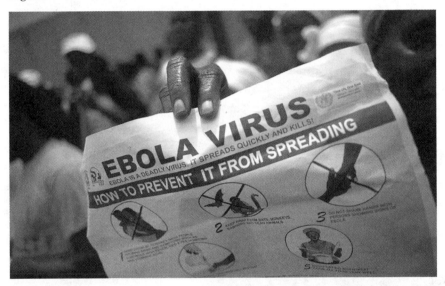

A public health effort aimed at handwashing to stop Ebolavirus spread in Liberia, 2014.

Many local Guineans were scared by the rapid response. From their perspective, sick loved ones were being whisked away into foreign-built isolation chambers, where they would often die without ever seeing family again. People who passed away from the disease were also denied traditional burial practices, as they were immediately zipped into body bags and buried. Aid workers could not offer vaccines or a cure, which had been a welcome relief for other diseases, such as malaria. The result was a fear of foreign aid workers and a general reluctance to help, and in some cases, that fear turned to violence. In the village of Womey, in southwest Guinea, eight Ebola aid workers were massacred with clubs and machetes. Some Guineans even hid their sick family members in their homes rather than bring them to a hospital, as rumors and conspiracy theories about the foreigners spread, including one that suggested they were secretly harvesting organs.

Meanwhile, Guineans kept on dying from the disease. By June, 300 had succumbed to Ebola. The number grew to 3,000 by September. As the outbreak worsened, Guineans grew both increasingly afraid and convinced of the severity of the disease, eventually trusting the containment measures and treatment methods introduced by the foreign doctors. One of the mainstays of therapy was IV fluids, which helped with severe dehydration, as well as antibiotics for the secondary bacterial infections like peritonitis (infection of the inner abdomen).

Virus hunters focused on the circumstances of the patient zero in the 2013 outbreak, the two-year-old boy who first caught the disease. They discovered that he often played with other local children near a hollow tree that contained a colony of Angolan free-tailed bats. These bats were sometimes captured, killed, and eaten by the children. While not proven yet, the connection between bat carriers of the disease and human outbreaks was once again circumstantially reinforced by this occurrence. It also underscored the growing health concerns of eating bushmeat (see sidebar, page 34).

By the time virus hunters found the tree, where they hoped to capture and test some of the resident bats, they encountered only a burnt-out stump. The tree had caught fire sometime during the outbreak, intentionally or not. Villagers who saw it burn said that it "rained bats" while the fire raged.

Protective medical gear dries after being cleaned in Gueckedou, Guinea.

Dubbed the "Western African Ebola virus epidemic," the outbreak that may have begun with that bat colony in 2013 ultimately spread across three countries: Guinea, Liberia, and Sierra Leone. By the time it was finally contained in January 2016, it claimed 11,000 people after infecting 28,000, putting its mortality rate at about 40 percent. The 2013 to 2016 eruption in West Africa remains the worst Ebola outbreak in history. Entire villages were abandoned, and remain so now, as local populations have yet to recover. Furthermore, publicized cases of Ebola occurred in dozens of aid workers and travelers who brought the disease to Europe, the UK, and the US, causing fear and panic. As with so many newer outbreaks, Ebola had far-reaching consequences and was not just a problem in its place of origin.

◆ ◆ ◆

Scientists recently discovered the Ebola virus can hide in parts of your body even after you've recovered from an infection. It can remain in the uterus or the testicles for weeks, even months. Survivors also often have pockets of the virus that linger in their eyes. Survivors, however, also continue to produce antibodies to fight the virus. In 2017, researchers led by Anne Rimoin studied fourteen survivors of the 1976 Zaire Ebola outbreak and discovered that, decades later, they still carried antibodies that recognized at least one of Ebola's proteins. Four of them had antibodies that could still completely neutralize the virus.

As of December 2019, the US Food and Drug Administration (FDA) approved the first ever Ebola vaccine (ERVEBO), which is a single-dose vaccine specifically effective against the Zaire ebolavirus subtype. The other subtypes still have no cure or vaccine. We can isolate patients and treat them aggressively with antibiotics and IV fluids, but there is an element of luck to surviving the disease. It's known to appear suddenly and unexpectedly, then disappear again just as unexpectedly, and for no apparent reason. This makes it difficult for scientists to test vaccines or cures—because how do you do that when the disease has disappeared?

Ebola will almost certainly return again. In the meantime, doctors and scientists are doing their best to prepare for another outbreak. However, the virus's ultimate return poses a potential problem: evolution. So far the virus has been relatively slow-moving since it requires the exchange of bodily fluids to move between hosts. Compared to viruses that are airborne, this evolutionary quirk has helped contain Ebola outbreaks to specific locations. If the virus ever evolved and, say, became airborne, we'd have a global pandemic on our hands that would make COVID-19 look tame. And that might be the scariest thing about the very horrible Ebola virus, the part that keeps scientists up at night: It evolves quickly. For example, in the Guinean outbreak of Zaire ebolavirus that began in 2013, the virus gained a mutation called A82V, making it four times more infective and leading to an uptick in cases. The virus will certainly continue to evolve in future outbreaks as well. Scientists are doing all they can to stay ahead of it.

GERM THEORY
From Miasma to Microscopes

Impact: A groundbreaking concept that proved certain diseases are caused by an invasion of microbes invisible to the naked eye

When: 1860

Who: Louis Pasteur

What Happened Next: A revolutionary change in public health and the practice of medicine that would forever alter life expectancies and human history

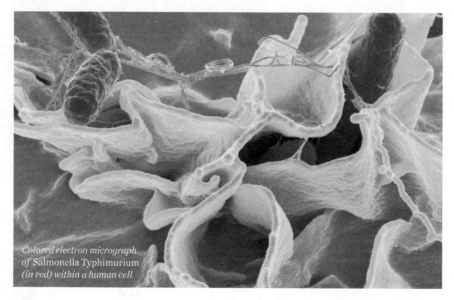

Colored electron micrograph of Salmonella Typhimurium *(in red) within a human cell*

For many millennia, humans blamed sickness on anything but germs. When boils, lethal amounts of diarrhea, and vomiting blood brought whole cities to their knees, responsibility was cast on comets, eclipses, earthquakes, demons, gods, or witchcraft. Powers far larger than ourselves were forever the reasons behind outbreaks of disease. But it turned out that our lens was too broad. We needed to constrict our visual field down from the heavens to that which we couldn't actually see, no matter how hard we squinted.

It has been a long and arduous path to simply discovering germs, let alone proving how they cause disease. Yet we are relatively good at internalizing the science that assists our survival, much of it ingrained in us since birth. We know, for example, it's a bad idea to lick a piece of raw chicken. We're told from a young age to cover our mouths when we cough and to avoid sneezing on people. The sight of a filthy portable toilet makes us queasy at best. Why? Because we know that what we cannot see with the naked eye for sure can sicken us—and possibly even kill us.

Germs.

The word means microscopic agents that can cause disease, including bacteria, fungi, protozoa, and nonliving pathogens, like prions and viruses. For centuries, these microorganisms went undetected, but they've always been with us, causing sickness and death since the dawn of the human race.

Before we discovered them, we tried to figure out why we got sick. There was no shortage of myths and theories to explain away our illnesses and anoint false prophets of protection. In ancient Rome, the goddess Febris was said to protect people from fever and malaria, with three temples built in her honor. The Bible's Old Testament refers to plagues being cast down by an angry God. "The wrath of the Lord was kindled against the people, and the Lord smote the people with a very great plague," reads Numbers 11:33. In Leviticus, there are instructions on cleanliness, quarantine, isolation, and food handling that would prevent pork tapeworm infections, for example. In ancient India, disease in the body and mind were attributed to demons that had to be driven away with charms. In 1348, the heart of the Middle Ages, King Philip VI of France was told by his physicians that the bubonic plague was the result of "a major conjunction of three planets in Aquarius," which could lead to "a great pestilence in the air." Particularly susceptible were "those following a bad lifestyle, with too much exercise, sex and bathing." In 1550s Venice, cases of syphilis that couldn't be cured were blamed on witches, at which point exorcism, or at least investigation by the Holy Office, was ordered.

A plague doctor, 17th century

Physicians and philosophers had other theories, as well. In the fifth century BCE, Hippocrates wrote in his texts that stagnant water and marshlands could cause malaria, diarrhea, and respiratory illnesses. For more than a millennium afterward, this concept seeped into the professional and public consciousness about disease, for better and worse.

Enter the miasma theory.

Miasma (from Greek for "pollution") was also called "bad air" or "night air" and thought to be nasty, disease-causing vapors that arose from rotting organic material. Mists arising from smelly swamps were similarly feared. The miasma concept was pervasive in Western culture from the time of Hippocrates all the way to the nineteenth century. The idea was that putrefied particles (*miasmata*) would rise up and ruin water, air, and nearby food, sickening people in its wake. Supposedly, you could identify the presence of miasma vapors by the telltale stink of rot. Classic areas at risk for miasmas were fetid, damp

Hippocrates (460–370 BCE) is often credited as the founder of modern medicine.

environments like swamps or filthy rivers full of garbage and excrement, including stinking city sewers inadequate for large populations. In fact, these were reasonable assumptions, given that excrement and rivers of garbage *do* harbor dangerous pathogens—just not in the form of "miasmatic" vapors.

In the first century BCE, ancient Roman architect Vitruvius noted that if "the poisonous breath of creatures of the marshes to be wafted into the bodies of the inhabitants, they will make the site unhealthy." The Chinese had their version of miasma as well, called *zhangqi*, or "pestilence vapors." These were often blamed for the colds and influenza (and sometimes malaria and dysentery) that hit southern China during the Sui (seventh century CE) and Tang (seventh through tenth centuries CE) dynasties. The casualties of disease in southern China from zhangqi slowed their development compared to the population in the north, which flourished for a good part of Chinese history. In the 1700s, the term *malaria* was coined from the Italian words for "bad air," and it was commonly thought that night air brought on the fevers that classically resulted from an infection.

Florence Nightingale, the famous nurse from the days of the Crimean War (1853–56) was initially a firm proponent of the miasma theory (see page 40).

She did not believe that contagious microbes were the issue—cleanliness was. Miasmas simply need to be kept at bay. Ever a proponent of hygiene, she believed that the idea that a contagion could cause the spread of disease was an "excuse for non-exertion to prevent its recurrence" and that "[because of] neglect of sanitary arrangements, epidemics attacked whole masses of people." Nightingale wasn't totally wrong, since sanitation and cleanliness *did* reduce infections. But the idea of contagious, microscopic sources of infection still wasn't widely accepted. Thomas Southwood Smith, a nineteenth-century physician and sanitary reformer, even rebuked the idea: "To assume the method of propagation by touch . . . and to overlook that by the corruption of air, is at once to increase the real danger, from exposure to noxious effluvia."

But the miasma theory led to some real gains. Problems with stagnant, smelly water led to sanitary reforms in England. The Great Stink occurred in the summer of 1858 when hot weather intensified the stench of rotting human excrement and industrial runoff in the River Thames. Talk about hellish: Overrun cesspits oozed methane, which occasionally caught fire and exploded. Three outbreaks of cholera preceded the summer, killing thousands of Londoners. Joseph Bazalgette, a civil engineer, designed new sewer systems, pumping stations, and improved drainage that ended up significantly lowering the number of cholera cases.

An anti-typhus illustration warning against miasma, or bad air, in 1885.

Other ideas competed with the miasma theory. One that emerged in the mid-1800s was the zymotic disease theory, which predicted that organisms, or microzymas (from the Greek word *mikros*, "small," and *zumoûn*, "to ferment"), caused disease. It seems a step in the right direction, except these microzymas were also thought to be minute constituents of all things, including rocks, plants, and animals. One of the proponents of this theory, French scientist Antoine Béchamp, a rival of Louis Pasteur, believed these microzymes created cells and organisms. If the native host was "unfavorable," then the microzyms would create pathogenic microbes. Despite making breakthroughs in organic chemistry

Papier d'Armenie, or Armenian paper, was infused with the resin of a specific tree bark and believed to cure cholera and croup, among other illnesses.

(producing dyes and an early version of an antibiotic), his zymotic disease theory did not take hold.

◆ ◆ ◆

An alternative disease theory that was highly entertaining but utterly wrong was spontaneous generation, or the idea that organisms arose from nonliving things. Examples include:

Oysters arise spontaneously from slime.

Westward winds create bookworms that will ruin libraries.

Anchovies come from sea-foam.

Maggots arise directly from spoiled meat.

Bees arise from cow carcasses (called *bugonia* in Byzantine texts).

In the early seventeenth century, Jan Baptista van Helmont asserted that full-grown mice could be produced in three weeks from a jar containing wheat and stuffed with dirty linen, and that if basil was crushed between two bricks, scorpions would arise.

But the idea of spontaneous generation started long before van Helmont. In the fifth to sixth century BCE, Greek philosophers speculated that certain elements came together to create creatures, such as fish arising when the sun reflected on water. In the fourth century BCE, Aristotle mentioned that in "instances of spontaneous generation some come from putrefying earth or vegetable matter, as is the case with a number of insects, while others are spontaneously generated in the inside of animals out of the secretions of their several organs."

But scientists like Francesco Redi, an Italian physician, biologist, and naturalist, worked to prove the fairy tale wrong. In 1668, he published an experiment that was both elegant and simple. He placed a dead snake, fish, eels, and milk-fed veal into containers and sealed them, then placed meat in other containers but left them open. Maggots only appeared on the decaying meat in the open flasks. When challenged over the lack of airflow as the reason for the results, Redi covered the flasks in fine Naples veil (or gauze), allowing air inside them. Still, no maggots appeared on the covered flasks. His experiment gave proof to an idea attributed to an English predecessor, physician William Harvey, who said: "*Omne vivum ex vivo*," Latin for "all life comes from life."

◆ ◆ ◆

More theories emerged. In the sixth century BCE, ancient Jain scriptures from India mention *nigodas,* or sub-microscopic creatures found throughout the world, including in plants and animals. In ancient Rome, scholar and writer Marcus Terentius Varro warned against swamps, because of "certain minute creatures which cannot be seen by the eyes, which float in the air and enter the body . . . and there cause serious disease." These creatures were referred to as *animaculae,* from the Latin "tiny animals." This was an impressive scientific deduction for the first century BCE, but bear in mind Varro also believed that cutting hair during a waxing moon made you bald. The Persian polymath Avicenna also speculated on the possibilities of microorganisms. But his ideas, like Varro's, were ignored in favor of the more popular miasma theory.

As time went on, scientists got closer and closer to the truth. In 1546, Italian physician Girolamo Fracastoro theorized that "seedlike" creatures could transfer infection from one body to the next, either by touch or through the air. In the second century, Galen, a Greek physician and philosopher whose views of humoral medicine dominated Western medical thought for over a thousand years, speculated about something he called the "seeds of fever." But the most famous discovery, the one that would change the trajectory of our understanding of germs, wouldn't happen until 1767.

It was made by the Dutch scientist Antonie van Leeuwenhoek, a contemporary of renowned painter Johannes Vermeer in Holland. Born to a basket maker and the daughter of a brewer, van Leeuwenhoek was a cloth merchant. In his forties, he started creating and grinding lenses that would eventually magnify objects up to three hundred times—ten times stronger than those traditionally found in microscopes at the time. Under van Leeuwenhoek's lenses, theory became truth.

A replica Leeuwenhoek microscope

He spied pond algae, blood cells, sperm cells, protists that looked like

What's in the Name

Bacteria, Virus, and Parasite

The word *bacteria* doesn't sound much like other words we know. The origin of bacteria is Greek, from the word *baktron*, meaning rod, staff, or cudgel, as the first observed bacteria were rod-shaped. The diminutive form is *bakterion*, or "little rod or staff." It first appeared in 1838, coined by German naturalist Christian Gottfried Ehrenberg.

Virus developed as a late Middle English word from the Latin *virus*, meaning slimy, malodorous, foul, or poisonous liquid, or poison sap. Not until the early 1700s was it used to describe an infectious agent that caused venereal disease. (In 1972, the same year that birthed HBO, Atari, and digital watches, the concept of the computer virus also came to life.)

Parasite comes from the Greek words *para*, or beside, and *sitos*, or grain, food, or bread. Put together, you get something that eats at the table of someone else, sometimes at their expense. It's a marvelous sixteenth-century word that also means a hanger-on and obsequious flatterer, or toady. The latter is a contraction of toadeater, the charlatan's assistant who apparently ate poisonous toads and yet miraculously remained alive due to the same charlatan's snake oil, which was naturally for sale nearby.

hairy bells, and from the plaque between his teeth, "little living animalcules, very prettily a-moving."

In short, van Leeuwenhoek found a new universe.

He wasn't the only one. Less powerful microscopes led to similar, albeit less diverse findings: English scientist, architect, and polymath Robert Hooke saw microscopic mold in 1665, and German polymath Athanasius Kircher had already seen "an innumerable multitude of worms" in vinegar and milk, which led him to suggest the plague was caused by microbes. (Kircher also wrote books on Egyptology and China, studied volcanoes, invented megaphones, and created magic lanterns—in addition to thinking armadillos were hybrids of turtles and porcupines.)

The tiny world torn open by powerful microscopes had an outsized impact on theories of disease. By the 1700s, physicians and scientists theorized that diseases such as smallpox were caused by microorganisms sometimes called "worms" or "poisonous insects" invisible to the naked eye.

In the unmagnified world, where people were vomiting, getting diarrhea too often, or suffering from rashes, swollen lymph nodes, and fevers that made them delirious, scientists were narrowing their vision by testing theories of infection in very specific ways.

A Hungarian obstetrician named Ignaz Semmelweis made a landmark finding in a maternity hospital in Vienna. As home births became less fashionable and so-called "lying-in" hospitals became the norm for childbirth in the Western world, puerperal fever (known as childbed fever) became rampant. Then as now, a woman would normally give birth attended by a physician or midwife. If the birth was difficult, forceps and manipulation of the baby might occur to help the mother. Days after the birth, some new mothers would suffer agonizing lower abdominal pains and fevers. The normally occurring postpartum discharge, or lochia, would become fetid. Shaking chills, or rigors, could get severe enough to chip teeth in chattering jaws. Nausea and vomiting would ensue, and often death.

Gram stain of Streptococcus pyogenes, *which can cause rheumatic and scarlet fever*

The cause? Septicemia, from bacteria introduced into the woman's uterus by dirty hands delivering the baby. But at the time, anything from miasmas to a poor emotional disposition of the mother were blamed. However, Semmelweis noticed that the mortality rate in the wards attended by midwives was 2 percent, far lower than the 16 percent of those attended by medical students. Those same medical students frequented autopsies between their patient care, where, notably, a pathologist died from sepsis after receiving a cut during one.

The connection was made. Midwives and their students washed their hands—the medical students did not. Semmelweis then had all the medical students wash their hands in chlorinated lime solution, and the mortality rate plummeted to 2 percent. Similar findings were found in Boston by Oliver Wendell Holmes Sr., the famous poet/polymath who hung out with Ralph Waldo Emerson. Unfortunately, Holmes's published findings were ridiculed. This was a time when surgeons' operating frocks were stiff with blood and pus, a sign of a surgeon so good he performed case after case without washing in between.

An American obstetrician, Charles Meigs, was furious at Holmes's assertions. "Doctors are gentlemen, and gentlemen's hands are clean," he said.

Only they were not. Around the same time, English physician John Snow was theorizing that miasma was a joke of an idea, writing in an 1849 essay called "On the Mode of Communication of Cholera" that the disease was contracted from "the excretions of the sick [diarrhea] at once suggest themselves as containing some material which being accidentally swallowed," and caused the illness. In other words, he was describing the fecal–oral route of contagion, which sounds only slightly better than: People poop dangerous germs and then other people inadvertently eat the pooped-out germs and get sick, then

French microbiologist Louis Pasteur

poop and infect more people. Snow even recommended water be boiled before drinking. This was during a time when people knowingly drank from water that was mixed with garbage and excrement. Boiling water was cutting edge, hence no one did it (see Cholera, page 282).

By 1860, another landmark experiment had occurred on Louis Pasteur's table. Pasteur, who pioneered pasteurization, the process of cooking foods and liquids to a point where spoiling or dangerous bacteria are killed, had done research to show that childbed fever was indeed due to bacteria. He collected pinpricks of blood from feverish postpartum women, then cultured the blood in warm chicken bouillon. Under a microscope, he found chains of cells growing in the broth that resembled "little tangled packets" that he described as resembling a string of pearls. He had likely discovered *Streptococcus pyogenes*. (*Streptococcus* was Greek for "twist or chain of berries," and *pyogenes* meant "pus-maker.")

But that wasn't Pasteur's defining experiment. What he is best known for happened when he took a glass flask of broth, boiled it to sterilize it, then heated and pulled the glass neck into an S-shape. This way, when air entered the neck, any particulates from the air would get stuck in the S-bend, and not in the flask of soup. When he did this, the broth remained clear. But when he left the flask open,

Are Viruses Really Alive?

The search for viruses was a bigger challenge than identifying much larger bacteria and parasites that could be seen under microscopes. The viruses behind illnesses such as the flu, MERS, COVID-19, chicken pox, herpes, polio, and HIV aren't like bacteria in significant ways. Primarily, they aren't cells, but they have genomic material like DNA and RNA. They can be killed with certain disinfectants.

Tobacco leaf infected with tobacco mosaic virus

In 1886, German scientist Adolf Mayer noticed a disease that caused a mottling of tobacco plant leaves and spread from plant to plant like a bacterial infection. A few years later, Dmitri Ivanovsky, a Russian botanist, passed tobacco leaf juices from infected plants through a Chamberland filter, which blocks bacteria-size particles, and discovered that the filtered liquid still infected plants. So whatever was infecting the tobacco leaves was even tinier. (A tobacco mosaic virus is about one-tenth the length of an *E. coli* bacterium, which is about one-third the length of a human red blood cell. Some viruses like polio are even smaller, a hundredth the size of a bacterium.)

In 1935, Wendell M. Stanley, an American biochemist and virologist who would go on to win a Nobel Prize in chemistry, crystallized the infective agent in the tobacco mosaic disease, and after dissolving it back to liquid, the contents were still able to infect plants. The finding was astounding. Here was a thing that infected living organisms, could be stored as a crystal on a shelf, then be brought back to life and once again be infectious. By itself it wouldn't grow. Rather it was a machine-like package of genomic material (DNA or RNA) enclosed in proteins. However, it needed cells to infect, to "brainwash" by reprogramming them to make more viruses before moving on, sometimes killing them in the process. (Other viruses never quite leave their hijacked cell.) It's estimated that 100,000 pieces of our genes come from viral DNA that never left.

What Stanley found was a tiny organism that reproduced, but not in a way people thought organisms replicated at the time. Could it die? Yes. If left out in the wrong environment, or exposed to certain chemicals and disinfectants, it falls to pieces and dies.

Did it eat? Did it have a metabolism? Did it breathe? No, not by itself. Some have used the metaphor of a seed—full of potential, but not autonomous until the conditions are right. And yet its virological goal is not just to exist, but to make more of itself. Is that not a reproductive lust for life, after all?

the broth grew cloudy and full of bacteria. When he made the S-bend again, but then tipped the flask so that liquid entered the S-bend, the broth got contaminated by the trapped particles and grew cloudy. The conclusion? Spontaneous generation as a theory was thoroughly debunked.

◆ ◆ ◆

By the late 1800s, the miasma theory was shunted aside for the budding science of bacteriology. German physician and microbiologist Robert Koch put the miasma theory to rest for good with his work on anthrax. Koch would take *Bacillus anthracis* from an infected sheep, cultivate it in a growth medium (like a broth), isolate the bacteria, then infect a mouse with it, which would then show signs of anthrax infection. He repeated this on twenty generations of mice (not lazy, Koch was). He then introduced his now legendary four postulates regarding how disease can be caused by a specific organism (see Plague, page 50). And being very not lazy, he created stains so bacteria could be seen better under the microscope, which led to the discovery that bacteria caused tuberculosis and cholera, leading in turn to the creation of new mediums for growing bacteria. (Koch also helped his assistant Julius Richard Petri create the Petri dish.)

Today, the fields of bacteriology, virology, and parasitology make impactful discoveries seemingly by the day. Molecular biology has laid bare many of the inner workings of these beastly germs in ways that Koch, Pasteur, and van Leeuwenhoek never imagined.

THE PLAGUE
Patient ZERO
Wong Chut King

Cause: *Yersinia pestis* (bacteria)

Symptoms: In the bubonic form, symptoms include swollen lymph nodes, fever, gangrene, and death

Where: San Francisco, United States

When: 1900

Transmission: Via infected flea bites or handling plague-infested animals and via infectious droplets coughed by an infected person or animal

Little-Known Fact: The plague is still around and prevalent in parts of the world, including Africa. An average of seven new plague cases occur annually in the US.

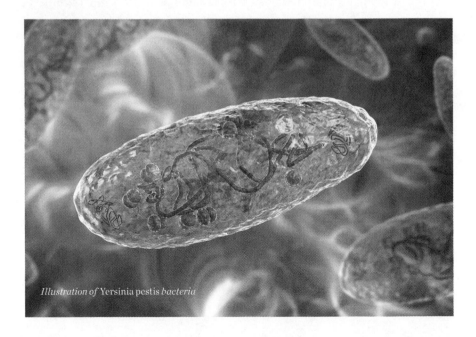

Illustration of Yersinia pestis *bacteria*

The Black Death, the Great Plague, the Pestilence, the bubonic plague.
The word *plague* tends to conjure visions of colossal pandemics that left mountains of dead in their wake, as well as ancient public health disasters from when the world was dangerous and filthy. That was largely true during the medieval period, when the plague killed tens of millions of people, with some estimates of up to half the European population. Then there were the realities of life during plague times, with bodies blackened by the disease and covered with abscesses flowing with disease-ridden humors. The era was also known by the bizarre, beaky masks worn by those treating plague victims that became something of a creepy specter of death itself. The expression that survives from medieval times—"Avoid it like the plague"—still holds heavy meaning and intention when used today.

The Black Death in Europe between 1347 and 1351 was one of the most infamous infectious disease outbreaks in recorded history, and it left its putrid mark on the entire known world. At its height during the fourteenth-century plague pandemic, approximately 30 to 60 percent of Europe's population was decimated; it took nearly two hundred years for that population to return to pre-plague levels. And it wasn't even the first plague pandemic. That distinction goes to the Plague of Justinian, which hit the Byzantine Empire in 541, killing a staggering quarter of the world's population.

Since the plague is such an ancient-feeling disease, seemingly more of a biblical horror story than a reality, you'd probably think it had been banished to the dark and distant past. But you'd be wrong. According to the Centers for Disease Control (CDC), the plague is still alive and making people unwell in the world, popping up now and then mostly in rural Africa. But also in the US: The CDC estimates the US gets seven

A protective hood worn by those treating plague patients.

new cases a year. In 2020, there were plague cases in California, Colorado, Arizona, and New Mexico. The bad old plague is very much still with us.

◆ ◆ ◆

Caused by the bacterium *Yersinia pestis*, the plague is a zoonotic disease, meaning it jumps from animals to humans. Rodents, such as the brown rat and the marmot, are reservoirs for the bacteria, meaning they are the animals in which the plague lives naturally and survives over time, even without any human infections occurring. In recent cases, people have gotten infected with the plague after consuming marmot meat in Mongolia. But the way humans usually get the plague is via bites from an intermediary, or vector—something that delivers an infectious pathogen into a living organism. In this case, the flea.

Plague-infested fleas are usually perfectly happy sticking with rats as a source of their blood meal. However, once that rat dies of the plague, that flea will abandon ship, so to speak, and if a human is nearby, that's a tasty enough substitution. When the *Yersinia pestis* bacterium is inside a flea, it has a brilliant trick: It blocks the upper digestive system of the flea. The flea starves, biting more frequently in desperation before it vomits replicating plague bacteria and blood into its host while feeding, thereby infecting a new host—the human.

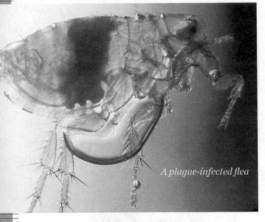
A plague-infected flea

Once inside a fresh host, the plague bacteria spreads inside the human lymphatic system, taking hold in lymph nodes and multiplying there until the tissue swells and becomes painful. These swellings, called buboes, from the Greek word for "groin," are also where the *bubonic* part of bubonic plague arises. The buboes can grow as large as an apple and usually occur in the armpits, groin, and neck. Occasionally, they erode right through the skin and ooze with syrupy pus, a yellowy mixture of live and dead bacteria as well as the remnants of infection-fighting cells.

When the plague spreads to the blood, there is a cascade effect of clotting and inflamed blood vessels, causing the tips of the nose, fingers, and feet to blacken with gangrene and die off—hence the moniker Black Death. Malaise then overtakes the patient, along with high fevers and chills, as the bacteria and toxins invade the rest of the body. Coma and death follow.

This is the classic bubonic plague. But the plague can manifest in two other ways: septicemic, which spreads via blood to bring fevers, shock, and bleeding to the skin and limbs, eventually causing tissues to turn black and die; and pneumonic, in which the lungs are infected. In the latter, the disease is spread from person to person via droplets and aerosols, the technical terms for what we spew out while coughing. In every plague pandemic, all three types of

Plague-related gangrene

the disease have been present, including the one that finally and inevitably hit the United States more than a century ago.

♦ ♦ ♦

The landing of *Yersinia pestis* on American shores can be traced back to 1900. At the turn of the last century, the plague was doing its level best to get a foothold on American soil, traveling mostly by boat. In November 1899, fleas from plague-ridden rats infected passengers and crew on ships docking in New York City. By January of 1900, the plague had also arrived in Port Townsend, at the mouth of the Puget Sound in Washington state. But in both cases, the infection wasn't able to make it onto the mainland. However, farther south in California the plague literally jumped ship and landed squarely in the city of San Francisco.

From an infectious-diseases standpoint, the turn of the last century was a tumultuous but exciting time. People still believed in miasma, in tainted air that could sicken people, or in ptomaine, a nonexistent microscopic disease-spreading boogeyman related to putrefying animal or vegetable matter once believed to spread illness. Antibiotics had yet to become the array of choices they are today. And yet microbiologists at the turn of the twentieth century were making groundbreaking discoveries about bacterial sources of illness, such as anthrax, tuberculosis, and cholera. They understood that containing contagious sources via isolation or quarantine could conceivably stop epidemics from becoming pandemics. Simple enough. But in a city that was loath to shut down any part of its industry for a disease that was dubbed "A Phantom" by the *San Francisco Chronicle*, what happened was far uglier and more complicated.

In 1900, San Francisco was a thriving port town on the West Coast, the "Paris of the Pacific," its denizens' pockets often lined with mining wealth or money made from ironworks and shipbuilding. The city's motto, "Gold in peace, iron in war," reflected its ambitious, take-no-prisoners approach to keep its economy surging. But the city's breakneck growth and bustling industries came with downsides: stinking sewage issues, polluted bay water . . . and rats. Not only did rats live in the city, but some would say they also led a life of cosmopolitan adventure, hopping on and off ships that crossed oceans in all directions.

At that time in 1900, the third plague pandemic, known as the Modern Plague, was surprisingly still in full force. It first erupted in China's Yunnan province in 1855, going on to kill millions of people and impacting every inhabited continent on Earth. In 1899, plague victims were found on the *Nippon Maru*, a Japanese shipping vessel that landed in San Francisco. Yet another Japanese vessel, as well as a British ship arriving from Brazil, had *Yersinia pestis* stowing away on board via sickened or dead crewmembers. Maritime quarantines kept them from entering San Francisco. The US was safe—or so it thought.

The chief quarantine officer of the Marine Hospital Service, Joseph J. Kinyoun, did his best to stem the tide. He ordered all ships entering San Francisco's port from the plague-infected lands of China, Hawai'i, Japan, and Australia to fly a yellow flag of warning. But the yellow flags were hotly contested by local officials, who believed they were bad for business, thus bad for San Francisco. The city's newspapers touted San Francisco as "plague proof." So the flags were ordered to be taken down.

That January, the steamship SS *Australia* arrived in port bearing goods from Honolulu and was found to be clear of the plague in crew and passengers. (Honolulu had been dealing with its own plague infections and related catastrophes, having accidentally burned down its entire Chinatown in an effort to control the spread.) The SS *Australia*'s cargo was unloaded near the sewer egress of San Francisco's Chinatown. How the infected rat came ashore will never be known, but what was seen soon after the SS *Australia* arrived in port was an increase in the number of dead rats in the area, as well as the sickening of one particular man.

◆ ◆ ◆

A resident of Chinatown for sixteen years, Wong Chut King was a forty-one-year-old bachelor who owned a lumber yard and lived in the basement of a reportedly run-down tenement. With the abundance of brothels and sex-trafficked Chinese women in the area, when Wong grew ill in early February, the malady was assumed to be a sexually transmitted infection such as gonorrhea, or perhaps typhus. A local healer thought it was a bladder problem. But Wong grew steadily worse, with fevers, vomiting, and profuse diarrhea. When he fell into a coma, he was brought to a shop that sold coffins as a way to keep him from healthy people.

A month after he first grew ill, Wong Chut King was dead.

On March 6, an autopsy was performed on Wong. The city's bacteriologist, Wilfred Kellogg, took samples from Wong's swollen lymph nodes before his body was wrapped in fabric doused with bichloride of mercury, placed in a lead coffin sprinkled with lime, and sent off to be cremated. Peering through the microscope, Kellogg believed he saw small, squat, sausage-shaped bacteria. Under the right stain, they would appear to look like plump safety pins. Without staining, most bacteria are transparent under a microscope. You can see them, but it's not easy. (Staining allows certain chemical dyes to bind to different parts of the bacteria, making them visually stand out under a microscope.) In the case of plague bacteria, special dyes resulted in a characteristic

Street scene, Chinatown, San Francisco, mid- to late nineteenth century

purplish "bipolar staining," meaning both ends of the sausage shape took up dye—which was why they resembled safety pins.

The safety pins meant it could be plague bacilli. Which meant the next step was quarantine.

Only this wouldn't be the usual quarantine of a ship or a single house. On March 7, in a display of Sinophobia, San Francisco city officials quarantined the entire fifteen-square-block area of Chinatown, sectioning it off with ropes, and within them the roughly thirteen thousand Chinese residents living there as

Whipping the Plague

The sheer horror of the plague in Europe forced the deeply Christian society to grapple with the unanswerable "why" of it all. As Giovanni Boccaccio, the Italian humanist, wrote in *The Decameron*, the plague seemed to have been "sent down upon mankind for our correction by the just wrath of God." Others looked to fanatical interpretations of Christian scripture in an effort to appease God through brutal practices of atonement for the sins of mankind.

A sect of religious zealots called the Brotherhood of the Cross sprang up in pockets around Europe. Members of the group would march through villages, towns, and cities, stripped to the waist, flagellating themselves with knotted scourges while chanting prayers of repentance. They believed these rites of self-harm would atone for the sins of the world and perhaps inspire God's mercy to withdraw the plague.

Although officially condemned by the Catholic Church, the sect was briefly—and enormously—popular while the plague ravaged the countryside. Some groups of flagellants numbered in the thousands and marched through Europe together, literally and figuratively whipping themselves into a spiritual frenzy. The largest recorded procession reportedly had more than 10,000 participants. An eyewitness described the arrival of a Flagellant sect in London in 1349, as the country remained firmly in the plague's grip.

> Thrice they would all cast themselves on the ground in this sort of procession, stretching out their hands like the arms of a cross. The singing would go on and the one who was in the rear of those thus prostrate acting first, each of them in turn would step over the others and give one stroke with his scourge to the man lying under him.

It is said that every night they performed the same penance.

While the movement was initially egalitarian in approach, with clergy, laypeople, men, women, and even children taking part, as it grew in size, opportunistic leaders layered the Brotherhood with a variety of cultish sensibilities and regulations. It transitioned from a brutal, if pitiable, attempt to inspire God's mercy into something altogether more disturbing. The Brotherhood became exclusively for men; they went to excessive lengths to avoid even accidental contact with women. Potential candidates to the Brotherhood had to confess all sins since the age of seven and flagellate themselves for thirty-three and a half consecutive days, one for each year of Christ's life on Earth. They also vowed to not bathe during that time, nor shave, sleep in a bed, or change their clothes. On top of all this, they were required to pay an entrance fee. That this cultlike business model was successful demonstrates the spiritual turmoil in Europe during the plague.

With this in mind, perhaps it's not so hard to understand the appeal of the Flagellants: For a brief period, their public displays of penance may have brought some spiritual relief to those who watched. Desperate and grief-stricken from losing so many friends and family to a horrific, entirely mysterious disease, mass public self-flagellation as repentance offered an opportunity for catharsis. Observers would sometimes scream, cry, or pray alongside the Flagellants to purge their own spiritual anguish.

As the Flagellant movement grew in number and power, its leaders began to make heretical claims, drawing the ire of the Catholic Church. The Flagellants doubted the sacraments and claimed to have the power to absolve each other of sins. They also granted themselves the ability to conduct exorcisms and to work miracles. (All of these claims were considered heretical.)

Pope Clement VI officially condemned the movement in 1349, prohibiting any further processions. He sent condemnation letters to the bishops of France, Germany, Poland, Sweden, and England. Denounced by clergy across Europe, the movement fell out of favor almost as quickly as it rose to prominence. By 1350, just a year later, the Brotherhood of the Cross was nearly defunct.

An illustration of Flagellants from the Nuremberg Chronicle, *published in the late fifteenth century.*

well—not a single one of whom was warned in advance. Beyond Wong, no further evidence of the plague had been confirmed.

White people were ordered out of the quarantined area because it was assumed—without evidence—that only Chinese people were responsible for and susceptible to the infection. All this for a single case of an unconfirmed disease. It was not simply an unprecedented and ignorant use of authority, but an ultimately useless one. After all, ropes aren't particularly effective at stopping rats. And worse, scientists had known since 1897 that fleas and rats could spread the plague, but leaders in San Francisco nevertheless chose to believe that poor living conditions in Chinatown was reason enough to isolate its entire population after a single case of plague.

Chinatown residents cooking their meal under quarantine during the Bubonic plague epidemic, San Francisco, 1900–1904.

The anti-Chinese intolerance that led to the decision to quarantine all of Chinatown had deep roots in California at the time. The Chinese had been referred to as "moral lepers" by San Francisco's chief health officer in 1870, contributing to the idea that Chinatown was a breeding ground for disease and contagion. But for the Chinese living there, it was near impossible to settle anywhere else, even with the rents set high by landlords (including Mayor James Phelan). And once Chinese renters took over a place, the maintenance of the building was thrust upon them. Given the steep rents and pricey upkeep, tenants had to live in congested spaces to share rental costs. The sewers that serviced the area were also in deplorable condition. When all these factors were mixed together in public discourse, the Chinese population was considered little more than a scourge themselves. Mayor Phelan referred to Chinese immigrants as "sullen . . . slaves to the opium habit . . . blackmailers and assassins" and that they "breed disease." With such open hatred and fear, it was no surprise the quarantine happened as it did.

Meanwhile, the scientists went to work. At the time, the best way to confirm if someone or something had the plague was to take a bacterial sample from an infected bubo, inject it into a healthy animal, wait to see if the animal sickened or died, then test it for plague bacilli. These steps were drawn from the postulates of Koch, who would later go on to win the Nobel Prize and in whose laboratory the city's chief quarantine officer, Kinyoun, had studied. Kellogg, the city's bacteriologist, extracted fluid from Wong's buboes and injected the samples into a lab rat, two guinea pigs, and a monkey in Kinyoun's lab. Then they waited.

As they did, the inhabitants of Chinatown grew furious that their neighborhood was cut off from the rest of San Francisco without their input, putting huge numbers of residents out of work. (An unintended consequence was that wealthier San Franciscans were reportedly miffed when their maids, launderers, and cooks were suddenly unavailable.)

In an effort to clean Chinatown, sulfur-infused fumigants clouded the streets. Chlorinated lime, a bleaching agent, was sprayed on the ground, gutters, cellars, and buildings, leaving behind telltale white smears and a harsh odor. Kinyoun was ridiculed, the newspapers characterizing the entire effort as a farce. One newspaper, the *San Francisco Call*, described the situation as a "Plague Fake," and the *San Francisco Bulletin* published this poem:

> *Have you heard of the deadly bacillus,*
> *Scourge of a populous land,*
> *Bacillus that threatens to kill us,*
> *When found in a Chinaman's gland?*
> . . . .
>
> *Well the monkey is living and thriving,*
> *The guinea pigs seem to be well,*
> *And the Health Board is vainly contriving*
> *Excuses for having raised the deuce.*

By March 9, only two days after the animals had been injected, none were sick or dead. In the meantime, no one else had succumbed to the plague. The quarantine was lifted on March 10 and the ropes were removed as San Franciscans breathed a sigh of relief and resumed business as usual.

The relief would last a day.

On March 11, one of Kinyoun's rats and two of his guinea pigs were dead from the plague. On March 13, the monkey succumbed as well.

What followed was a chaotic swirl of denial, racism, fear, and the incessant march of infection upon the inhabitants of the city. California governor Henry Gage refuted the idea that his state had been infected by the plague. He refused to believe the scientific evidence, going so far as to accuse Kinyoun of planting plague bacilli in cadavers to falsify the outbreak. Mayor Phelan later sent out hasty telegrams to the mayors of fifty cities stating that there was "no further danger" in San Francisco. The newspapers backed him up. Even when a federal commission of investigators confirmed the plague's existence in the city, the ongoing battle between safety and public health continued to be laced with overt xenophobia.

As the days passed in March, three more Chinatown residents died of plague. The city's health board tried to squash news of the outbreak. Home inspections were performed, high-risk houses sanitized, and sewers disinfected with bichloride of mercury. Rats were killed when found. Foul-smelling sulfur fumigations continued. Certificates of inoculation or proof of health were demanded if Chinese residents tried to leave the city. At ferries and railroad stations to

A fire set in Honolulu's Chinatown in 1900 was meant to burn the house of a suspected plague victim, but consumed most of the neighborhood.

and from San Francisco, Chinese people—and only Chinese people— were refused passage and turned back. Residents of Chinatown were loath to report any illnesses or give up their dead for fear of autopsy and cremation. In one case, a dead resident was allegedly propped up at a card game to evade detection by health officials.

In May, a plague vaccine was ordered to be given to Chinatown residents. If they refused, stronger measures were threatened according to notices posted in the area. Doctors and health workers gathered the vaccine doses and syringes, ready to inoculate as many Chinese Americans as possible. They would be using the Haffkine plague vaccine, which consisted of heat-killed bacteria in broth that theoretically included plague toxins.

Developed by Waldemar Haffkine, a bacteriologist who left Russia to work at the Institut Pasteur, the vaccine was first used in a Bombay prison. The side effects of being "Haffkinized" were terrible—headaches and high fevers, plus pain and severe swelling at the injection site. Recipients were usually ill for days. Despite anecdotal reports of its efficacy, the vaccine was experimental—there was no firm evidence it worked. So there was reason for Chinese American residents of San Francisco to be wary, and few of them allowed themselves to be vaccinated.

Mayor Phelan, in a continuing pattern of inflammatory statements, felt that the white residents should be kept away from the "filthy" Chinese Americans, who were a "constant menace to public health" and "bred disease." Phelan would later run for Senate touting the not-so-subtle slogan "Keep California White." Phelan also advocated renewing the Chinese Exclusion Act of 1882, which would become permanent in 1902, only to be later dismantled in pieces. It wasn't completely undone until 1965.

As the 1900 plague outbreak expanded in San Francisco, Chinese American residents began fighting back—and for good reason. Though the plague was indeed in Chinatown, it wasn't nearly as widespread as believed. The number of cases was low, their legitimacy debated. Residents successfully battled through legal and diplomatic means to stop the mass inoculation program and the forced quarantine in *Wong Wai v. Williamson*. However, the San Francisco Board of Health began planning to remove residents of Chinese descent en masse to another location, and then raze Chinatown. A second arbitrary quarantine in June was fought in federal court, the Chinese American plaintiffs again winning in *Jew Ho v. Williamson*, the judge ruling that the actions of health officials were carried out "with an evil eye and an unequal hand." All the while the scientists had to fight politicians who wanted desperately to pretend the plague wasn't happening, which led to a lack of funding to fight the disease.

By spring 1901, a more effective campaign was launched to clean out homes in Chinatown, which, combined with the eradication of rats, made a dent in the epidemic. In 1904, after 119 plague deaths mostly confined to Chinatown, another outbreak occurred in California, killing seventy-eight mostly white victims this time. Soon after, the epidemic would finally be over. But those dark times and callous expressions of xenophobia and racism wouldn't be forgotten, nor would the epidemic's first victim on American soil, Wong Chut King.

Ring-a-Round the Rosie

Ring-a-round the rosie,
Pocket full of posies,
Ashes, ashes,
We all fall down!

In the United States, many of us remember singing this rhyme as children. We'd clasp hands, dancing in a circle around the "rosie," then fall down in a fit of giggles at the end. Only as we grew older did some of us learn the story behind this singsong poem.

The "rings" were the red marks of the plague on skin.

The "pocket full of posies" were flowers whose fragrance hid the stench of death.

"Ashes, ashes" signified the mass burning of plague victims. Alternately, "achoo" or sneezing sounds substituted the "ashes" line, which may have indicated that sneezing or coughing were the final symptoms of victims.

Finally, "we all fall down" referred to how the plague killed so many, how there was no escape, even for children.

Horrible, right? Except none of it is true.

The poem was indeed a singsong childhood rhyme, and not just in North America. There are Maori, Indian, British, German, Italian, and Serbian and Croatian versions. Many of them date back to the nineteenth century. But it wasn't until after World War II that the plague popped up as an explanation behind the seemingly sweet rhyme. There is no mention of the plague in any discussion of the rhyme in the nineteenth century. Further, the "rings," sneezing, or coughing aren't even hallmarks of the bubonic plague.

So what does it really mean? "Ring a Ring o' Roses," as it's known in England, was likely born there. A "rosie" may come from *rosier*, or "rose tree" in French. There are theories that the song is pagan in origin, with the sneezing being a superstitious action. The falling down could refer to curtsying.

But as with many songs and stories handed down via oral traditions, we'll likely never know. In the meantime, children around the world can continue ringing around the rosie without the macabre implications.

AUTOPSY
From Humoral Theory to Grave Robbing

Impact: Enormous, laying the essential groundwork for understanding the causes and effects of diseases within the body

---

When: As early as about 300 BCE, but not widely performed until the sixteenth century

---

Who: The Greek teacher Herophilus (335–280 BCE) wrote the first known treatise on autopsy

---

What Happened Next: Intermittent autopsies on executed criminals and vivisections

---

Little-Known Fact: Michelangelo's autopsy work and personal history of kidney stones led him to paint God inside a huge kidney-shaped cloak on the ceiling of the Sistine Chapel.

A strong man, sixty years of age, was bit in the wrist by a rabid dog. . . .

Three months had elapsed after the bite ... he began to be agitated with unusual and surprising timidity, trembling with fear at every little noise, regarding every stranger as a betrayer . . .

Soon after these symptoms arose he was seized with a dread of light and water, and was brought into the hospital, where he lived two days. His attempt to swallow water occasioned contortions of the body . . .

His incredible fear—his aversion to water—and the difficulty and uneasiness he felt on beginning to swallow it, continued to the time of his death.

His body was opened on the 21st of May 1727.

So begins a case written by Giovanni Battista Morgagni in his famed 1761 work *The Seats and Causes of Diseases Investigated by Anatomy*. Morgagni is considered a founding father of modern pathology, the study of the causes and effects of diseases.

Pathology is how we know that syphilis can tear open your aorta or cancer can chew away at bone. But one part of pathology often overlooked nowadays has historically been one of the most influential methods of understanding living and dying, and the sicknesses that connected them.

The autopsy.

A dead body carries with it a unique story. Not the kind that says whether they went to college or played baseball or voted for a detested politician. It's a deeper story that goes to the sinews of your being, your genetic history and lifestyle choices. The inevitability of your fate. The story, in some ways, was written at conception. Only with autopsy, it's up to someone else to unravel the plot twists.

Autopsies have been happening for millennia in pursuit of answers to how we lived and why we died. They can be windows into why certain bodies succumb to certain infections and diseases, but not others. They seek the answer to why an orchestra of an organism—vessels humming with blood, neurons firing just so, gases sliding in and out of our tissues— changes its synchronous tune to one of

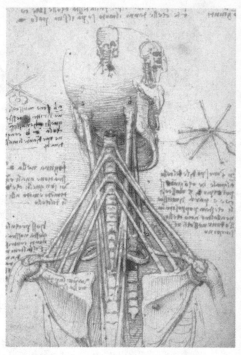

A neck anatomy drawing by Leonardo da Vinci, ca. 1515

chaotic destruction. And ultimately, autopsies are done so we can learn how not to succumb to the same ailments ourselves.

The path to looking inward began not with humans, but animals. Dating back to 3500 BCE in Babylon, the reading of animal entrails was considered a way to see the future. Called haruspicy, this type of divination involved examining entrails of slain animals, in particular the liver (considered the seat of the soul), which was thought to contain omens and messages from the divine. For example, the ancient Babylonians believed that if the left-sided vein in the liver was defective, it meant that your enemies would suffer defeat.

The side effect of such beliefs was learning what was and wasn't normal anatomy in animals, a foundational approach in future anatomical study on humans. In Deuteronomy, Jews were following the dictate that "Thou shalt not eat of anything that dyeth of itself." This meant that, in the sacrifice and butchering of animals, attention was to be paid to the minute details of the organs and overall healthiness of an animal's innards, so as to avoid disease arising from them.

Other theories and beliefs rose above mere mortality. In eighth century BCE, Homer's *The Iliad* describes a pestilence that took down the Greeks, supposedly as an answer to a prayer made to Apollo. And yet a certain amount of anatomical understanding must have existed. Bodily organs had names, though their function was still a wonder. Warriors knew keenly of how to deliver wounds that would kill. The spear of a hunter, a cook's knife, the sacrificial blade—all could discern the different inner organs of animals, and surely its holder could see similarities in humans. But no dissections or autopsies were recorded from that time.

In the fourth and fifth centuries BCE, the Greek physician Hippocrates so entranced the world with humoral theory that it would become a medical mainstay for more than a millennium. According to the basic theory, blood, yellow bile, black bile, and phlegm, as well as cold, wet, dry, and hot elements could lead to illness and death if they fell out of balance. And yet, looking *inside* the body remained less important. All the information a practitioner gleaned was through what came out (vomit, urine, feces, and all things that swelled up or were coughed, bled, or sweated out). It would take nearly another millennium before pathology became a driving force—one that showed the cause and effect of disease upon the body, pushing humoral theory to the sidelines.

By the first century BCE, ancient Egyptians had perfected the technique of mummification through the removal of brains and organs before preserving the bodies with natron, a drying agent, and then storing them in canopic jars. Those who prepared the dead certainly had vast opportunities to see normal and abnormal anatomy after death. But no doubt being members of a lower class, what enlightenment they had would likely not have seeped upward to those who could pass it on through writing and more intensive medical study. Information in the medical-rich Ebers Papyrus (1550 BCE) and Edwin Smith Papyrus (ca. 1660 BCE), a pair of ancient writings, concentrated on herbal, magical remedies as well as treatment for traumatic injuries, again with no discussion of internal examination. Problems related to internal organs were still relegated to the realm of magic.

A replica of a canopic jar made in the image of Egyptian King Tut.

Autopsy

How It's Done

Autopsies today are systematically performed to determine the cause of death, beginning with the backstory of the deceased. There are two basic types of autopsies—hospital-based and forensic. The former elucidates the cause of death from a medical standpoint; the latter considers the cause of death from a possible criminal act.

Time of death, where and with whom the decedent was, and illnesses are all gleaned before the autopsy begins. Ideally, the autopsy will occur within twenty-four hours of death, and begin by taking body measurements and weight. Other identifying characteristics are noted, such as eye and hair color (natural or artificial), skin tone, and ethnicity. Depending on the type of autopsy, certain details are more important for the forensic cases, such as evidence of gunpowder, type of clothing, and stain samples, as well as hair and nail samples. UV light may be employed to look for body fluid residues. In hospital-based autopsies, care is taken to note things like incisions, IV sites, and scars. Sometimes X-rays are made if broken bones are suspected. If visual findings are remarkable, photos will also be taken.

Next, a block is placed beneath the torso, and usually a Y-shaped incision is made from the shoulders, meeting in the middle of the chest, then downward to the lower abdomen. Skin and soft tissue are teased away from the chest wall before using an oscillating bone saw to remove the rib cage. Then the internal organs are observed in situ, or where they sit. The organs are then removed, sometimes as a whole (called "en bloc"), after which each organ is dissected off one by one,

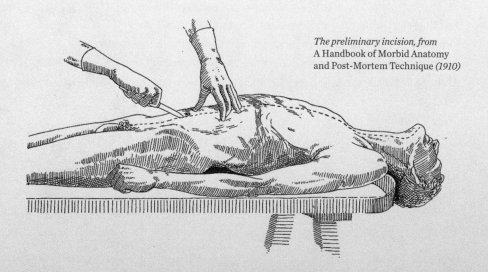

The preliminary incision, from
A Handbook of Morbid Anatomy
and Post-Mortem Technique *(1910)*

measured, and weighed. In hospital cases, solid organs like the liver are cut into sections for later microscopic evaluation. Hollow organs, like the stomach, are opened and fixed in formalin, then examined for abnormalities. Cultures can be collected if needed, and in some cases, tissue swabs are taken from the stomach, urine, blood, gallbladder, vagina, anus, cuts or bites, and even eyeballs (the potassium level in the vitreous humor can help estimate the time since death).

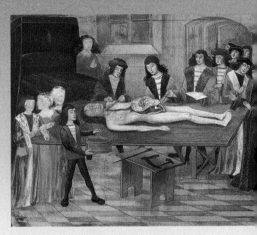

Dissection scene from Guy de Chauliac's Chirurgia Magna, *1363*

If the brain must be examined, a cut is made from ear to ear around the pointed back, of the head, the skin peeled forward and back, and then the bone of the skull sawn off with that same oscillating saw. The brain is always fixed in a preservative—like formalin—to be examined later, because it's too soft to work with (texture-wise, think Jell-O before fixative, firm avocado after). The limbs are usually not examined by cutting them open.

Once done, the diener (morgue worker, from the German word meaning "servant") finishes the work. The organs are returned (wrapped in plastic bags) or if not (as is usual in hospital autopsies, unless requested for religious or family reasons), the body is stuffed with cotton wool or other material. The skin is stitched closed with a typical "baseball" stitch. The scalp incision is quite well hidden and hair is preserved, which is helpful for certain funeral arrangements.

All told the process takes only two to four hours, but final results of microscopy and lab testing may take several weeks. After the autopsy is complete the body is released to the funeral home, where makeup and clothes are used to allow the deceased to be appropriately viewed in an open casket funeral. As for cost, it's not cheap: A privately ordered autopsy can run upward of $3,000 to $5,000. After all, the dead don't reveal their secrets to just anyone.

One final note: Pathologists and dieners must protect themselves with personal protective equipment, even as far as using chain mail to guard against fatal cuts from powerful bone-cutting saws. Pathologists and dieners have been known to contract lethal strep infections, tuberculosis, blastomycosis, HIV, hepatitis, and even rabies during autopsies.

Corpses may be dead, but it doesn't mean they can't still kill someone else.

But human dissections *did* happen before the Common Era. In fact, during the third century in Alexandria, Greece, the famous teacher Herophilus (335–280 BCE) wrote a treatise on anatomy after dissecting condemned criminals. Later, he was accused of being a butcher for dissecting men while they were still alive, called vivisection, as part of their punishment, though this is debatable. Herophilus discovered the prostate and the duodenum (Greek for "twelve fingers," which is the length of that portion of the intestine). However, Herophilus didn't really tackle *why* organs and bodies fell ill—he was still trying to figure out the basics of *how* bodies worked.

A contemporary of his, Erasistratus, took matters a step further and began to note changes that occurred with illness, and that could be seen in the body after death. For example, after a snake bite, the liver was soft and mushy, though when someone died from dropsy (bodily swelling, most often from congestive heart failure) the liver was rock hard. Though Herophilus and Erasistratus made great strides in anatomy, they did so by performing autopsies in addition to hundreds of vivisections. Though Erasistratus denounced humoral theory, he still didn't quite understand how the circulatory system worked.

Autopsies did not appear often in historical records for some time after Erasistratus, as Roman rule forbade human dissection by the second century BCE. But they did occasionally make it into the history books. In 44 BCE, Julius Caesar was autopsied after his assassination, which confirmed that his twenty-three stab wounds did, in fact, kill him. One in particular pierced his aorta (which made the other twenty-two stabbings a bit unnecessary).

Elsewhere, in ancient India, Sushruta (ca. 600 BCE) recommended human dissection, though it didn't appear to happen frequently. The first dissection in China was performed in 16 CE, and in Japan it happened in 456 CE and was related to the suicide of Princess Takukete (it was noted she had a stone in her abdominal cavity).

Galen (130–200 CE), a philosopher, physician, and surgeon to gladiators, was unable to dissect humans, as it was forbidden in the Roman Empire at the time. He performed dissections on Barbary apes and pigs instead, his understanding of humans coming from their living injuries, or from collecting skeleton remains he'd find after being washed from a grave during a flood. None of this prevented the larger force of his work. Galen took Hippocrates's, views of the humors and doubled down, such that his humoral theory of the body would prevail almost unquestioned for another millennium and then some.

An anatomy lesson depicted in Mondino de Luzzi's Anathomai corporis humani *(1316), one of the first books on dissection.*

The generally unfavorable view of anatomical study and autopsy lasted for quite some time. An early Christian author, Tertullian (160–230 CE), and Saint Augustine (354–430 CE) both decried dissection as a moral crime against humanity. Saint Augustine spoke of how anatomists "with their cruel zeal for science they have dissected the bodies of the dead, and sometimes of sick persons, who have died under their knives, and have inhumanely pried into the secrets of the human body to learn the nature of disease and its exact seat, and how it might be cured!"

And yet history is punctuated with important work that happened in spite of general mores against dissection, particularly in the realm of pandemics. In 543 CE, physicians performed dissections on plague victims in Byzantium. Arabic physician and poet Ibn Zuhr (also called Avenzoar, 1095–1152) performed illicit dissections (the Quran forbids mutilation of the dead) and discovered an infectious case of scabies, a parasite that burrows into the skin and causes intense itching. It was a "small" finding, but the discovery itself ran completely against the orthodoxy of humoral theory. A scabies mite, after all, doesn't parasitize you because you've got an overabundance of phlegmy humors.

By 1231, Holy Roman Emperor Frederick II allowed at least two executed criminals to be used for anatomical teaching purposes—a huge stride against the anti-dissection biases of the time. Seemingly in opposition to this, Pope Boniface VIII outlawed the dismemberment of corpses for the ease of transportation in 1299, specifically the boiling of bodies in order to remove flesh from bones. Technically, this wasn't anti-autopsy, and autopsy itself wasn't outlawed, though it was infrequent.

In 1300s China, autopsy was making headway. Song Ci, a physician, judge, and forensic scientist wrote *The Washing Away of Wrongs*, the first ever book on forensic science. In it, Ci explained how to conduct medicolegal evaluations, or finding out times and causes of death, the nature of poisonings, decomposition, and how to determine if a suicide was faked.

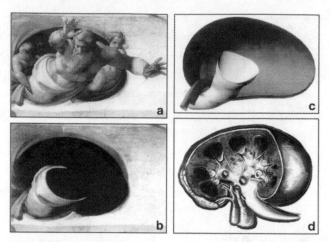

Michelangelo's depiction of God on the ceiling of the Sistine Chapel is based on his anatomic understanding of, and obsession with, the human kidney.

As the fourteenth century progressed, more autopsies happened and were even authorized by the Pope to investigate the victims of bubonic plague, particularly in Padua in 1363. But they were not frequent—perhaps a few per year. Nevertheless, investigations of the dead to understand infectious disease indeed occurred during a time most consider a backward era.

Dissections also touched the art world in Renaissance Italy. Leonardo da Vinci dissected a 100-year-old man in 1506. By 1513, he'd dissected thirty corpses, and his anatomic understanding is clear in his work. Michelangelo's experience in autopsies began with his participation in public dissections. He was so obsessed with the kidney (he suffered from kidney stones) that it ended up on the ceiling of the Sistine Chapel. The ruby-colored mantel of the Creator is in the shape of a gigantic, split-open kidney. Around the same time in Florence, Antonio Benivieni wrote the landmark *The Hidden Cause of Disease*, progressing away from pure anatomic study to the causes of diseases themselves. Stomach cancer and that aching old hip from osteoarthritis were among his discoveries.

◆ ◆ ◆

In North America around the same time, religion played a part in the first recorded autopsy there, which involved the dissection of deceased conjoined twins connected from the belly button to the middle of their chest. Born in Hispaniola (now the Dominican Republic) in 1533, Joana and Melchiora Ballestero both died eight days after birth. Two baptisms were performed out of an abundance of caution because it wasn't clear if Joana and Melchiora had one or two souls. Sometimes one twin would sleep while the other was awake, or one cried when the other was content. An autopsy showed that each twin had her own liver and gallbladder. Since some believed the liver to be the seat of the soul, the twins were thought to have two souls. (This was much to the chagrin of the father, who wasn't happy to pay for the two baptisms of his "double monster" child.)

Andreas Vesalius, from his dissection treatise De humani corporis fabrica *(1534)*

In the sixteenth century, Flemish anatomist Andreas Vesalius made his mark by writing one of the most important works of anatomy, *De Humani Corporis Fabrica Libri Septem* (*On the Fabric of the Human Body*). Before this work, however, he had to learn how to dissect a body. Vesalius was such a zealous student he even stole a corpse from a roadside gibbet (gallows) and puzzled the bones back together at home. His efforts showed him that Galen's work on animals was no substitute for human dissection. Autopsy (Greek, from *autos*, "self," and *optos*, "sight, to see," meaning "eyewitness" or "to see for oneself") was becoming more grounded in observation-based anatomy, as opposed to the authority of voices that laid ground rules centuries before him. Medical schools soon began to dissect cadavers routinely.

In 1576, Guillaume de Baillou performed an autopsy on a seven-year-old boy who died during the Paris diphtheria epidemic. His finding? The telltale false membrane covering the airways, suffocating victims. These early discoveries would pave the way to breakthroughs. Doctors never see cases of diphtheria today because we have an effective vaccine, which is usually combined with a tetanus shot.

In 1699, the Republic of Lucca (a historic state in Italy) recommended autopsy to help understand and limit the spread of tuberculosis. Tornius, Benivieni, Bonet, Dryander, Estienne, Colombo, Fabricius, Fallopia (of the fallopian tube), Eustachio (of the Eustachian tube), and many others contributed to the vast reservoirs of knowledge that needed to be filled.

By the eighteenth century, one pathologist took the postmortem investigation of disease to new heights. Giovanni Battista Morgagni (1682–1771) built on the foundational work of Théophile Bonet's pathology. Morgagni painstakingly performed and recorded autopsies on more than 600 corpses to inform the public on how disease manifested within the body. He also did something quite different at the time—he tied autopsies to the victims' illnesses, such as strokes, heart failure, and pneumonias. And he performed autopsies on people from all walks of life, including bishops, merchants, nuns, and thieves. He recognized that their illnesses provided a clinical story that paired with the pathology of the autopsy—the beginning of what healthcare professionals call "clinical-pathological correlation." It seems an obvious and small thing, but it was a huge step in the realm of pathology.

◆ ◆ ◆

The study of pathology would make great leaps and steady strides in the next two centuries from the addition of histology, or the study of microscopic tissues, thanks largely to practitioners such as Marie François Xavier Bichat (1771–1802). Now the macro (gross anatomy) could work together with the micro (as in microscopic) to unearth more revelations. Bichat declared that in medicine, we ought "to dissect in anatomy, experiment in physiology, follow the disease and make the necropsy in medicine." A pair of German anatomists, Rudolf Virchow and Carl von Rokitansky, would continue the passion that preceded them; Rokitansky in particular performed an estimated 30,000 autopsies in his career.

During the Civil War, William Hammond gathered "all specimens of morbid anatomy" to further medical understanding of disease and trauma, including various bullet wounds, scurvy, and infectious diseases such as typhoid fever and dysentery. Death via kick from a horse was included, as well as gangrene of the face due to excessive use of the commonly prescribed toxic mercury-containing medicine, calomel. Illustrations and descriptions were compiled into volumes of *The Medical and Surgical History of the War of the Rebellion, 1861–65*, which included mystery ailments such as "army itch," which no one would ever want to get.

Anatomical illustration from The Medical and Surgical History of the War of the Rebellion *(1883)*

At the turn of the twentieth century, William Osler, a Canadian physician and one of the founders of modern medicine, recommended autopsy in medical education. Though study of anatomy via cadavers had been going on for a few centuries, autopsies were still not a regular part of medical education. Afterward, autopsies increasingly became a part of hospital care and medical schools to verify diagnoses, for research, and to aid in teaching. The hospital autopsy rate went from 10 percent in 1919 to 50 percent in the 1940s. There were warnings about physicians needing autopsies so as not to "bury their mistakes." But after World War II, autopsy numbers declined for various reasons, despite a recommendation from the Joint Commission on Accreditation of Hospitals, established in 1951 to improve quality of care, that hospitals aim for a minimum autopsy rate of 20 percent. Nowadays, outside of epidemics and pandemics, autopsies are still used to verify mysterious diagnoses, unexpected deaths, and for forensic purposes when deaths are a criminal matter. But they are far rarer now that diagnostic techniques have improved vastly.

When it comes to novel infectious diseases, however, autopsies are still in demand as a source of critically important information. Sometimes, the information comes years after the pandemic has ended. In 2005, the entire genome of the 1918 H1N1 pandemic influenza virus was sequenced thanks to samples saved from prior autopsies (see 1918 Influenza, page 224). The first autopsy of MERS victims in 2014—more than two full years after the first MERS case occurred—provided insight into how the disease progressed and was a sad reminder of lost opportunities to understand the disease earlier.

Relatively recent diseases such as hantavirus, Ebola, Legionnaires' disease, microsporidiosis, Lassa fever, and leptospirosis were all better understood thanks to pathologists and autopsy.

Remember that poor gentleman at the opening of the chapter who died of rabies back in 1727? He was autopsied by Morgagni and found to have swollen blood vessels in his brain and inflamed nerves, both findings typical in rabies encephalitis. Our understanding of this disease and countless others started with those who dared to observe the dead and even put themselves at risk—against the mores of the governments, religions, society, and contemporaries—to glean the truth.

Sometimes, the dead really do speak. Time and time again, those voices have posthumously saved the living.

Body Snatchers

Political cartoon satirizing the "burking" of the constitution in the name of Catholic emancipation, ca. 1688

In the recent history of medicine, the study of anatomy could simply not occur without corpses. In the last few centuries, the inclusion of gross anatomy classes was a mark of a good medical school. The problem was, there just weren't enough corpses to go around.

Anatomists relied on executed criminals on both sides of the Atlantic, but even then it wasn't enough. However, they were able to pay a pretty penny for a corpse, and hence an illicit trade boomed throughout the last three centuries, leading to rampant body snatching and grave robbing.

Fresh graves were often robbed by "resurrectionists" within hours of burial. Grave robbing became such a problem that families hired guards for their newly buried. Sometimes, they employed mortsafes (sturdy metal cages over the gravesites) and watchtowers, particularly in Scotland. Coffins were locked and made of metal. Occasionally wood and metal coffin collars attached the corpse's neck to the floor of the coffin. In 1878, one clever inventor created a coffin "torpedo" that would blast a robber with lead pellets. "Mort," or dead houses, were employed in churchyards so bodies could decompose before burial, as resurrectionists would only steal fresh corpses.

"Mortsafes" protected the newly dead from graverobbing.

In 1788, the so-called Doctors' Riot broke out in New York City, a response to the stealing of corpses by medical students. Many of the victims were enslaved Black people, as well as poor and immigrant residents. Better rules were crafted to protect the dead, but the poor, recent immigrants, and Black people were often the target of body snatching, which continued well into the early 1900s.

Sometimes the snatching took on an even more sinister turn. William Burke and William Hare, a pair of Scots, didn't wait for people to die to sell them to a local anatomist, Dr. Robert Knox. They killed sixteen people in the early 1800s in Scotland by intoxicating them, pinching their noses, and sitting on their chest to smother them to death without leaving a mark. The method of murder became commonly known by people and the press as "burking."

In an open display of payback, Burke was later convicted of murder, hanged, and had his body publicly dissected in an anatomy theater. His bones are still on display in the University of Edinburgh Anatomical Museum, a testament to the ghoulish and dishonorable practice long gone but not forgotten.

MAD COW DISEASE
Patient ZERO
Cow No. 133

Cause: Bovine Spongiform Encephalopathy, BSE (prion)

Symptoms (in animals): Arched back, poor coordination, weight loss, panic, aggression

Symptoms (in humans): Depression, memory loss and behavioral changes, tingling and numbness, incoordination, shaking limbs

Where: Great Britain

When: December 1984

Transmission: Cows eating fellow cows (as fed to them by humans)

Little-Known Fact: Four cases of human BSE have occurred due to blood transfusion.

I t started with cow No. 133.

In December 1984 on a farm in southern England, a cow was sick, her back oddly arched, and she'd lost weight. Just before Christmas, veterinarian David Bee visited her and was perplexed by her symptoms. She'd developed other problems as well—her head trembled and she'd lost coordination, as if forever inebriated. By early February 1985, cow No. 133 was dead from her strange illness.

Dr. Bee considered the possibilities. Had fungus contaminated the feed? An internal parasite? Mercury poisoning? Lead ingestion? Testing showed none of those as the culprit.

Then six more cows in the herd got sick. They too suffered from terrible coordination, were wasting away, seemed panicked, and sometimes got inexplicably aggressive. When they all died, one of them, cow No. 142, was autopsied. The pathologist looked at slices of the cow's brain under the microscope and noticed something odd: The brain tissue was stained pink and there were white holes everywhere. It looked like a slice of sponge. No. 142 had something called "spongiform encephalopathy," a brain disease named for its spongy, porous structure. It looked a lot like the brain of sheep that suffered from a disease called scrapie.

Humans have been aware of scrapie since 1732. It sometimes afflicted entire sheep herds spontaneously or would spread from sheep to sheep. The affected animals would get a humpback shape, lose weight, walk jerkily, chew bizarrely and repetitively, and have an insatiable desire to rub and scrape (hence the name) against fences, resulting in bare patches of skin. But scrapie only happened in sheep, and the disease took its time to manifest. It was also 100 percent fatal. Though it seemed capable of passing from one generation to the next, possibly through infected placentas, it never jumped to other species.

Twenty-five years earlier, scientists discovered that scrapie resembled two human diseases, kuru and Creutzfeldt-Jakob disease (CJD). CJD had puzzled doctors for decades. It struck after age fifty with symptoms of memory loss, apathy, depression, difficulty walking, and jerking muscles, among others. Within five months, it was fatal. Occasionally it ran in families, at other times randomly showing up to kill a single victim.

But what clinched the fate of cow No.133 was the link between CJD, scrapie, *and* kuru.

In Papua New Guinea, members of the Fore people (pronounced FORE-ay) suffered from a neurodegenerative disease that resembled CJD and dated back to around 1900. The word *kuru* came from the Fore word *kuria* or *guria*, which means to shiver and shake from fever or cold, a hallmark of the disease. Like CJD, those suffering from kuru would experience bouts of strong emotions, ups and downs, often zigzagging between melancholy and inappropriate bouts of laughter. For this, kuru is also called "the laughing disease."

People with kuru typically feel fine for anywhere from five to thirteen years. Then for a year their walking becomes unsteady and they have difficulty pronouncing words. Terrible tremors arrive next and walking is no longer possible. This is followed by the emotional rollercoastering and weird laughing spells. In the final stage, eating, drinking, and swallowing become difficult. Eventually kuru victims can't sit up without support and finally become unresponsive. Death inevitably follows.

Kuru patients at a hospital in Okapa, Papua New Guinea

Outsiders initially believed psychosomatic behavior explained kuru, the symptoms attributed to a series of cultural tics; the Fore themselves wondered if it was from magical sources, like ghosts. In 1957, a particular habit of the Fore tribe finally drew attention: They practiced cannibalism.

As a funerary practice to return life force back to the villagers, bodies of the dead were prepared by women and children. Those sick from dysentery or Hansen's disease (leprosy) were not consumed. Men preferred to eat flesh, but often ate none at all—and traditionally not the flesh of women. Women and children consumed what the men left, according to anthropologist Shirley Lindenbaum:

> "Opening the chest and belly, they avoided rupturing the gallbladder, whose bitter content would ruin the meat. After severing the head, they fractured the skull to remove the brain. Meat, viscera, and brain were all eaten. Marrow was sucked from cracked bones, and sometimes the pulverized bones themselves were cooked and eaten with green vegetables. In North Foré (sic) but not in the south, the corpse was buried for several days, then exhumed and eaten when the flesh had 'ripened' and the maggots could be cooked as a separate delicacy."

In the late 1950s and early 1960s the cannibalism–kuru connection was made thanks to the help of research by Nobel laureate Dr. Daniel Carleton Gajdusek, among others. Government patrols in Papua New Guinea subsequently outlawed endocannibalism. Brain tissue in Fore victims of kuru showed protein plaques with a moth-eaten appearance in their neurons. But the key that unlocked the mystery was how much their brain tissue looked like that of sheep with scrapie and humans with Creutzfeldt-Jakob disease. All three are classic for what is termed a spongiform encephalopathy: "spongelike" because that's how

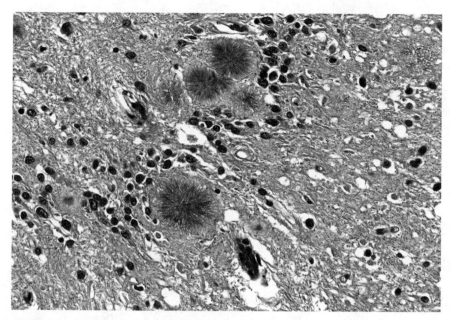

A brain tissue sample showing amyloid plaques

the brain appears under a microscope, and "encephalopathy" for a brain disease that alters brain function. But the latent period of the kuru meant that villagers continued to come down with the disease years and decades after cannibalism was outlawed. Mercifully, cases declined rapidly and the last fatality from kuru was in 2005.

The similarities between kuru and CJD and what was happening to English cows in 1984 was profound, but the cause of the illness remained elusive. Scientists considered it a "slow virus," as the disease seemed to manifest from nowhere. They also knew that injecting brain fluid from scrapie-infected sheep into healthy sheep caused disease. In the late 1960s, the same principle was proven with brain material from humans who died of kuru and CJD, which was shown to infect the brains of healthy chimpanzees. Complicating matters was how tough the infectious agent appeared—it couldn't be killed by heat or alcohol. Formalin (a formaldehyde-like chemical) couldn't hurt it. Ionizing and UV radiation, as well as harsh detergents, didn't work. Even autoclaves, which used high temperatures and crushing pressures that normally inactivate hard-to-kill spores, were ineffective. So it wasn't a virus or bacteria. Whatever it was it seemed nearly indestructible, a super pathogen invisible to microscopes and immune to all conventional methods of killing it. The pathogen also had no DNA or RNA, could infect other victims, and with deeply strange symptoms, would kill its host 100 percent of the time.

◆ ◆ ◆

A new theory eventually emerged.

Perhaps the cows weren't afflicted with a virus, bacteria, protozoa, fungi, or any other living thing that typically infected humans. Maybe it was something simpler that could infect people without using nucleic acids like DNA or RNA? (Which was outlandish, by the way.) All known pathogens at the time were believed to have DNA or RNA—the nucleic acids containing the plans that allowed them to replicate and readily infect others. Perhaps what was disabling and killing the cows had no master plan, no nucleic acid "brain." What if "it" was simpler, only a protein? They are, after all, the amino acid structures of which every living thing is made. And they have lots of jobs in organisms, making up parts of every single cell, though they can't reproduce on their own.

Human cells operate like little factories. The DNA and RNA are the bosses that tell the factory workers to make parts, like fingernail protein or hair or muscle fibers. Now imagine the hair protein—a thoughtless piece of amino acids—deciding to go rogue and make its own hair protein without the boss, kind of like a monstrous Cousin Itt. The idea was biologically absurd. Almost.

Even the possibility of it was like a tsunami barreling toward the central dogma of biology. A protein can't be a living thing that also infects and kills, can it? Without a "brain" of nucleic acid, without a survival plan, how could it live? For perspective, James Watson and Francis Crick had only just discovered the correct DNA helix structure in 1953, a decade or so before the kuru–cannibalism connection was made in Papua New Guinea.

It was not until 1982 that American neurologist Stanley Prusiner bravely declared the etiology, or cause, in this string

Illustration of a human prion protein

of diseases was an infectious particle, a proteinaceous infectious particle, or prion. Prusiner's declaration set off a firestorm. Other researchers were convinced a different, traditionally infectious agent was responsible. The announcement put Prusiner's university tenure at stake, and he lost his government funding.

In the end he'd have the last word—because he was right. Prions are real, and Prusiner went on to win a Nobel Prize for his work.

Though they feel straight out of science fiction, prions are proteins very similar to those we have in our own brains. Right now. The key difference is that healthy, normal prion proteins in our brains are folded in a certain way. When those proteins come into contact with an infectious isoform (which has minor changes in their amino acid chain that make the protein fold in a different 3D shape), they change the normal protein into the new shape—making it an infectious isoform of prion protein that can now convert another normal protein into an infectious one, and so on, and so on. Where a virus takes over a cell to replicate itself, prions make more prions through, well, conversion. They're not unlike Star Trek's infamous Borg, who physically and mentally assimilated new members into their roving collective.

More insidious perhaps, abnormal prion proteins don't elicit an immune response, so a body doesn't know to fight them off. They hide so effectively because the protein chain itself is encoded within the chromosomes. They're *supposed* to be a part of our bodies—just not in that conformation, or shape. And when those misfolded prions start increasing, they clump together and form plaques, which in turn become islands of protein that build up in our neurons. The body won't destroy them because they're our own proteins. But those abnormal prion proteins are toxic to neurons. Soon enough, brain cells start dying. The little holes start showing up—the "spongy" texture seen under a microscope. And the damage is irreversible, inevitable, and fatal.

In figuring out what happened to cow No. 133, prions naturally emerged as a possibility. Doctors and scientists considered the seeming inability of scrapie to jump species. But what if it could? Or perhaps the cow disease started like a random case of CJD, appearing sporadically but rarely in a population? Or, just maybe, it got passed to other cattle the way kuru did among the Fore people?

Cow cannibalism.

The symptoms: loss of coordination, wasting away, occasional aggression. The shorthand diagnosis: "mad cow" disease.

The push in modern farming to produce beef as quickly as possible meant nourishing growing calves and cattle with protein supplements. In the US, much of those protein supplements come from vegetable sources, like soybean meal. So it made sense that "mad cow" disease never hit American cattle like it did those in the UK. After all, soybeans grow lush and easily in North America, but not across the Atlantic, where they often turned to protein sources that were byproducts of the meat production process: bone, spines, and other unused portions of slaughtered cows and sheep. It was science fiction come to life, a bovine *Soylent Green* (or *The Matrix*, if you like).

By 1986, two years after cow No. 133 became ill, the disease had a name: bovine spongiform encephalopathy, or BSE. Two years later, 421 cattle had died from BSE. By March of that year, it was clear that the source was what Dr. Gajdusek called "high-tech cannibalism." By July 1988, the British government banned the use of bonemeal and meat in cattle and sheep feed. A year later, Britain banned the use of bovine entrails, including spines and brains, for human consumption. The US was so worried about a potential zoonotic (animal-to-human) spillover they banned the import of any livestock, including goats, sheep, cows, and bison, from countries harboring BSE in their animals or beef products.

British Agriculture minister John Gummer and his daughter ate hamburgers in an attempt to show that the nation's beef supply was safe.

And yet the question lingered: If this disease made a species jump to cows, possibly from a scrapie-infected sheep ground up into cattle feed, could it jump species to infect humans? British authorities stated that British beef was safe. There hadn't been any evidence to show humans were affected yet, and panic needed to be quelled—there was an enormous beef industry to protect. To "prove" England's beef was safe, the British Minister of Agriculture, John Gummer, had his four-year-old daughter eat a juicy hamburger before the cameras.

But all was not safe. In 1990, a house cat and several zoo animals (five types of antelopes, all fed with commercial cattle feed) came down with a disease that looked just like BSE. British pet food manufacturers had already removed offal (animal entrails and organs) from their food sources, yet it felt not soon enough to scientists, who held their breath and wondered if humans had managed to escape the danger. By 1993, 120,000 British cattle had died from BSE. By the mid-90s, Britain finally banned the feeding of meat and bonemeal to animals, as well as its inclusion in fertilizer.

With that, perhaps enough had been done. Maybe the BSE nightmare was finally over.

◆ ◆ ◆

In August 1994, a teen named Stephen Churchill got into a car accident in Wiltshire, England. He'd driven his family's Ford Fiesta across a median and into an oncoming army truck. Luckily he and his passenger weren't hurt, but Stephen couldn't explain why the accident occurred. It just . . . happened.

It was an odd turn of events for a bright, generally happy student who did well in school. Until then, Churchill had been a normal teen—healthy, robust, a good appetite. And yes, he'd eaten his share of hamburgers, sausages, and beef growing up. In the months after the accident, however, his parents saw a sudden change in him. A Royal Air Force cadet, Churchill became depressed and fell so far behind in school he was unable to continue his studies. In November, his mother noticed that, after taking her son out for a meal, he couldn't remember it afterward. By December, Churchill had difficulty walking and couldn't write normally. Hallucinations set in next. His psychiatric diagnosis of depression was questioned, and by March of 1995, he was undergoing neurological tests in London. At some point, a doctor wrote in his chart:

"C.J.D.?"

But it couldn't be. CJD, or Creutzfeldt-Jakob disease, struck down people over fifty. CJD symptoms were more like dementia, including memory loss, as opposed to the psychiatric symptoms Churchill was experiencing. His test results were different from classic CJD, too, and yet the similarities were too striking to ignore.

Churchill died in May 1995. Shortly afterward, several young people came down with identical symptoms. Among the differences from classic CJD were the pathological findings in the brain. The new cases showed tight bundles of protein plaques, called amyloid, surrounded by a halo of holes. With the pink tissue staining, they looked like little flowers and were called "florid plaques." A full year after the "C.J.D.?" noted in Churchill's chart, ten young people had died of the same illness. The House of Commons then issued a terrifying public statement: The victims had most likely died "from exposure to B.S.E."

The new disease was called Variant Creutzfeldt-Jakob disease, or vCJD. The dreaded jump from cows to humans had finally happened.

◆ ◆ ◆

All told, more than 170,000 cows were infected with BSE and some four million of them destroyed for safety reasons. That number was staggering—170,000 infected cows—because no one really knew just how many other cows had already entered Great Britain's food chain. Or how many had been exported all over the world. It was subsequently determined that eating just a quarter teaspoon of infected brain was enough to kill a cow, about the volume of two M&M'S. Since there seemed to be an incubation period, the first ten victims of vCJD could have been the start of a bona fide epidemic of neurodegenerative disease in the UK.

Thankfully, the epidemic never materialized.

But why? After all, the prions that caused it were practically indestructible. So much so that the Centers for Disease Control and Prevention (CDC) recom-

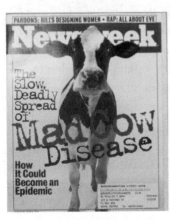

Fear of mad cow disease continued into the new century, with this 2001 Newsweek cover.

mends instruments that touch highly infected CJD tissues should be incinerated when possible. Normal sterilization techniques that allow the reuse of medical equipment, such as radiation, hot temperatures or pressures, or detergents, don't work. More than 500 accidental cases of CJD have been transmitted by surgical instruments, transplanted tissue, and contaminated human growth hormone.

We also know how prions work: One misfolded protein makes a normal one misfold, and on and on. It sounds like the perfect recipe for a neurological disaster on a global scale. One theory why it didn't happen points to a genetic component of susceptibility. Even with prions, we have small variations in our amino acid chain based on what we inherit from our parents. Nearly all cases of vCJD share one thing in common: They all have the same duplicate gene for prions that have the amino acid methionine in a very particular place on the protein. And people with these two duplicate genes are called methionine homozygotes.

Fatal Insomnia

It is as bad as it sounds.

Prions don't always kill by causing dementia, or making us stumble and become uncoordinated. Some prions kill us by denying us the things we need most.

Like sleep.

Fatal insomnia can be both familial (inherited through genes) or sporadic (by random mutation). But without question, it's a prion disease that appears to attack a part of the brain called the thalamus, which regulates sleeping and waking cycles.

First noted in an Italian man who died from a months-long stupor in 1765, the disease signals to the sufferer of what's to come: pupils become small, sweating begins, then trembling and incoordination. Women suddenly go into menopause and men become impotent. All these symptoms are signs that the thalamus is under attack, neurons are dying, and the brain is beginning to experience that Swiss-cheese texture.

Then the sleeplessness begins.

Within a year, a sufferer will die from a complete inability to sleep. Sedatives and sleep medications don't work. At best, they can get some drowsy "rest," but inevitably the sleeplessness takes its toll and the victim falls into a coma.

It's a devastating disease that causes a staggering amount of suffering. Luckily, it's quite rare. Only a few families are affected, but the disease can happen sporadically as well.

But why do some people with prion diseases get fatal insomnia, and others are stricken with classic Creutzfeldt-Jakob disease? Again, it comes back to our genes. Abnormal prion proteins aren't all identical. Within that protein chain, substitutions in single amino acids change how they cause disease, and these different amino acid substitutions can be coded in our DNA. But in many cases, a variety of symptoms and age of onset can occur depending on that explicit amino acid switch. A particular mutation can lead to Creutzfeldt-Jakob symptoms, or another prion disease called Gerstmann-Sträussler-Scheinker disease.

For example, patients with the abnormal prion protein have one abnormal gene, D178N. In addition to this one, having switched amino acid at the 129 position mean the difference between the two diseases. If you have the amino acid methionine there, you'll get fatal insomnia. If you have a valine at the 129 position, you'll have familial Creutzfeldt-Jakob disease.

Either way, it shows how incredibly delicate humans are. One or two amino acid switches on one of the more than twenty-thousand protein coding genes in humans can mean the difference between life and a fairly horrible death.

Not a lottery anyone would knowingly join.

Their normal prion proteins are more susceptible to the infectious folding from the bad prions. But there's a significant catch—40 percent of the Caucasian population are methionine homozygotes. So, again, why so few cases?

In surveys of archived British appendix tissues, one out of two thousand contained infectious prion. In appendixes from people with no symptoms, the prions could lurk elsewhere in the body, unbeknownst to people. Which is why there continues to be a ban on blood donations in the US from anyone who lived in the UK during those so-called "mad cow" years.

Prion diseases were initially thought to be caused by a slow virus because it seemed to take a long time between infection and disease manifestation. In some cases we're talking a few years, but in others, fifty. It's not clear exactly why the incubation period is so long. It could be related to the characteristics of the patient (apparently a large brain volume is protective) and the specific prion's tendency to aggregate. What we do know is that once the prion is "replicating" at high rates and brain tissue turns spongy, the symptoms start appearing. Thus, the UK's "mad cow" problem may not be over, as thousands of people could be harboring a prion that isn't ready to make itself known yet.

Overall, the number of vCJD cases has declined steeply. But they haven't quite gone away. The most recent US case was in 2012. An American victim succumbed to vCJD, his brain showing the typical florid plaques. He'd previously lived in countries that imported beef from Britain during the at-risk years. Like other patients, it took more than a year from the beginning of his symptoms before the diagnosis was found.

In the meantime, BSE keeps popping up sporadically in cows, mink, sheep, and humans. It has crossed species barriers time and again, even showing up spontaneously.

That little prion isn't just indestructible, it's apparently aiming for immortality, too.

ANATOMY OF AN OUTBREAK
Calling in the Public Health Cavalry

Impact: By systematically analyzing disease outbreaks, we can often identify and stop their spread, saving countless lives.

When: 1946

Who: The CDC, via their Epidemic Intelligence Service Officers, plus local and state agencies

What Happened Next: Outbreak investigations have added to a growing wealth of information on infectious diseases, as well as identified new pathogens.

Little-Known Fact: In 1981, 85 people were sickened by a Salmonella outbreak linked to contaminated marijuana.

I t was 1976. The year had been thus far packed with celebratory events for the bicentennial of the country's founding. On July 21, the Pennsylvania chapters of the American Legion, a thriving veterans group, opened their 58th annual convention. Given the special year, it was all too perfect to hold the four-day meeting in Philadelphia, the city where the Declaration of Independence was signed. Hosting the convention and festivities was Philadelphia's magnificent Bellevue-Stratford Hotel, with seven hundred rooms reserved for a portion of the four thousand Legionnaires flooding the area. Delegates from the Ladies Auxiliary and families of Legionnaires were also in attendance. A deluge of meetings and caucuses were planned, as well as plenty of opportunities for socializing, a dance, and even a small parade. For many in attendance, it was *the* social event of the year.

Legionella *bacteria growing on an agar plate and an American Legion logo.*

It was also the perfect setup for a new infectious disease to flourish.

The Bellevue-Stratford Hotel, built in 1904, was ideal for a convention. Its French Renaissance grandeur set the scene for the occasion, with beautiful light fixtures designed by Thomas Edison. The hotel also stayed comfortably cool on those warm July days thanks to the building's large air-conditioning units. Aside from the Bellevue-Stratford, Legionnaires were also staying in a half-dozen other hotels all over Philadelphia.

As usual, the convention was a hit. People mingled and drank together, caucused and voted together, and danced and reminisced together. Then the Legionnaires went home, and the hotel personnel and restaurants all over town who served them continued business as usual.

Retired Air Force Captain Ray Brennan returned from the convention feeling tired, likely from all the excitement and merrymaking. But three days later, he started having chest pain and was short of breath. Brennan already had heart trouble, but this time he also had a fever. At first he resisted getting medical help, but he eventually made his way to his local hospital where his condition worsened, his lungs congested with frothy, bloody secretions. When he died, the doctors blamed a heart attack.

The Bellevue-Stratford Hotel, Philadelphia, 1976

Three days later, another Legionnaire died from similar symptoms. Then three more died. By August 1, a week after the convention ended, six more Legionnaires were dead. Word spread that Legionnaires were dying of an illness with symptoms that included shortness of breath, chest pain, fever, and lung congestion. It took a few more days before doctors and Legion officials could coordinate with state authorities. By the next week, those first few cases had blossomed into 130, with twenty-five more deaths. One was a bus driver who attended the convention, another was a man who delivered food to the Bellevue-Stratford the week of the convention. There were other symptoms: abdominal pain, vomiting, and diarrhea. Sometimes the afflicted would become delirious or have problems pronouncing words. In other cases, the center portion of an infected lung would simply die, leaving a gaping hole.

Clearly, something lethal was taking out not just Legionnaires but those who'd worked around the convention.

This is where the first step of the outbreak investigation began. More than 150 federal and state scientists got to work, establishing a phone line that took calls at the rate of 400 an hour. In Harrisburg, Pennsylvania's state capital, a "war room" was set up. This was 1976, so a map with pushpins showing locations of cases was used—red for deaths and yellow for just the reported cases of what they were calling "Legion Disease." A group of investigators contacted hundreds of hospitals in the state searching for more cases, gathering information on sick patients, and then comparing their illnesses to others. Connections started to be made: the fever, the pneumonia, the convention. Data was graphed as it came in, and those graphs changed daily. The investigative team decided on a case definition: anyone who had a cough, fever, and chest X-ray that showed pneumonia; plus, they had to have attended the convention *or* entered the Bellevue-Stratford Hotel between July 1 and the onset of the illness.

The work being done in Harrisburg didn't stay there. Pennsylvania medical societies and local hospitals were notified of the outbreak in hopes of collecting more data and finding additional cases. Hospital-based nurses did daily rounds to report new illnesses and recorded each date of onset of the illness, the clinical description, and the patient's association with the convention. News outlets reported on the outbreak and urged the public to share info on new illnesses.

Hotel guests and employees as well as Legionnaires were surveyed by phone and peppered with questions. "When did you arrive? When did you leave? Did you go to the dance? To the go-getter's breakfast? Did you have contact with pigs?" Emergency rooms and hospital admission records were also checked.

Meanwhile in state laboratories, blood and lung tissue samples from victims were being analyzed. Air, water, dirt, and other materials from the hotel rooms were

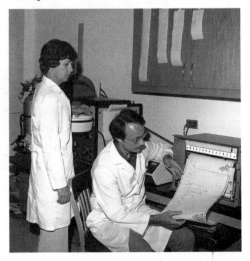

Crunching data on the chromatography profile of Legionella *bacteria.*

collected. Rodents and birds were trapped and tested. Researchers took samples of infected tissue to perform the usual gram stains, the process of dyeing bacteria to help distinguish their shape and affinity to particular dyes, in order to classify their species. However, the gram stains didn't help find an obvious culprit. Without seeing lots of bacteria on the gram stains, scientists narrowed their search—and began to suspect that it had to be a virus.

The prior winter, swine flu had struck in Fort Dix, New Jersey (see page 230), so researchers considered if somehow that was the cause. But none of the affected cases had been exposed to pigs, and animal lab testing on chick embryos, mice, and cell cultures showed the diseased tissue didn't have swine flu, or any influenza for that matter. Another hypothesis considered a type of zoonosis called ornithosis, which involves catching an infection from bird droppings or by touching birds. But this, too, was disproved by lab tests.

Also considered were: tularemia, mycoplasma, Marburg virus, Lassa fever, spiroplasma, plague, and typhoid—but none were the Legionnaire killers.

In comparing people with "Legion Disease" to those who didn't catch it, investigators discerned several key facts. Food or drink didn't seem to be the culprit. Nor were any particular restaurants or other hotels implicated. Also, weirdly, hotel employees weren't getting sick, and neither were family members or spouses of sick and dying Legionnaires.

However, one commonality was clear: All the stricken had, at some point, been in the lobby of the Bellevue-Stratford, or at least in front of it during the parade. And while investigators had identified a common link, the actual cause of the sickness remained elusive. Meanwhile, weeks passed and the outbreak ended almost as swiftly as it began. Clearly it wasn't spreading, but there was also no way to learn how to stop it from happening again.

Without a definitive cause for the outbreak, wild rumors began circulating, including those who blamed the sickness on everything from toxic gases to heavy metals and nickel carbonyl intoxication. (The idea that communists or the pharmaceutical industry were secretly murdering veterans was even floated.) There were accusations that the whole investigation had been botched, with mixed-up samples or a general lack of them. People got angry at the CDC for not providing immediate answers, and politicians naturally got in on the act and began pointing fingers, as did various media outlets.

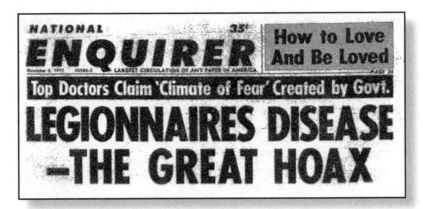

NATIONAL
ENQUIRER
35¢
LARGEST CIRCULATION OF ANY PAPER IN AMERICA
How to Love
And Be Loved
Top Doctors Claim 'Climate of Fear' Created by Govt.
LEGIONNAIRES DISEASE
—THE GREAT HOAX

◆ ◆ ◆

Most of us learn the details of these sorts of investigations through the news. Perhaps it's about a bad batch of romaine lettuce that is already making people sick. Sometimes, we hear about the health consequences long after harm is done, such as a warning on a box of tampons about the risk of toxic shock syndrome. Often we're completely unaware of how outbreak investigations made things safer for us. For example, grocery produce-misting machines keep our veggies crisp without giving shoppers pneumonia. (Yes, there was a pneumonia-causing veggie-mister outbreak; see page 92.)

But few of us really need to understand the rules of outbreak investigations and how they happen—until we're the ones who are sick or just bought recalled strawberries. The first thing to know: Time is of the essence. There is the obvious urgency of preventing more illnesses and deaths, but also (and always) pressure to wrap up the investigation from the various parties involved, who often have financial and legal concerns.

And while the clock ticks loudly, the investigation still has to be systematically done with rigor, following multiple basic steps already seen in the Philadelphia outbreak above, many of which often happen simultaneously. Once an epidemic is confirmed to exist and the diagnosis of illness is also confirmed, then the illness has to be clearly defined. For example, if investigating an outbreak of *E. coli*, someone with symptoms of diarrhea during a specified time period and a positive laboratory test for *E. coli* may be considered a confirmed case. Additional cases of diarrheal illness that lack laboratory confirmation may be defined as "probable" or "possible" depending on other factors.

The next step is to determine if the rate of infection is above the usual number of illnesses at that time and place, called the background rate. For example, since 1993 the background rate of measles has been less than 200 cases a year in the entire US. So, if cases of measles start popping up in a community, that's cause for alarm because the associated background rate is so low. This step includes descriptive epidemiology—including demographics like gender, age, and race/ethnicity, but also information such as occupation, recent travel, and vaccination records. When those grocery store misters were giving people pneumonia, for example, it helped to know that more women than men were getting sick, and those women were doing more of the grocery shopping. In another outbreak that was caused by salmonella, investigators noticed that those affected tended to be younger adults than the usual demographic that catches salmonella. Also, most of the sick were exposed to marijuana—and that's how they learned that contaminated weed was the source of the outbreak.

Next, a hypothesis is formulated on how the outbreak occurred. When it's a disease they already know, past lessons help. But not always. If the disease is brand-new or the method of infection is novel—like the marijuana-salmonella outbreak—and the cause can't be found by way of the initial hypothesis, investigators have to look closer. Home refrigerators can get inspected and interviews are ongoing, so the information keeps coming as the investigation uncovers new things. Sometimes people who've been infected are brought together by the investigators to talk and discover that elusive, common thread on their own.

The hypothesis is tested in various ways. One method is to compare the cases to controls, another is by looking at a population cross section to see if the hypothesis holds up. If it doesn't, you come up with a new hypothesis. Testing field samples can help zero in on the source. Once identified, the source needs to be controlled, which entails public announcements of food recalls, press releases, and making sure all infected sources, like food, are pulled from the shelves. The whole report is finally put together, and the final step is maintaining surveillance so that the numbers actually go down and the outbreak can be declared over.

◆ ◆ ◆

By fall 1976, the deaths in Philadelphia had mercifully ended, but the research had not. The cause of the outbreak was still a stubborn mystery. One CDC scientist, Joseph McDade, was still at work on the outbreak. McDade wasn't just a microbiologist, but a rickettsiologist. He specialized in the bacteria belonging to the genus *Rickettsia*, responsible for diseases like typhus and Rocky Mountain spotted fever. McDade was tasked with ascertaining if the sickened Legionnaires had Q fever, (once classified as a rickettsial disease), even though many experts were convinced the "Legion" disease was viral, not bacterial.

Which made sense; after all, the tissue samples didn't show anything on a traditional gram stain, and nothing grew in the petri dishes used to culture bacteria. When injected with solutions made from tissues of diseased patients, specialized lab mice used to confirm bacterial infections didn't get sick. So it had to be a virus, *had* to be.

Investigators tried another method for finding viruses—using chicken eggs to grow the pathogen. Because they didn't want random bacteria infecting the samples, the eggs were always treated with antibiotics. But again, nothing seemed to turn up.

During the holiday season in December 1976, McDade continued to work. He enjoyed the quiet time between Christmas and New Year's, when he decided to visit the lab to "tidy up loose ends." While there, he took one last peek at some slides related to the Philadelphia outbreak he'd already reviewed. On one slide of infected guinea pig tissue—unlike special lab mice, guinea pigs displayed classic symptoms of the Legion disease—McDade saw some scattered bacilli. This was no big deal—a few random bacteria here or there wasn't illuminating. But in one corner of the slide, he saw several clusters of bacilli.

That was something.

Regrouping, McDade repeated the experiments on the chicken eggs but this time left out the antibiotics. And the cultures reacted to the serum of disease survivors. Which meant he finally had the pathogen responsible for the Legion disease. Subsequently, McDade stained the cultures with the Gimenez stain, which was often used in rickettsia research, since those bacteria never pick up the usual gram stain very well. Using a bright-pink fuchsin stain for the bacteria and a malachite-green counterstain for the background, the bacilli were now visible.

McDade had found the bug.

It turned out that those lab mice that never got sick had probably somehow picked up an immunity to the new disease—now officially called Legionnaires' disease. And the bacterium, named *Legionella pneumophila* (from the Greek "lung loving"), had been discovered.

Legionella was hard to grow in a lab, which was why McDade and others had so much trouble discovering it. The bacteria prefer high levels of cysteine, iron

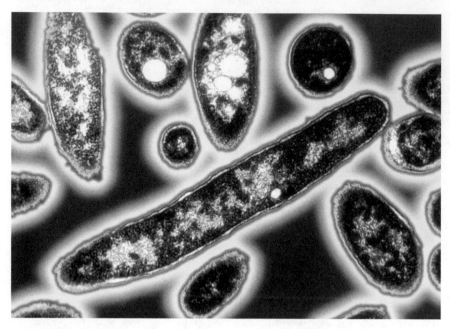

Legionella Pneumophilia *bacteria*

supplements, low oxygen, low sodium, and activated charcoal in lieu of the usual bacterial growth media. They also like higher-than-body temperatures to grow. Some refer to it as "pond scum" based on its required growth environment. And where can you find pond-scum-like warm temperatures with lots of iron?

Air-conditioning towers.

The cooling–heating exchange found in these units often have a biofilm growing on their surface—the so-called pond scum—and *Legionella* love it. The bacteria are also found in hot springs, as well as areas where nuclear reactors spew their hot water runoff. What likely happened at the Bellevue-Stratford was that bacteria from the air-conditioning towers were spewing off infected droplets into the air, and older Legionnaires with weakened immune systems were inhaling them and getting sick. It explained why the hotel staff didn't get ill—they'd probably already been exposed to the bacteria previously, so they were immune. Furthermore, it's theorized that infected water from drip pans beneath the air-conditioning towers were sprayed by the powerful fans of the units, resulting in droplets outside the building which got sucked into the lobby—both areas that were associated with infection.

Thanks to the outbreak investigation and countless hours of toil by "shoe leather" epidemiologists in the field, plus the tireless work of scientists in their labs and the Harrisburg war room, they finally had a handle on their new disease.

Only it wasn't so new after all. Once they identified *Legionella*, the investigators dug through tissue samples from other similar but mysterious illnesses. It turned out that *Legionella pneumophila* was responsible for Pontiac fever, a disease that sickened 144 people in Pontiac, Michigan, in 1968 at the county health department. The source was likely a defective air-conditioning unit that also sickened the lab guinea pigs. There had been another outbreak at a Hormel meatpacking plant in 1957, as well as an outbreak in a Washington, DC, mental hospital in 1965.

Like many other diseases freshly "discovered," *Legionella* had in fact been with us for far longer than we'd realized.

Now that we know much more about the disease, when new outbreaks do happen, it's far easier to identify the cause. Remember the grocery store water misters that caused pneumonia? In early 1990, the misters had sickened several people, mostly women shoppers, and gave them—you guessed it—*Legionella*. We now know how to prevent *Legionella* growth by keeping water systems clean, free of biofilm, and decontaminated. We have good tests to look for the disease via a simple urine sample, and excellent antibiotics to save people's lives when they do catch it.

All of this is because of the logical and critically important steps of an outbreak investigation, combined with the tenaciousness of those who refuse to stop looking for answers.

Biblical Plague: A Historical Outbreak Investigation

For biblical scholars, historians, and epidemiologists, the Bible's ten plagues have long been a matter of interest and secular debate. Beginning with: Did they really happen? And if so, when? And how did the wrath of God occur in a scientifically plausible way?

The story in Exodus begins when the Pharaoh taunts God by saying he "does not know Yahweh" (the Hebrew name for God), to which God responds: "The Egyptians shall know that I am the LORD." God then gets busy flinging plagues upon the Egyptians.

At that point in human development, plagues might have seemed like manifestations of divine wrath, but when viewed through a modern scientific lens, the ecological events described in Exodus actually connect to each other in an intriguingly logical progression, one that would have inevitably led to an outbreak of infectious diseases among a given population. In the end, the reason for those plagues, if they indeed happened, might've been climate change, as some researchers such as John S. Marr, Curtis D. Malloy, and Stephen Mortlock have suggested.

The Egyptian Pharaoh being attacked by flies in the Bible's fourth plague

With the staff that is in my hands I will strike the water of the Nile, and it will be changed into blood. The fish in the Nile will die, and the river will stink and the Egyptians will not be able to drink its water. (Exodus 7:17–18)

In the first plague, the Nile River's water turned to "blood," a phenomenon now attributed to a red algal bloom that would have made the Nile appear red. In deoxygenating the water, fish would have died in great quantities, their decomposing bodies further contaminating the river, which could've led to the second plague, where frogs overran Egyptian settlements along the Nile.

Let my people go, so that they may worship me. If you refuse to let them go, I will plague your whole country with frogs. The Nile will teem with frogs. They will come up into your palace and your bedroom and onto your bed, into the houses of your officials and on your people, and into your ovens and kneading troughs. The frogs will go up on you and your people and all your officials. (Exodus 8:1–4)

A red algal bloom

Frogs, initially freed from predation by the large fish who died in the algal bloom, would have quickly bred in great quantities. As the river water turned increasingly toxic, however, the frogs would have tried their luck farther inland, overrunning Egyptian towns and villages by the sheer size of their overbred population. But those frogs would have soon died in huge quantities, too. Their quests for new food sources would have largely failed, and their decomposing bodies would have further contaminated local water sources, creating an ideal environment for plagues three and four.

The third plague, when "all the dust throughout the land of Egypt became lice," likely described the arrival of biting midges in great numbers. Midge larvae feed on the microorganisms abundant in decomposing material, such as all the dead fish and frogs that would've littered the land in this scenario. Also, the midge population could thrive with the sudden absence of one of their primary predators: frogs. Midges bite both animals and humans with vengeance and regularly transmit disease. The midge would in turn become the biological vector for the fifth plague. But we get ahead of ourselves.

Plague number four: a swarm of flies. With contaminated water supplies, conditions were ripe for fly species to overpopulate. As such, modern scholars attribute the fourth plague to the stable fly, which focuses its energies on biting cattle (hence its name), but bites humans, too. Cattle heavily infested with stable flies become anemic and have lower milk yields.

Egyptian livestock would be hit again by the fifth plague, described as "the hand of the LORD will bring a terrible plague on your livestock in the field—on your horses and donkeys and camels and on your cattle and sheep and goats." Contemporary science identifies two diseases transferred by biting midges (see plague number three) as the likely culprits. African horse sickness and bluetongue, neither of which infect humans but have dire mortality rates in animals, could have easily spread among livestock thanks to the explosive growth of midges that year.

> Then the LORD said to Moses and Aaron, "Take handfuls of soot from a furnace and have Moses toss it into the air in the presence of Pharaoh. It will become fine dust over the whole land of Egypt, and festering boils will break out on men and animals throughout the land." (Exodus 9:8–9)

The sixth plague in our ecological scenario was made possible by the zoonotic transfer of disease to humans, when God declared that "festering boils will break out on men and animals throughout the land." Owing to rapidly declining meat and milk supplies (see all previous plagues), the consumption of contaminated meat and milk was all but guaranteed for a huge chunk of the Egyptian population. And it would unleash an outbreak of what is now believed to have likely been glanders.

A zoonotic infectious disease caused by the bacterium *Burkholderia mallei*, glanders is highly contagious and typically transmitted from direct contact with infected animals and/or through fly bites. (We can assume there were an abundance of stable flies biting sickened animals and then humans as well.) In animals, glanders strikes horses, donkeys, mules, and goats in particular. It begins as a respiratory infection and then progresses to the lymphatic system, where it gets deadly.

In humans, glanders symptoms begin with the formation of nodular lesions in the lungs, as well as ulcers in the upper respiratory tract. Coughing, fever, and sneezing are all signs of an acute infection. If the disease advances to the chronic form, tumors (boils) develop inside the nose and around other parts of the body. These tumors eventually ulcerate as well, leading to death within a few months. If someone survives the disease, they end up a carrier of it as well.

The seventh, eighth, and ninth plagues are not as connected to the linear narrative of ecological progression from contaminated Nile water to an outbreak of disease among local populations. Rather, these plagues were random, unfortunate events: an unusually large hailstorm whipping through the region (seventh), causing significant damage to crops, which were then further destroyed by a plague of locusts passing through (eighth). The ninth plague was likely an exaggerated reference to some sort of unusual astronomical event—an eclipse perhaps, or the passing of a comet.

Then the LORD said to Moses, "Stretch out your hand toward the sky so that darkness will spread over Egypt—darkness that can be felt." So Moses stretched out his hand toward the sky, and total darkness covered all Egypt for three days. No one could see anyone else or leave his place for three days. (Exodus 10:21–23)

But the tenth and final plague, the most devastating of the lot, brings us right back to another possible outbreak of infectious disease: "Every firstborn son in Egypt will die, from the firstborn son of Pharaoh, who sits on the throne, to the firstborn of the slave girl, who is at her hand mill, and all the firstborn of the cattle as well. There will be loud wailing throughout Egypt—worse than there has ever been or ever will be again." As presented in Exodus, the event takes on the qualities of the supernatural, implying the progression of an outbreak that would have been spread over a few weeks takes place in a single night. But it's hard to knock a good story.

Modern researchers have proposed a more convincing cause of death from the tenth plague than the mercurial will of God: mycotoxins. They are toxic compounds produced by mold that can grow on a variety of stored foods, grains in particular. In warm, damp conditions, mold can flourish on stored food, as that long-overdue visit to the back of your food pantry will confirm. With all of their standard food supplies devastated by a rapid series of plagues over the course of a year, most Egyptians would be starving and desperate. And just like you only reach for suspect food in the back of your pantry when you're out of all other options, so would Egyptians turn to their oldest foodstuffs in their least-used granaries to feed themselves. They likely suspected that the food was risky to eat, but starvation would've forced their hands.

And if the Egyptians were left only with mycotoxin-infected grains, it actually makes sense that firstborn males were those who died first. In a traditional, deeply patriarchal society, the firstborn male is going to be the first to eat when food is scarce. The proclamation of firstborn cattle succumbing to the same fate can be similarly explained—the most dominant cattle would have had the greatest access to whatever food was available.

The cumulative effect of all of this eating of infected grain would've made a local outbreak of mycotoxin poisoning more likely, not dissimilar from how ergotism (see page 2) spreads. A slew of mycotoxin deaths would not have happened on a single night, but they could have occurred in rapid succession over a period of just a few days. It wouldn't have been much of a leap for a storyteller to condense the timeline to a single night. But then again, it's fairly impossible to do a scientifically rigorous outbreak investigation for something that occurred thousands of years ago. However, the lesson is clear: Outbreaks of disease are not always what they seem.

YELLOW FEVER
Patient ZERO
Kate Bionda (Memphis outbreak, 1878)

Cause: Yellow fever virus

Symptoms: Fever; jaundice; abdominal pain; vomiting; dark urine; bleeding from the mouth, nose, eyes, or stomach

Where: Worldwide, with recent outbreaks in Angola, Democratic Republic of Congo, and China

When: Ongoing, but peaked in the late nineteenth century

Transmission: The bite of the *Aedes aegypti* mosquito

Little-Known Fact: A yellow fever outbreak shut down Philadelphia while it served as the capital of the US in 1793, causing a constitutional crisis.

"Lucille died at ten o'clock Tuesday night, after such suffering as I hope never again to witness . . . the poor girl's screams might be heard for half a square at times and I had to exert my utmost strength to hold her in bed. Jaundice was marked the skin being a bright yellow hue: tongue and lips dark, cracked and blood oozing from the mouth and nose . . . To me the most terrible and terrifying feature was the 'black vomit' which I never before witnessed. By Tuesday it was as black as ink and would be ejected with terrific force. I had my face and hands spattered but had to stand by and hold her. Well it is too terrible to write any more about it."
—W. E. GEORGE OF MEMPHIS, IN A LETTER DESCRIBING THE DEATH OF HIS NIECE FROM YELLOW FEVER

n the summer of 1878, Memphis, Tennessee, was not a pleasant place to be. The humid and heavy heat arrived early that year in the bustling city of 47,000 located on a bluff overlooking the Mississippi River. The heat was accompanied by a drought, a particularly galling problem in Memphis. Unusual for a city of its size in the last quarter of the nineteenth century, Memphis still did not have running water or piped sewage, so it relied entirely on the river and rain cisterns to collect water. Sewage and garbage removal were persistent problems,

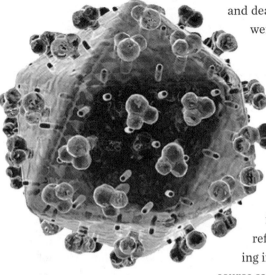

3D illustration of a yellow fever virus particle

and dead animals rotting on the street were a common sight.

Residents would normally dump their garbage and the contents of their privies into the Gayoso Bayou, a five-mile-long (8 km) creek that drained into the Mississippi River. In bad drought years like 1878, the Gayoso Bayou would grow stagnant, creating a putrid mix of dead animals, refuse, and human waste, all cooking in the summer heat. And then of course came the mosquitoes.

In the midst of Memphis's stench and squalor, thirty-four-year-old Kate Bionda operated an Italian restaurant and snack shop in the Pinch District, the city's first commercial district, largely populated by Irish immigrants who had fled the Potato Famine (1845–52). At the time, the neighborhood had a reputation for gambling, sex work, and carousing. Bionda's business primarily catered to riverboat workers—and there were many. Memphis was a hub for one of the major cotton-producing regions of the world, perfectly positioned on major American trade routes, with the Mississippi River and three railroad lines converging on it.

On the evening of August 1, a deckhand named William Warren stepped off a visiting steamer and ate at Bionda's restaurant. The next morning Warren got sick and was admitted to a quarantine hospital, where he would die three days later from one of the most dreaded diseases of the nineteenth century: yellow fever.

A viral disease transmitted by *Aedes aegypti* mosquitoes, yellow fever was named in recognition of one of its primary symptoms—jaundice, or a yellowing of the skin and eyes. An advanced symptom was black-colored vomit from internal bleeding. Once bitten by an infected mosquito, the virus travels via the blood to the lymph nodes where it replicates. Yellow fever varies in severity—most who catch it endure a few mild symptoms, such as aches and pains, fever, nausea, and dizziness for several days before it recedes. However, if the disease progresses,

A four-part representation of the progression of yellow fever

it attacks the entire body. The jaundice and vomiting kick in, along with delirium and internal hemorrhaging, with a brutal mortality rate. Between 20 to 50 percent of people who exhibit this level of infection will die—gruesomely—from the disease. An infected person can bleed from their nose, gums, and uterus and pass blood-filled stools. They will throw up a mixture of stomach contents, acid, and blood as "black vomit," (sometimes know by the Spanish term "vomito negro"). The heart becomes inflamed and beats erratically. The kidneys cease to work, and the inability to produce urine heralds death. A terminal victim will eventually fall comatose as their organs fail, causing death within seven to ten days of catching the disease. Perhaps the one saving grace of yellow fever is lifelong immunity if you survive.

The disease is endemic in Africa, where it spreads by bites from infected mosquitoes. We don't know when yellow fever first began infecting people, but some theories suggest the disease-causing virus originated with African monkeys, mutating at some point before it was able to spread when mosquitoes bit humans. As global trade networks began to mature and stretch across the Atlantic Ocean in the seventeenth century, yellow fever outbreaks became increasingly common in American summers, carried to its shores by sailors, traders, soldiers, and the enslaved.

Yellow fever thrives in hot and humid environments with stagnant water where mosquitoes flourish and multiply. While the vector of the disease (mosquitoes) remained unknown until the early 1900s, in the US it was often considered a disease that hailed from the South. Fear of an outbreak in major

US southern cities in the nineteenth century was perennial as the heat and humidity of the summer arrived. Considered strange at the time, the disease seemed to crop up with little warning during the summer, killing many people before receding again as the weather turned cold. Without an understanding of how it was spread but with a full understanding of its ravenous symptoms and high mortality rate, outbreaks of yellow fever in America would often spur a full-scale panic.

So when deckhand William Warren visited Kate Bionda's snack shop in Memphis and died from the disease, city officials never put his case on record, not wanting to create just such a panic. Instead, they held on to a feeble hope that Warren's was an isolated infection.

It was not.

Bionda herself caught yellow fever and died on August 13. The city's board of health was then forced to act, reporting the case, anointing her the official patient zero for the Memphis outbreak. Her body was burned within five hours of her death. The Bionda's restaurant and house were blockaded and all adjacent buildings were disinfected. The Memphis newspaper, the *Daily Appeal*, in Pollyannaish prose, attempted to stem the rising panic, writing, "The sad case of Mrs. Bionda, who left two little children and a grief-stricken husband, does not prove necessarily that others will follow. There is no need of a panic or stampede."

But that's exactly what happened. Every person with enough money and resources to do so left the city as soon as they could. Many residents remembered the last time yellow fever broke out in Memphis, five years earlier in 1873, claim-ing 5,000 lives. Within five days of the news of Bionda's death, 25,000 Memphians, a full 50 percent of its population, had fled the city in an exodus by train, carriage, and boat.

"On any road leading out of Memphis was a procession of wagons piled high with beds, trunks, small furniture, carrying also women and children. Beside walked men, some riotous, with the wild excitement, others moody and silent from anxiety and dread," wrote an Episcopal nun involved in relief efforts.

Meanwhile, yellow fever ripped through those left behind. Although Bionda was the first Memphis resident to die from it, she probably wasn't the first local to catch it. It was likely circulating on the outskirts of town by late July, as it trav-eled up the Mississippi River from New Orleans, embroiled in its own yellow fever epidemic at the time. By September, Memphis had been reduced to 19,000 residents from a starting population of 47,000.

Of those 19,000 people, 17,000 would catch the disease, and 5,150 would die from it.

◆ ◆ ◆

The first recorded yellow fever epidemic happened in Barbados in 1647, where it was called the "Barbados distemper." In response to that outbreak, Massachusetts governor John Winthrop established America's first ship-quarantine regulations to help protect the Massachusetts colony from the spread of disease from Caribbean trade routes. About 5,000 people died from yellow fever in the Barbados outbreak. The next year it spread from Barbados to Cuba, Saint Kitts, Guadeloupe, and the Yucatán Peninsula, spawning epidemics wherever it appeared.

And for the next 250 years, it was one of the most dreaded diseases in the world. Yellow fever made a perennial appearance in port cities during the warm weather of summer as it spread along maturing trade routes that connected Africa, Europe, the Caribbean, and the Americas. (Asia, curiously, has to this day managed to avoid an outbreak of yellow fever.)

Philadelphia was hit particularly hard in 1793, when it was serving as both the state and nation's capital. Refugees from a Caribbean outbreak of yellow fever fled to Philadelphia in the late summer of that year. Unfortunately, they brought the disease with them. What began as a handful of infections close to the Delaware River soon spread throughout the city, launching a full-scale

How Mosquitoes Defeated Napoleon

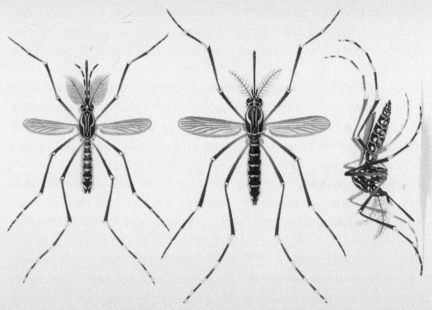

Aedes aegypti *mosquitoes: Males (left) do not bite for blood; only females (center and right) do and thus transmit disease.*

Napoleon's plans for global domination were hampered on multiple occasions by disease (see Typhus, page 161), most famously during his failed invasion of Russia. The general at one point also had great ambitions for an American empire. Before the Louisiana Purchase, France owned extensive North American territory, stretching from New Orleans north through the Midwest up to Canada, and Napoleon had plans to expand the French holdings even further if not for a rebellion by enslaved residents in French-occupied Haiti.

In 1801, a Black Haitian leader, Toussaint Louverture, led a successful revolt in Haiti, taking control of the island and prompting Napoleon to send an army to the Caribbean island to restore French dominion. Within just a few months of their arrival in Saint-Domingue, however, and well before the army could rout the rebels, a yellow fever epidemic ripped through the French troops, killing 27,000 of them. Only a handful of survivors remained to tell the tale.

Thoroughly defeated by the mosquitoes, Napoleon's army withdrew, surrendering what would later be known as Haiti to its own people. Napoleon was reportedly so disillusioned by the whole affair that he gave up entirely on the Americas, selling French territory—for a bargain—to Thomas Jefferson in the famous Louisiana Purchase that doubled the size of the fledgling United States.

panic. Almost 50 percent of Philadelphia's population fled town as the death toll reached up to one hundred people per day.

Among the refugees were George Washington, then serving as the country's first president, along with most of his cabinet, an unusual situation that created one of the nation's first constitutional crises. With no president in town, and with so many other congressmen absent, a regular meeting of Congress was impossible.

Alexander Hamilton suggested holding congressional sessions in another city, but this was specifically prohibited in the Constitution. The reason: English monarchs were notorious for moving Parliament to remote locations when they needed a vote to go in a certain direction. So the still-fledgling US government came to an abrupt halt while yellow fever ravaged Philadelphia for months.

Hamilton himself was a victim. Both he and his wife, Eliza, caught it. They sent their children away to stay with family in Albany, New York, while they convalesced in Philadelphia from the disease. Both recovered, but their attempt to reunite with their children was thwarted by armed guards charged with preventing Philadelphia traffic from from entering Albany. The Hamiltons were ultimately allowed to enter Albany, but not until their clothing and luggage had been burned, their servants disinfected, and their carriage abandoned. Furthermore, all five of Albany's physicians had to examine the Hamiltons and agree they were in good health before they were finally allowed to cross the Hudson and enter the city.

With the federal government in disarray and much of the state and local government having fled the city, management of the crisis came down to the city's mayor, Matthew Clarkson, and a group of dedicated volunteers, many of them free Blacks recruited for the effort by Dr. Benjamin Rush, a leading physician of the era who served as a surgeon general of the Continental Army and was also a signer of the Declaration of Independence. Erroneous medical thinking at the time had concluded that Black people had some innate immunity to yellow fever. They didn't, but a complex array of socioeconomic factors had contributed to some yellow fever outbreaks disproportionately killing white people and creating an illusion of Black immunity. Rush, a committed abolitionist, had ties to the Free African Society in Philadelphia, who agreed to supply workers.

At the time, the prevailing theory of disease was that it spread by bad air, so a variety of recommendations were issued to help "purify" the air, such as setting bonfires and firing guns. The smells of woodsmoke and gunpowder were every-where that summer. Citizens were also encouraged to breathe through cloths soaked with camphor or vinegar.

Hamilton wrote to the College of Physicians of Philadelphia to recommend the same course of treatment he believed resulted in his return to health: cold baths, glasses of brandy, and quinine enemas. But the truth was Hamilton's immune system simply conquered the disease, without the help of the surely unpleasant quinine enemas.

Benjamin Rush chimed in with a dramatically different approach. Since victims vomited blood, Rush concluded that the body needed to rid itself of blood during the disease. So he recommended copious bloodletting, as in drain-ing four-fifths of the body's blood supply. For good measure, Rush also offered his "ten and ten" cure, which was ten grains of highly toxic mercury combined with ten doses of a powdered jalap root, which promoted defecation. The com-bination of the two will in fact poison you and result in copious vomiting and diarrhea, both of which were symptoms yellow fever sufferers needed no help expressing.

The Philadelphia outbreak of yellow fever ultimately was not defeated by mercury or jalap root. In the end, it was the weather. When a cold front moved in with freezing temperatures at the end of October 1793, it killed off huge numbers of mosquitoes, substantially dropping case numbers. Within a few more weeks, the disease had vanished (at least in the eyes of Philadelphians unaware of how it spread in the first place). A full one-tenth of the city's population—between 4,000 and 5,000 people—died of the disease.

◆ ◆ ◆

As bad as Philadelphia's 1793 outbreak was, what happened in Memphis in 1878 was worse. Yellow fever first arrived in America that year via trade ships traveling from Havana, Cuba. In New Orleans, where it struck first, some 20,000 people were infected. Not until October, when cooler weather finally began to kill off some of the mosquitoes, did the outbreak relent, leaving approximately 4,000 people dead in its wake, with a mortality rate around 20 percent. The death toll would've been even worse if not for the fashionable trend among wealthy Southerners to leave the city for second homes elsewhere during the hot sum-mers. Because of that, a full one-third of the city's population wasn't in town during the outbreak.

In 1793, carriages in Philadelphia picked up those dying and dead from yellow fever.

After New Orleans, the virus traveled up the Mississippi River, hitting Vicksburg, Mississippi, before arriving in Memphis and blowing up. It would eventually spread up the Mississippi and Ohio River valleys, infecting at least 100,000 people, killing 20 percent of them, a vicious mortality rate that nevertheless paled in comparison to the nearly 70 percent the disease killed in Memphis the summer of 1878.

If you were to visit Memphis that summer, the scene would've been barely indistinguishable from a medieval village stricken with the plague in the fourteenth century. The city was unnaturally quiet and deserted, save a few people darting quickly about on essential business, their mouths covered with sponges. Coffins were stacked along Main Street. The homes of victims were marked with yellow pieces of cardboard, some with the words "Coffin Needed" scrawled on them, along with approximate dimensions of the body. The heat hovered near 100 degrees (37 degrees Celsius), which would've made the stench unbearable. Despite the heat, windows were boarded shut and citizens were advised to keep fires burning inside their homes in a misguided effort to "cleanse the air." A daily wagon patrolled the streets with undertakers issuing the dreaded call to "Bring out your dead."

Such was daily existence for the 19,000 remaining residents in Memphis for three horrid, drawn-out months, until the first frost of the year arrived in October, and with it the death of the mosquitoes that were transporting the disease. Whole families were wiped out. In one tragic example, a widow named Mrs. Flack, along with all seven of her children from age three to twenty-eight, died.

Yellow fever didn't spare the wealthy residents who attempted to flee the city, either. One particularly gruesome but sadly common scene in the country-side surrounding Memphis was a boarded-up house with rotting corpses found inside when the outbreak finally subsided. In one example, the Angevine family fled Memphis for their 4,000-acre country estate in Grenada, Mississippi, at the beginning of the outbreak. Mr. Angevine, a Memphis attorney, wasn't about to take chances with the disease that had claimed his wife the previous year. He boarded up his country house and prayed for their preservation from the disease.

Yellow fever found them anyway, infecting each member of the family until none could help the others. When an old family servant, a former enslaved man who stayed on as an employee after the Civil War, came down from Memphis, he discovered a tragic scene. Upon encountering a locked and boarded-up house with no signs of life, the servant pried open the shutters and broke through one of the windows. The stench of decomposition, heightened by the enormously warm summer, came rushing through the broken windowpanes.

Knowing what he would find, he proceeded anyway. One by one in the darkened rooms of the boarded-up house, he found the bodies of the Angevine family in various stages of decomposition. However, the last room he visited held a surprise. Nine-year-old Lena Angevine, the youngest member of the family, was lying on the floor with flies buzzing around her head. But her body had not yet begun to decompose. The servant carried her out of the house and placed a piece of raw bacon on her mouth. He watched, amazed, as the child began to suck on the bacon, the first bit of food she must have had in days. On the very brink of death, she was found just in time, and over the next few weeks, she would recover fully from the disease.

An ambulance used for yellow fever patients in New Orleans, 1905

US Army physician Dr. Walter Reed, who discovered that mosquitoes transmit yellow fever

In a poetic turn of fate, Lena, now immune to yellow fever, became a nurse later in life. In 1900, she answered an advertisement calling for nurses immune to the disease to help US Army doctor Walter Reed investigate yellow fever in Cuba during the Spanish-American War. In the conflict, more American troops were dying from yellow fever than in battle, and the US government pinned its hopes on Reed to solve the mysteries of the disease. With Lena at his side in Cuba, Reed first proved conclusively that the mosquito (specifically the *Aedes aegypti*) was the vector of yellow fever. After the discovery, cities began to make concerted efforts to eradicate mosquitoes, leading to a plummet in yellow fever outbreaks, with the last major American epidemic of the disease occurring in 1905.

The viral nature of yellow fever was discovered in the 1930s, and a vaccine followed the same decade after a major breakthrough by South African virologist and physician Max Theiler, who would go on to win the Nobel Prize for his discovery. (As of this writing, Theiler is the only person to win a Nobel Prize for a virus vaccine.) While the disease was effectively eliminated from the developed world, it's still endemic in Africa, South America, and several Caribbean islands where mosquitoes persist in large quantities and immunization and mosquito eradication programs have lagged due to lack of resources and rapid urbanization.

Memphis, meanwhile, significantly ramped up sanitization efforts after the 1878 outbreak subsided with the arrival of cold weather. While the cleanliness efforts were aimed at a misidentified source of the disease (medical thinking at the time centered on germs as the agents of transfer), the cumulative effect was positive since sanitary streets and clean drinking water are less friendly to mosquitoes.

As it turned out, not allowing animal corpses to rot in the street has a variety of positive benefits.

COVID-19

The Origins of SARS-CoV-2 and the Politicization of Plagues

Impact: The US leads the world in COVID-19 infections and deaths as of early 2021, withdrew from the World Health Organization (WHO) in July 2020, rejoined in January 2021

When: December 2019

Where: Possibly Wuhan, China

Who: Still under investigation by the WHO, Chinese authorities, and international scientists

What Happened Next: Obfuscation and blustering by world leaders who attempted to play politics with the WHO

Little-Known Fact: The animal that hosts a coronavirus most similar (96.2%) to SARS-CoV-2 is the intermediate horseshoe bat, *Rhinolophus affinis*.

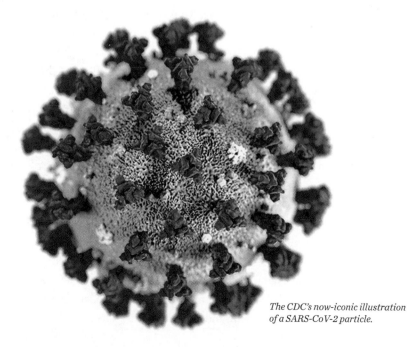

The CDC's now-iconic illustration of a SARS-CoV-2 particle.

As of January 2021, COVID-19 infected more than 90 million people worldwide and killed nearly 2 million. At the time, the US led the world in cases with more than 22 million, with the virus claiming more than 400,000 lives. For a stark comparison, the 2019–20 flu season in the US killed an estimated 24,000 to 62,000 people. Across a number of waves in the US—or one long one, if you prefer—ICU beds became scarce and healthcare workers and systems strained to keep up with the increasing and seemingly relentless rise in cases. The novel coronavirus that became known simply as COVID to the general public has tested doctors, healthcare facilities, scientists, leaders, and everyone else in ways that were both unexpected and shocking. COVID-19 harkened back to the deadly 1918 influenza pandemic, with some haunting similarities to the HIV global epidemic that continues today.

At the beginning of the pandemic, a flurry of questions arrived with the SARS-CoV-2 virus. What was it? How bad was the disease itself? How did it spread? Was there a test for it? Why didn't we have it yet? And of course the all-consuming one to which we still don't have an answer: Where did it really come from? This relative unknown was nearly as frightening as the virus itself. And yet most of us had no idea what was really coming.

In a world where we can receive information in the milliseconds it takes a tweet to travel the online universe, what we received between December 2019 through the beginning of March 2020 didn't come quickly or accurately enough. But data and details of something as big as an epidemic and emerging pandemic never arrive directly to the general public. They are filtered through complex webs and chains of people, organizations, and governments. And the information is not simple data spewed out by an emotionless computer—when it comes to disease outbreaks, narrative comes first, statistics second.

The story of the COVID-19 pandemic began in late 2019. And its underlying narrative, perhaps more than that of any other previous pandemic, is a tale of how pandemics are, and have likely always been, politicized. That includes the infectious disease origin stories—the core "patient zero" discoveries epidemiologists strive to uncover so they can build a fact-based understanding of a new pandemic. After all, more than one in ten deaths across the globe are caused by infectious diseases—and that's in non-pandemic times. Suffice to say these diseases, novel or otherwise, don't happen in isolation.

As is widely known, a new type of pneumonia appeared in December 2019 in Wuhan, China, confounding physicians. Though Chinese authorities

An entrance to the Huanan Seafood Market in Wuhan, China, theorized to be the source of the COVID-19 outbreak.

would officially claim that the first patient with COVID-19 was diagnosed on December 8, other sources cite December 1, or as early as November 17. By December 26, two companies (BGI Genomics and Vision Medicals, both based in China) were already attempting to sequence samples from sick patients. Sequencing the virus would reveal if it was novel or old, and the virus type. Also, if it was similar to other viruses doctors might already know how to treat, and how to prevent their spread. Knowing a new virus's genome would further mean being able to create testing kits and vaccines, as well as allow for tracing its origins and evolution, both steps crucial to stopping it.

According to the Xinhua News Agency, the state-run press agency controlled by the People's Republic of China, on December 27 critical care physician Zhang Jixian filed a report in her Hubei hospital about three family members suffering from the same pneumonia of unknown cause, speculating that "It is unlikely that all three members of a family caught the same disease at the same time unless it is an infectious disease." More cases appeared beyond Wuhan, the capital of Hubei Province, prompting the Wuhan Municipal Health Commission to investigate. However, *Caixin Global*, a privately owned news outlet, reported later that the two companies—BGI Genomics and Vision Medicals—knew by December 29 that it was a novel coronavirus. The genome had also been sequenced, though the magnitude of the finding was not understood at the time.

On December 30, medical institutions in Wuhan were given an urgent notice by the Wuhan Municipal Health Commission. Cases of pneumonia appeared linked to the city's Huanan Seafood Wholesale Market, where not just fresh seafood was sold but live animals, such as badgers, snakes, turtledoves, giant salamanders, hedgehogs, foxes, raccoon, and sika deer. Bats, common suspects in zoonoses, when diseases jump from animal to human (see Zoonosis, page 12), were absent. The message was leaked online.

Dr. Ai Fen, head of the emergency department at Wuhan Central Hospital, recorded and then sent a clip of a CT scan showing an infected patient's lungs to a former classmate at Tongji Hospital. Ai later saw a lab report with a positive SARS test result, which was erroneous (the virus would eventually be identified as SARS-CoV-2, not SARS-CoV-1, the cause of the SARS outbreak in 2002). Ai circled the result in red and sent it to classmates and colleagues at her hospital.

That same day, ophthalmologist Dr. Li Wenliang sent a message to several medical school classmates about "seven SARS cases from the Huanan Fruit and Seafood Market." Another urgent notice by the Wuhan Municipal Health Commission went out later in the day urging providers to track cases and report them. This message, too, was quickly leaked online.

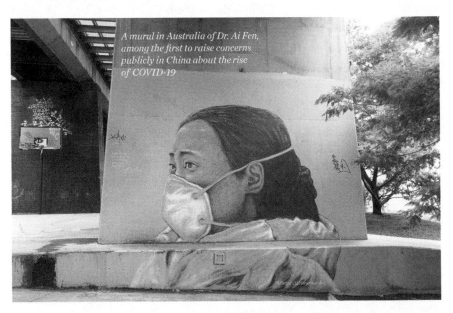
A mural in Australia of Dr. Ai Fen, among the first to raise concerns publicly in China about the rise of COVID-19

武汉市公安局 武昌分局 中南路街派出所
训诫书
武公（中）字（ 20200103 ）
被训诫人 李文亮   性别 男   出生年月 19861012
身份证号码或及号码：
现住址（户籍所在地） 武汉市：

工作单位： 武汉市中心医院
违法行为（时间、地点、参与人、人数、反映何问题、后果等）：
2019年 12月30 日在微信群 "武汉大学临床 04级" 发表有
关华南水果海鲜市场确诊 7例 SARS 的不属实的言论。

现在依法对你在互联网上发表不属实的言论的违法问题提供告
示和训诫。你的行为是严重扰乱了社会秩序。你的行为已超出了法律
所允许的范围，违反了《中华人民共和国治安管理处罚法》的有关
规定，是一种违法行为！
公安机关希望你积极配合工作，听从民警的规劝，至此中止违
法行为。你能做到吗？
答：能

我们希望你冷静下来好好反思，并郑重告诫你：如果你固执己
见，不思悔改，继续进行违法活动，你将会受到法律的制裁！你听
明白了吗？
答：明白

被训诫人：李文亮   2020 年 1 月 3
训诫人：胡桂芳 徐炼   工作单位：

The letter Dr. Li and others were required to sign by the Chinese authorities after trying to bring attention to the COVID-19 outbreak.

On December 31, *Yicai*, the financial arm of the state-owned news outlet Shanghai Media Group, confirmed the leaks and published a story about the outbreak in Wuhan. One minute before midnight, a user of the US-based listserv ProMED (Program for Monitoring Emerging Diseases) posted the *Yicai* article.

And that is how the WHO learned of the outbreak—not from the Chinese government.

The next morning, January 1, the Huanan Seafood Wholesale Market was shut down. A team from Beijing-based Chinese Center for Disease Control and Prevention (China CDC) collected 515 samples from the market. It was the right idea but too late: Vendors reported that mask-wearing workers had already sprayed down the market with disinfectant two days prior, greatly reducing the chances of acquiring quality samples for testing. The China CDC would return on January 12 to take seventy more samples, specifically from the wild animal vendor stalls. But no samples were taken directly from the actual animals at the market, which made it impossible to prove that *if* the COVID-19 pandemic truly originated from the market, it came from a human patient zero or an original animal spillover source. (For comparison, when the Middle Eastern respiratory syndrome virus, MERS-CoV, was found to be linked to camels, it was confirmed by direct samples from Egyptian slaughterhouses.)

That same day, on January 1, Dr. Ai, Dr. Li, and six others were investigated by the Wuhan Municipal Public Security Bureau for "spreading rumors" and creating a "negative social influence." Days later, Dr. Ai and Dr. Li were severely reprimanded by their hospitals as well, and Li was required to post an online letter declaring that his "false statement" had "severely disturbed social order." Ai was also forbidden to speak to anyone about the pneumonia cases, though she later regretted that her silence cost lives because she didn't "keep screaming." The Hubei Health Commission then ordered genomic companies to destroy all samples from the outbreak and stop testing, despite them already having crucial answers in hand. The Chinese state machine was now in full swing to control the outbreak narrative.

◆ ◆ ◆

By January 2, there were officially forty-four people with the strange new pneumonia, some with ties to the Huanan market. But about a third of the victims had no direct connections to the market, making the Huanan source theory questionable. Eleven of the forty-four were extremely ill.

On January 3, the China CDC finished identifying and sequencing the novel coronavirus, despite yet another entity (the Wuhan Institute of Virology) having done so the day before. But the China CDC did not announce the finding. Officially, they would claim that the sequencing was finished on January 7. What did those few days mean? Acting on the genomic sequence of a pathogen days into an outbreak can result in a meaningful jump on getting control of it. It can also lead to a quicker understanding of its source, development of effective testing kits, and discovery of pharmaceutical treatments.

This time, the more powerful Chinese National Health Commission "ordered institutions not to publish any information related to the unknown disease, and ordered labs to transfer any samples they had to designated testing institutions, or to destroy them," according to *Caixin Global*, whose journalist saw the report. Nevertheless, professor Zhang Yong-Zhen of Fudan University in Shanghai, a famed virologist, sequenced the virus and submitted it to the GenBank database. The latter is a publicly available National Institutes of Health genetic sequence database that includes international databanks and enabled the sequence to be shared openly worldwide. He waited for the sequence to be reviewed and notified the Chinese National Health Commission that the virus was like SARS, and likely very infectious via respiratory passages. "We recommend taking preventive measures in public areas," his notice read.

At this point, the WHO, the US, and China were well aware that an outbreak was underway, though the timeline of the origin was being methodically obscured by the Chinese government. The WHO issued a formal statement on January 5 stating that, "Based on the preliminary information from the Chinese investigation team, no evidence of significant human-to-human transmission and no health care worker infections have been reported." The WHO further advised "against the application of any travel or trade restrictions on China" at the time.

On January 7, an infected sixty-nine-year-old patient underwent surgery at Wuhan Union Hospital, and would spread COVID-19 to fourteen medical workers, making him the first "superspreader." Chinese authorities did not let medical professionals know about the infections until fourteen days later. The *Wall Street Journal* reported on January 8 that Chinese scientists had genetically identified a new coronavirus based on the Wuhan pneumonia, deeply embarrassing both the WHO and the China CDC. The WHO's executive director of its Health Emergencies Programme, Michael Ryan, went on record with his frustration: "The fact is, we're two to three weeks into an event, we don't have a laboratory diagnosis, we don't have an age, sex, or geographic distribution, we don't have an epi curve." These are all basic data points necessary early on in an outbreak investigation (see Anatomy of an Outbreak, page 87). On January 9, the WHO continued to not recommend specific measures for travelers, or travel restrictions related to China. On January 10, the WHO recommended *against* entry screening

Dr. Michael Ryan, director of WHO's Health Emergencies Programme

for travelers, stating that "there is no significant human-to-human transmission, and no infections among health care workers have occurred."

By January 11, China had its first death, a man who had frequented the Huanan market related to a growing number of COVID-19 cases. That day, on behalf of Professor Zhang and his team, Australian virologist Edward C. Holmes published the novel coronavirus sequence publicly on virological.org, a site used by international researchers, while noting that the sequence had also been sent to GenBank. Though the China CDC had planned on sharing the sequence with the world, they'd apparently wanted to do so on their own terms, and to have their own China CDC scientists awarded the credit first. The delay was possibly due to an internal competition within the system and the desire to control the narrative of how the sequence was obtained. The China CDC responded to the push for getting the genome out by quickly sharing its sequences for the novel virus on an influenza sharing platform, GISAID (Global Initiative on Sharing All Influenza Data). The China CDC shuttered Zhang's lab the next day in retaliation. Approximately 600 more people were infected that week, a roughly threefold increase.

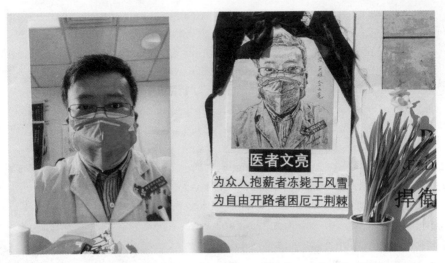

医者文亮

为众人抱薪者冻毙于风雪
为自由开路者困厄于荆棘

Dr. Li, an early whistleblower on Chinese government efforts to hide the severity of the COVID-19 outbreak, would later succumb to the virus.

The next day, on January 12, Dr. Li—the whistleblower who leaked early information about the outbreak online—became ill with the new virus. The WHO again claimed that the cases were all linked to Wuhan's Huanan Seafood Wholesale Market, and that "there is no infection among healthcare workers, and no clear evidence of human to human transmission." Finally, on January 13, when two experts from Taiwan's CDC visited Wuhan to investigate the outbreak, they were told by a health official that "limited human-to-human transmission cannot be excluded," a step away from denying any transmission was occurring, but still a non-admission that obfuscated a terrible truth. One Taiwan expert, Chuang Yin-Ching, was there to listen to the Chinese presentation about the outbreak in Wuhan and interpreted the words with more clarity: "That means human-to-human transmission absolutely."

Thailand soon reported its first case, a traveler from Wuhan. On January 14, the WHO tweeted that there was "no clear evidence of human-to-human transmission of the novel #coronavirus (2019-nCov)," even though the acting head of WHO's emerging diseases units announced in a press conference that "it is certainly possible that there is limited human-to-human transmission." Lunar New Year celebrations began on January 18, including huge numbers of travelers within China and, of course, in and out of Wuhan.

By January 20, new cases were being seen in the US, Japan, and South Korea.

In the US, the first case was a thirty-year-old man in Washington state who traveled to America from Wuhan. On January 26, the China CDC would report large quantities of the novel coronavirus in thirty-three out of the 585 samples from the Wuhan market. Thirty-one of them came from the wildlife vendor areas.

◆ ◆ ◆

On January 23, Chinese authorities cut off Wuhan and its eleven million inhabitants from the rest of the world. The draconian measures at first meant no public or private transportation at all, and people were allowed out of their homes every two days to get food and supplies. But soon, no one was allowed out, and couriers delivered necessities. Temperature monitoring was widespread and bleach foot baths were used before entering some areas. Door-to-door health checks sought out anyone with symptoms, exposures, or suspected of having the novel coronavirus. They were taken (sometimes forcibly) to temporary quarantine and isolation centers. The streets were empty. Most stores and all schools were shut down. Hong Kong–based *South China Morning Post* reported that a boy with disabilities died when his caretaker family members were removed for quarantine.

Even while this happened, an intense discussion was underway among WHO Emergency Committee members over whether or not to declare a global emergency. According to the *New York Times*, diplomats and health officials noted that China had lobbied them not to do so, their concern being the declaration would indicate that the WHO had no confidence in China's ability to investigate and control its own outbreak. The global emergency was not declared. Instead the WHO's director general, Tedros Adhanom Ghebreyesus, publicly praised China's leader, Xi Jinping, and the Chinese pneumonia surveillance system despite its failure to spot the outbreak.

The WHO was treading a careful diplomatic path. Funding by a collection of nations covers the vast majority of its budget. Between 2018 and 2019, the US was the WHO's largest donor, contributing more than $850 million toward its budget, compared to China's $89 million. The agency, created after World War II and operating under the authority of the United Nations, was deeply underfunded and couldn't continue to function without its top donors. Also, despite the cooperation of multiple nations who strive to meet the WHO's guidelines on global health, the WHO does not have the authority to enforce international regulations, nor can it independently investigate epidemics within a given country.

Dr. Tedros Adhanom Ghebreyesus, director-general of the World Health Organization, with Chinese President Xi Jingping, January 2020

As a global power that increasingly rivals the US, China is not a country the WHO wishes to upset. Tedros also had a reputation for preferring not to overtly criticize member countries. Lawrence Gostin, Professor of Global Health Law at Georgetown University, commented that "His strategy is to coax China to transparency and international cooperation rather than criticizing the government." Given their limited power and authority, the WHO naturally sought as much cooperation as possible. Since access was tantamount to success, showing countries in a positive light was a useful strategy, or had been. Tedros and the WHO also did not want to draw the ire of Chinese authorities, thereby getting Chinese scientists in further trouble.

But by the end of January, the US restricted travel from China. Official infections were at 9,800 worldwide, with 213 deaths. Tedros and Xi met in Beijing on January 29. The WHO were subsequently allowed to send an international group of scientists to investigate the outbreak. The WHO requested that China share biological material, an indication they had not yet done so. Tedros remarked on this diplomatically, stating that "We appreciate . . . the transparency they have demonstrated, including sharing data and genetic sequence of the virus." On January 30, the WHO reconvened its Emergency Committee. Despite the positive comments from Tedros the day before—and against the advice of the Chinese delegate—the committee finally declared a PHEIC—a Public Health Emergency of International Concern. Tedros softened the announcement by saying, "Let me be clear: This declaration is not a vote of no confidence in China," adding, "The WHO doesn't recommend limiting trade and movement," knowing that any recommendation affecting either would infuriate leaders around the world.

Not limiting trade or movement made the virus go global by February. It and the disease it caused now had names: SARS-CoV-2 and COVID-19 (from **CO**rona**VI**rus **D**isease 20**19**), a designation that eliminated the blame-filled, origin-based "Wuhan Virus" that had been gaining traction until then.

On February 3, a Japanese cruise ship was quarantined and eventually reported 712 infected passengers. Dr. Li died of COVID-19 on February 7. The virus then exploded and overwhelmed Italy. A WHO mission to China sent to understand the initial outbreak agreed not to investigate China's early response, or a possible animal source. At a time when every early moment was crucial to finding the original source of the virus, the trail for COVID-19 had already gone cold under the opaque shroud of China's state control.

The lack of transparency on the part of the Chinese government led to conspiracy theories, one of which emerged in late January and claimed that the virus was leaked (either accidentally or on purpose) from a biosafety level 4 lab in Wuhan. Rumors included that it was human-made. Seemingly in retaliation and to point blame at anywhere else but China, a counter rumor emerged in early February that the virus was of US origin, a biological weapon unleashed on Chinese soil.

Despite the disease being renamed COVID-19 by early February, many US officials continued referring to it as the "Wuhan virus" when discussing the infection, a clear rebuttal to the US-origin conspiracy theories. US President Donald Trump then upped the ante by referring to it as the "Chinese virus" on March 16, crossing out "Corona" in his speech notes and replacing it with "Chinese." Despite criticism of the term being xenophobic and racist, anti-Asian sentiment spread in the US and through media sources. It was no surprise when anti-Asian hate incidents and crimes flourished in the months that followed, echoing what happened during the SARS outbreak in 2002, when racial tensions against Asian Canadians flared in Toronto.

But the US finger-pointing at China and the WHO for botching the initial messaging and control of the pandemic was clearly done in part to deflect blame from the US government's own missteps in allowing the infection to take hold on US soil. The seriousness of the outbreak in China was downplayed by the US at first, as if the virus couldn't possibly explode outside of China the way it already

An image of Donald Trump's speech from March 19, 2020, with the word "Corona" crossed out and replaced with "Chinese."

had. By the time testing kits were available in March, the virus was already spreading widely in the US, and still federal officials considered the pathogen to be only as dangerous as a regular flu virus—a grave mistake. Trump complained in early April, tweeting, "The W.H.O. really blew it. For some reason, funded largely by the United States, yet very China centric. . . . Fortunately I rejected their advice on keeping our borders open to China early on. Why did they give us such faulty recommendation?" And so began the US narrative machine, led largely by Trump and those politically loyal to him, including various conservative media outlets like Fox News.

◆ ◆ ◆

US social media accounts quickly became rife with claims that disease outbreaks, such as COVID-19, only happen during major election years. A viral meme circulated, pointing out the pattern: SARS 2004, Avian 2008, Swine 2010, MERS 2012, Ebola 2014, Zika 2016, Ebola 2018, Corona 2020. The claim was swiftly debunked, given that the years for SARS, Ebola, swine flu, and Zika were wrong—and the other diseases like avian flu and MERS had no meaningful bearing on US elections.

Meanwhile, the US Centers for Disease Control (CDC) was criticized for its initial steps to address the pandemic. The CDC's first COVID-19 test kits were faulty, related to contaminated reagents—likely a side effect of their rushed manufacturing. The CDC also flip-flopped on the necessity of public mask wearing, even as understanding grew that the virus could spread in asymptomatic and pre-symptomatic people. All the while their recommendations were sidelined by the White House and those allied with it. A traditionally nonpolitical entity, the CDC had to walk a fine line between policymakers and their own scientists to guard their reputation as a neutral health authority. But their recommendations were repeatedly downplayed or subjected to forced edits by the Trump administration, which controlled the CDC via the US Department of Health and Human Services (HHS). As a result, their other public health recommendations, such as social distancing and shuttering churches, schools, and businesses, were met with partisan fury.

Naomi Oreskes, a science historian from Harvard University, noted, "The exploitation of uncertainty of science is a very nefarious intellectual strategy, but it's effective because people are disinclined to act anyway."

Drugs can also become political fodder. When Trump learned in March 2019 of a small, flawed study that showed a benefit against COVID-19 from a well-known rheumatological drug called hydroxychloroquine, he pushed for it relentlessly as a treatment, even claiming to take a "preventative" course of it himself. At the time, there were no known treatments for COVID-19 and deaths were mounting, along with desperation. Still, politics outshone science. On April 4, "speaking on gut instinct," Trump authorized the US government to purchase and stockpile 29 million hydroxychloroquine pills. The FDA issued an emergency use authorization, which was later revoked on June 15. After multiple carefully conducted studies, hydroxychloroquine was determined not to be the saving grace for which leaders like Trump had hoped. (Notably, when Trump later became ill with COVID in October, his doctors did *not* prescribe him hydroxychloroquine, nor did they inject disinfectant or hit him with powerful UV light, two other recommendations that Trump floated during a news conference on April 23.)

On May 19, the WHO adopted a landmark resolution to work with organizations and countries to identify past zoonotic sources of outbreaks and how they were introduced to human populations—in particular, COVID-19. China eventually cosponsored the resolution, as it would reflect poorly on them if they didn't cooperate. After submitting a list of demands related to the WHO resolution, Trump threatened to leave the organization if they were not met. It was a move to force Tedros's hand, put pressure on China, and reinforce US influence within the WHO, but it backfired—Tedros did not negotiate. The day the resolution was delivered at WHO headquarters in Geneva, Trump withdrew the US from the organization, formally severing ties in July. Months later, Tedros would remark on Twitter, "Governments should focus on tackling the virus and avoid politicisation."

◆ ◆ ◆

Politics in the realm of disease is not new. Outbreaks and pandemics have been tooled and weaponized by political figures and parties throughout history. Often, political allegiance speaks more to how the public agrees or disagrees with specific health issues.

In 2020, the COVID-19 pandemic played an enormous role in the US presidential election. It started early in the year when Trump heard about the virus. In March, as it began readily spreading through the country, he knowingly downplayed it. Throughout the year, between early February and the end of October, Trump said forty times that the pandemic would simply go away. ("One day, like a miracle, it will disappear." "It would go away without the vaccine." "Now, the virus . . . a lot of people think that goes away in April with the heat.") But by summer, Trump's assurances were proven wrong, as COVID-19 cases and deaths continued without abating. Throughout the pandemic, Trump disagreed with Drs. Anthony Fauci and Deborah Birx, well-known public health authorities and lead members of his own White House Coronavirus Task Force, on testing, masks, and the aggressiveness of the US response.

Meanwhile his presidential opponent, Joe Biden, wore masks at events and quickly shifted to virtual campaigning. Trump and his circle within the White House shunned the wearing of facial masks and derided Biden repeatedly for wearing one. This occurred while the vast majority of Americans supported the wearing of masks. Biden's pandemic approach became "truth over lies, and science over fiction." And the truth was backed up in the numbers. By July, the US led the world not in its healthcare response, but in the number of COVID-19 deaths. The lack of testing capability, poor contact-tracing efforts, lagging shutdowns of schools and gatherings, and premature reopening of state shutdowns all played a part.

In early October, Amy Coney Barrett was sworn in as a US Supreme Court judge at the White House, where many celebrants were seen in close quarters indoor without masks. A high point of Trump's presidency, the event would soon be sullied by the news that it was a COVID-19 superspreader event, resulting in the

Trump removes his mask while still infectious with COVID-19, upon returning to the White House after receiving treatment for the disease.

infection of forty-eight people, including US senators, Congressional representatives, media personnel, and White House staff. Trump and First Lady Melania Trump were infected, resulting in Trump's hospitalization for three days.

Trump brushed off the illness, yanking off his mask immediately upon returning to the White House despite likely still being contagious.

A Reuters/Ipsos poll at that time revealed that 67 percent of registered voters agreed that if Trump had taken the virus more seriously, he wouldn't have gotten infected. With less than a month to Election Day, AP-NORC polls showed 65 percent of Americans believed the president had not taken the coronavirus outbreak seriously enough. In August, he had only a 31 percent approval related to his handling of the pandemic, a drop from 44 percent in March. For voters whose number one election concern was the pandemic, 81 percent of them voted for Biden.

Shortly before the election, the Chairwoman of the Republican National Committee, Ronna McDaniel, said, ". . . if he loses, it's going to be because of COVID." Trump's former homeland security advisor, Tom Bossert, noted, "He's always been better at controlling the narrative than the levers of government."

On November 7, 2020, Trump officially lost his reelection bid to Biden.

◆ ◆ ◆

The search for the origins of SARS-CoV-2 continues. But if there's one thing to be learned from the way the COVID-19 pandemic unfolded, it's this: It is perhaps the most highly politicized pandemic in history. But pathogens do not have political motives. They take advantage of gaps in time, of slowed efforts and human in-fighting from politics, of marginalized and poor communities, never stopping as they evolve and maneuver around and through us. Facts and science alone do not fully drive the outcome in pandemics. Looking back to the beginning, to the mistakes, to the expected spillover, can help save lives the next time.

Politicization of Diseases

GERM THEORY

Germ theory espoused contagion as the means of spreading diseases, and quarantines were often a means of controlling them. Anti-contagionists, who believed in the miasma and "filth" theory of disease, considered quarantines and the "cordon sanitaire" stifling to commerce and individuality and a sign of reactionary despotism. The anti-contagionist beliefs in associating filth with disease also meant that people of lower socioeconomic classes were labeled as dirty and disease-producing.

1976 US "SWINE" FLU VACCINE

On the cusp of a reelection bid, US President Gerald Ford pushed to order a nationwide vaccination campaign against a novel H1N1 "swine" flu virus. The subsequent campaign was ambitious and rushed, and the flu season turned out to be mild. The vaccine used was live attenuated (see page 248), which took less time to produce than a killed vaccine, but possibly resulted in a small number of Guillain-Barré syndrome cases, a neurologic disease with a superficial resemblance to polio (see page 262). Ford would lose his reelection bid, but the American public would not forget the fear stoked by rushed federal public health actions.

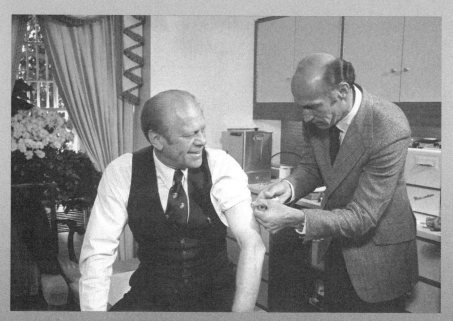

President Gerald Ford photographed receiving a swine flu inoculation in 1976.

HARM REDUCTION VIA NEEDLE EXCHANGE PROGRAMS

Needle exchange programs (NEPs), in which used syringes from non-prescribed drug use are exchanged for sterile ones, aims to prevent the sharing of needles and thus deadly diseases such as HIV, hepatitis C, and hepatitis B. The data is abundantly clear that NEPs reduce risk of transmission and save taxpayer money in the long run. But many lawmakers worry that it promotes non-prescribed drug use and crime. (It doesn't.) In 1988, Congress prohibited federal funding of such programs. This was reversed in 2016—except that no funds can be used to purchase syringes. As of 2021, there are only about 185 NEPs in thirty-eight states, despite all the evidence of the good they do.

ZIKA VIRUS

When Zika began infecting people in 2016, the goals of preventing its spread and the teratogenic effects on the unborn seemed straightforward. Eliminate mosquitos (the Zika virus vector). Use mosquito repellents to keep from being bitten. Don't travel to areas with Zika. And avoid getting pregnant if you must be in such an area.

What this plan completely neglected was the critical need for political sway within an epidemic. Politics—and policy—are necessary for public health. Those at highest risk for Zika were disproportionately marginalized and poor people in Brazil, especially women. They were at risk for sexual abuse, with no options to move to safer areas, and did not have access to healthcare, contraception, or abortion. In other words, much of the burden to fight Zika was placed on lower-income women without the means or choices to help themselves.

YELLOW FEVER

When the infamous yellow fever epidemic of 1793 struck Philadelphia (see Yellow Fever, page 100), beliefs about the origin of the scourge ran along party lines: Democratic-Republicans (a single party at the time) believed that the source of the disease was local, related to unsanitary conditions. The Federalists (opponents of the Democratic-Republicans) blamed French immigrants from the Caribbean. Fearing revolutionary France, they wished to keep French immigrants at more than arm's length. Conversely, Democratic-Republicans appreciated the immigrants and their link to the lucrative West Indies.

Cures, albeit fruitless at the time, also fell along partisan lines. Founding Father and physician Benjamin Rush, who bravely stayed behind to treat victims of the epidemic, was a proponent of "heroic depletion therapy," which involved unsafe amounts of bloodletting and purging by consuming copious amounts of a toxic mercury compound called calomel. Rush's methods could be done by anyone, as opposed to the more elitist practices of the Federalists who espoused "stimulant" methods, such as wine, cold baths, and cinchona bark containing a relative of the hydroxychloroquine compound that was touted as a cure for COVID-19.

UKRAINE AND "SWINE FLU"

In 2009, Ukrainian prime minister Yulia Tymoshenko was running for president when the H1N1 "swine flu" popped up in the US and began spreading around the world. Contrary to the WHO recommendations, Tymoshenko fanned fears by pushing school closures, quarantines, and public gatherings like political rallies (after hers had already happened). The frightening pandemic was a phantom crisis created for a hero—Tymoshenko—to solve. In the end, Tymoshenko lost the election, but as her senior campaign advisor put it, "We won in the media."

Former Ukrainian prime minister Yulia Tymoshenko, who lost her election to become president in 2009 while overplaying the dangers of the swine flu.

EBOLA

The Ebola outbreak hit West Africa in 2013 but only arrived on American shores in 2014 with a scant handful of cases. The arrival coincided with the midterm elections. Overwhelmingly Republican politicians politicized the threat, calling for stricter immigration laws, closing the southern border, halting flights from affected countries, and criticizing the Obama administration for not doing enough—and Republicans reaped big electoral gains in that election year.

THE HAMBURG CHOLERA EPIDEMIC

In 1892, cholera struck Hamburg, a self-governing city under the shadow of a centralizing-leaning Prussian government. Chemist and public health leader Max von Pettenkofer believed that individuals were responsible for their health. He'd previously refused to have river water filtered to residents, or acknowledge that cholera deaths were doubling by the day. Robert Koch arrived to help, but his germ theory butted heads against von Pettenkofer's anti-contagionist ideas. Von Pettenkofer even went so far as to drink a glass of bouillon laced with cholera bacteria. He got sick, but survived—being either too healthy to succumb or previously inoculated.

But it was Koch who truly proved his point. He was able to lead Hamburg out of the scourge by providing a clean water supply, disinfecting homes, and supporting the hospitals. Ten thousand died due to von Pettenkofer's decisions. Soon after, businessmen were voted out of Hamburg in favor of scientifically minded politicians who favored health over profit, and German-unified Hamburg became self-governing no more.

SPREAD

HIV virus concept illustration

HIV
Patient ZERO
A Primate

Cause: Human immunodeficiency virus (HIV)

Symptoms: Initially fever, rash, body aches, and fatigue; when progressed to acquired immunodeficiency syndrome (AIDS), symptoms reflect diseases that invade the host, such as Kaposi's sarcoma, diarrheal infections, brain infections, and lymphoma, among many others

---

Impact: Markedly affected individual behavior, relationships, communities, and the economic growth of nations. Continues to be a major global health issue.

Where: The Democratic Republic of Congo, now worldwide

When: 1981

Transmission: Sexual, blood product transfusions, needle sharing and accidental needlesticks, pregnancy, childbirth, and breastfeeding

---

Little-Known Fact: The virus is more than one hundred years old and was likely circulating in the US for about a decade before it was recognized in the early 1980s.

In June 1981, doctors and public health officials took notice of a small article in the *Morbidity and Mortality Weekly Report,* a weekly Centers for Disease Control (CDC) publication that reports on infectious diseases. The case study by Michael Gottlieb, an immunologist at the UCLA Medical Center, highlighted five patients in Los Angeles with odd infections. They all had "PCP," a pneumonia caused by *Pneumocystis carinii* (nowadays called *Pneumocystis jirovecii*).

Pneumocystis is a yeast-like fungus that's everywhere. But most people don't inhale it and die of pneumonia, so it was strange to see a cluster of patients suffering from it out of the blue. In these patients, all men, it was replicating inside their airways, damaging the alveoli, or minute air sacs, where oxygen is absorbed from the air into the blood vessels. Without treatment, PCP would asphyxiate the men—and in some cases it did.

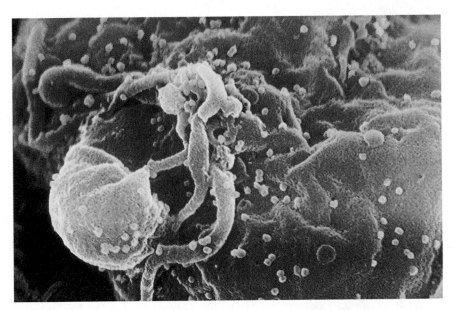

A scanning electron micrograph showing HIV-1 virus (in green) on a white blood cell called a lymphocyte

The men also had coughs and fevers, classic symptoms for any pneumonia. But it wasn't the only problem—all five also had a viral infection called CMV (cytomegalovirus, or one of the causes of mononucleosis, known as mono). Only in the LA patients it was also causing pneumonia and, in some cases, an inflamed esophagus. Four of the cases also had candidiasis, a yeast growing in creamy white plaques in their throats and esophagus. Yeast is delightful in bread and beer, but dreadful when it overtakes your organs. The CDC report was striking because these patients were being attacked by diseases that normally don't bother healthy people. And the attacks were happening as if the patients' immune systems weren't functioning.

All five also happened to be gay men. Who didn't know each other. And all were previously healthy.

Around the same time, reports emerged of young gay men in the US with Kaposi's sarcoma (KS), a skin condition with slightly raised, purplish marks that were usually seen in older men from Central Europe or the Mediterranean. These men didn't fit the demographic at all.

More and more cases emerged of gay men with PCP and KS, and not just in the US but Europe, too. In New York, a pediatric immunologist noticed similar diseases to PCP, as well as severe immune deficiencies—not in gay men, but in children whose mothers were sex workers or used IV drugs. His findings were not taken seriously by his peers. By the end of 1981, there were 337 reported cases of the disease worldwide. By New Year's Eve, 130 of them were dead.

More details came to light. Some of the men suffering from it reported having sex with other men who had PCP or KS. Many of those same men were now dying. It seemed clear that something infectious and deadly was attacking gay men and spreading among communities of them.

A CDC task force was created to begin gathering information on primarily gay men showing up with KS, PCP, and other symptoms like swollen lymph nodes, fatigue, weight loss, and fevers. In 1982 after the first case reports, the world accepted that an epidemic was on their hands, and the disease got a name: AIDS, or acquired immunodeficiency syndrome. (This was infinitely better than an earlier term, GRID, or Gay-Related Immune Deficiency, which only served to stigmatize the gay community inaccurately as the sole source and sufferers of the disease.)

Despite the flurry of data gathering and task forces formed to track and combat AIDS, the cause of the disease remained unknown. But new clues were beginning to emerge. Reports in the *MMWR* in July 1982 revealed that patients with hemophilia, who require regular blood product transfusions to replace a missing blood clotting factor, were dying of AIDS. Infants who had received blood transfusions were also coming down with AIDS-like symptoms.

The same *MMWR* report detailed an array of opportunistic infections showing up in a cluster of Haitian immigrants in Florida and New York. These Haitian men were heterosexual, and a mere fraction of them were IV drug users. (Oddly, their cases were glossed over for some time, much like the children born to mothers who were also drug users or sex workers.)

Whatever AIDS was—virus or bacteria, how it spread, what the cure was—it was clearly getting into the donor blood supply, which isn't a surprise in retrospect, as AIDS was then being seen in patients across the country and broadly around the world. In 1982, Brazil, Canada, Australia, and Italy reported their first cases. In San Francisco alone, the number of patients with AIDS went from a handful in 1981 at the dawn of the epidemic to more than eight hundred by the end of 1984, a shocking rise.

By 1983, researchers found that one particular infection-fighting cell—the CD4+ T lymphocyte—was low in patients with AIDS. This cell is an infection fighter responsible for recognizing pathogens and setting off an immune cascade

Dr. Anthony Fauci at a national AIDS conference, San Francisco, 1989

to help destroy them. It would be another year before scientists unlocked a valuable secret of the disease—that CD4$^+$ T cells were also how the HIV virus entered the body. This was why it made immune systems so weak. Without enough of these cells, certain infections were bypassing the body's innate immune response to even recognize them as foreign.

With the floodgates open, viruses, fungi, bacteria, and even cancers like KS and lymphomas, ran wild. Even with antibiotics and antifungal medications. Patients would weaken, lose weight, and barely recover before being attacked again and again by yet another opportunistic invader. Sometimes it was pneumonia, other times it was crippling diarrhea, delirium-inducing brain infections, or cancer. AIDS would always, inevitably, win. It quickly became known that a diagnosis of AIDS was a death sentence.

Between 1983 and 1984, three separate teams of scientists discovered the cause. One group called it HTLV-III; another named it AIDS-associated retrovirus, or ARV; the third group called it lymphadenopathy associated virus, or LAD.

They were all the same virus. It was renamed HIV, or human immunodeficiency virus. It turned out there were two major types, HIV-1 and HIV-2, the latter not nearly as globe-trotting, keeping mostly to West Africa. But inside and outside of sub-Saharan Africa, HIV-1 spread to pandemic proportions.

An inevitable tidal wave of information swelled as scientists learned more about the disease, including how it was transmitted and exactly how it caused illness. But they still didn't know how to stop it, nor could they answer the pervasive question of origin.

One place to start investigating was how the sick were connected. It was clear from the earliest case reports that the emergence and prevalence of HIV in the gay community pointed to a sexually transmitted infection (STI), particularly as those infected often reported having had sex with others who ended up with the hallmark diseases associated with AIDs, including KS and PCP.

◆ ◆ ◆

In 1984, a group of investigators focused on how a cluster of cases was woven together. They homed in on the very first patients in southern California, asking them about their sexual pasts—essentially contact tracing. It led to cases being linked in Los Angeles, San Francisco, New York City, Florida, Georgia, New Jersey, Texas, and Pennsylvania. Basically, all over the US map. They created a diagram that resembled a molecule, with thirty-nine circles representing the cases, each labeled with symptoms and locations, and connected to one another with lines. And at the center of this big, interconnected molecule was a single patient, labeled with an ominous "O."

Patient O. Not Patient Zero, Patient letter O.

The O indicated "Out[side]-of-California." Because Patient O had been linked to several cases in Southern California, where the study started, but was also connected to cases in New York and was not from the US. In the article, Patient O was estimated to have had 250 sexual partners per year in the three years leading up to the study. Patient O had also kept an address book containing memorable partners, and was able to name a whopping 72 of his 750 partners—a gold mine for contact tracers.

Patient O was rebranded as "Patient Zero" in Randy Shilts's book, *And the Band Played On*. Shilts noticed that some doctors and scientists had mistakenly called the so-called "index case" Patient Zero, confusing the letter O for zero. It made sense to call the originator Patient Zero, like Ground Zero—the point on the earth's surface closest to the detonation of a nuclear bomb, or the center of the destruction. In Shilts's book—noted for its flair, rumor, and hearsay—Patient Zero went from being an unnamed patient in scientific circles to a living person at whom society could now point a finger.

His name was Gaétan Dugas.

Dugas was a Canadian-born flight attendant. In Shilts's book, Dugas was portrayed as a man who carelessly gave AIDS to gay men across the US as a result of his voracious sexual appetite. This was helped in no small part by his good looks, confidence, and charm— reportedly, he would strut into a bar  and announce, "I am the prettiest one"—not to mention his disarming, sultry Québécois accent. Dugas was also portrayed as overtly malevolent in Shilts's book once he was made aware of his HIV status. In one infamous scene, Shilts describes Dugas having sex in a dark cubicle of a bathhouse. Afterward, he turns the lights on and eyes the KS lesions on his own body, saying, "Gay cancer." And then to the man with whom he just had sex, "Maybe you'll get it too."

But with the number zero burned into public consciousness and reinforced by Shilts's painting of Dugas as an infected gay sociopath, there was no thought to "Patient Negative One." In most everyone's minds, Dugas truly was the beginning. He was chapter one in the burgeoning story of HIV and AIDS, the world's greatest super-spreading villain.

Many things were left out of Shilts's book. On the page, Dugas is made out to be reckless, selfishly putting his sexual freedom first and risking the lives of others to satisfy an unquenchable hedonism. Meanwhile, off the pages in the real world, Dugas was vilified for being unabashedly gay and also unashamed of his sexuality. Anecdotes about Dugas not wanting to hurt a possible new lover were left out. As was his kindness, and his seclusion. A friend who spent a day out on a picnic with Dugas noted his isolation from the gay community. Dugas later gave his friend a heartfelt thank you for "a normal day." Dugas also contributed to local AIDS charities and became a resource for others with AIDS; he even flew to the CDC in Atlanta to donate blood samples. All these facts were left out of *And the Band Played On*, considered the authoritative book on the early AIDS crisis.

Dugas was also demonized as Patient Zero in a recklessly irresponsible *New York Post* headline, "The Man Who Gave Us AIDS." His face was broadcast across the country in an episode of *60 Minutes* entitled "Patient Zero." All this while the country wrestled with the fear and trauma of a lethal new virus that appeared to

The front page of the New York Post, *October 6, 1987; a glaring reminder of the ignorance and fear laid bare by the AIDS crisis.*

be targeting the gay community, which was rewarded with even more anti-gay bigotry thanks to the disease. Meanwhile, more and more members of gay communities were getting sick, suffering horribly, and dying at alarming rates.

So little was understood about AIDS when it first came to light that many leading virologists and researchers thought it possible that the disease originated with one man. In retrospect, that was a ridiculous notion, a point made by none other than American infectious disease expert Dr. Anthony Fauci, then a working physician and immunologist racing along with others to understand and treat AIDS. Further obscuring the truth was the belief that "We thought, based on very little data, that it was only about two years from infection to death," explained Fauci. It made sexual partners from encounters just before the complications of HIV set in seem like the ones who gave them HIV, which was unlikely. In truth, the disease usually takes a decade to show itself in healthy people, which removed Dugas from the possibility that he could have ever been the origin of the virus in America.

Like so many others with HIV and AIDS in the 1980s, Dugas was a victim of it, too. He died in 1984 at age thirty-one from complications related to the virus.

But if not Dugas, who was the true patient zero of HIV? Who gave it to *him*? Scientists needed to learn from where HIV came, and not for finger-pointing reasons. Surely there was a beginning to this story that would teach us about why HIV and AIDS took the world by storm in the 1980s, and why we had never seen anything like it before. And if we could find the beginning, it might illuminate how it could happen again, and how the virus works.

◆ ◆ ◆

In 1964, a Danish physician and surgeon named Grethe Rask began working in Zaire (now the Democratic Republic of the Congo). She left and then returned to the country in 1972. In 1975, she began to feel ill. Rask was constantly fatigued, lost weight, was plagued by diarrhea that wouldn't go away, and had swollen lymph nodes. By the time she went back to Denmark in 1977, she needed oxygen support and was being attacked by opportunistic infections like candidiasis and PCP. She died that year. Nine years later, Rask's preserved blood sample tested positive for HIV. She had likely contracted it at some point in those Zaire years, possibly from a breach in her protective equipment during surgery.

For the sake of argument, let's say she caught the virus in 1964, the first year she was in Zaire. At that time, Gaétan Dugas was twelve years old, which makes it basically impossible that he was patient zero of the HIV epidemic.

Other early cases came to light as well. In 1976, a Norwegian man named Arne Vidar Røed died from multiple lung infections and a mysterious neurological disease that caused dementia and muscle problems. Røed was known to have multiple encounters with sex workers in Cameroon around 1961. In 1969, a sixteen-year-old St. Louis teenager named Robert Rayford suffered from a baffling disease consisting of weakness, KS, and skin infections. The boy was quiet about why he was positive for chlamydia, and doctors suspected that he was abused as a child sex worker. He ended up dying of pneumonia that year. Twenty years later, in 1985, his tissue samples would react with antibodies against HIV.

So when did HIV actually arrive in the US?

AZT

HIV has become a chronic disease.

That may sound ominous, but it's actually a good thing. There was a time when HIV-positive status was tantamount to a terminal cancer diagnosis. Now, with treatment HIV can be managed to the point where life spans are not altered by the disease. And that is incredible progress. As of this book's publication, there are twenty-five medications for HIV, in six different classes.

But it all started with one medication: AZT, or azidothymidine, better known now as zidovudine.

Idovudine, or AZT, crystals, seen through polarized light.

AZT wasn't made to be an HIV med. It was actually created in 1964 by Burroughs, Wellcome & Company as an anti-cancer medication. However, it didn't seem to work in mice, so it was largely forgotten. In 1974, scientists did note that it appeared to work against the murine leukemia virus, a retrovirus in mice that caused cancer.

In the 1980s after HIV was discovered and the AIDS epidemic was spreading across the world, the search for new medications against it turned into a life-or-death race against the clock.

In 1984, Samuel Broder of the National Cancer Institute invited companies to submit drugs for screening via a test developed in his lab for possible activity against HIV. AZT was selected. Human clinical trials came swiftly thereafter, confirming that it suppressed the virus and reduced deaths. AZT worked by competing with thymidine and inhibiting the enzyme that helped HIV to replicate itself, something called reverse transcriptase. It was fast-tracked through the FDA for approval. Usually, these sorts of approvals took a decade. From the time it was discovered as a potential HIV drug to FDA approval was an astounding twenty months.

AZT helped turn the tide in the fight against HIV. It's not a perfect drug—it has a lot of side effects like nausea, vomiting, and headaches, and HIV can become resistant to it. But its discovery proved that even when medications don't work against the diseases they've been created to fight, they can still be heroes down the line.

AZT remains in the armamentarium against HIV and AIDS today.

Activist group ACT UP protests at FDA headquarters (below) in 1988 and in San Francisco (above).

In 2016, using a novel technique called "jackhammering," Michael Worobey and his team took an early genome "snapshot" by examining and sequencing thousands of archived serum samples. The samples belonged to volunteers who donated their blood for medical studies dating back to the late 1970s in

New York City and San Francisco. The new technique was necessary given the very fast rate of degradation of the RNA, and the approach filled in gaps by discovering overlapping sequences. It turned out that HIV was already genetically diverse in the 1970s and was linked to a preexisting epidemic elsewhere (see below). The particular subtype of HIV that infected so many gay men in the 1980s was HIV-1, group M, subtype B. (The study examined Dugas's blood sample and clearly found that there was "neither biological nor historical evidence that he was the primary case in the US or for subtype B as a whole.") Plus, most of the strains of HIV-1 in the US appeared to hail from one particular strain . . . that came from Haiti.

This statement—that HIV in the US arose from Haiti—is fraught with controversy for multiple reasons. In 1983, the CDC created a list of those who were at high risk of having HIV, including people who were homosexual, addicted to heroin, suffering from hemophilia, and now Haitians. The dreaded "4-H" meant that an entire group of people, including those born in Haiti and those naturalized in the US from Haiti, were all thrown into a bucket that was considered horrifically tainted. In 1990, the Food and Drug Administration (FDA) even banned blood donation from Haitian Americans who'd been naturalized for decades, no matter how low or nonexistent their actual HIV risk.

Swirling around this anti-Haitian sentiment was the fact that cases from Haitian immigrants were swept under the rug in the 1980s. Now, the genetic analysis emphasized HIV's link to Haiti and the fact that HIV is incredibly diverse genetically. The family tree of the disease is dizzying to behold. By looking closer at the phylogenetic, or evolutionary, relationships and archival viral RNA samples, M. Thomas P. Gilbert and his team were able to estimate that HIV-1 arrived in Haiti between 1962 and 1970, more specifically around 1966.

However, doctors in Haiti didn't encounter the first cases of AIDS until 1978. Furthermore, no banked blood samples from the 1970s have shown to be HIV positive. The timing didn't make sense, considering that the progression from HIV infection to AIDS in Haiti was faster than the US, where the median time from infection death without HIV medication is ten years; in Haiti, it was far shorter, at five to seven years. The accelerated progression and spread of HIV among many in Haiti was due to factors such as having concomitant infections like syphilis and tuberculosis, poor access to health care and clean water, and that relentless killer known as poverty.

In the first decade or so of the epidemic in Haiti, rates of infection were twice

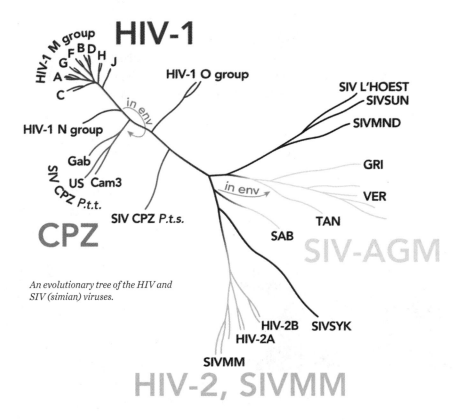

An evolutionary tree of the HIV and SIV (simian) viruses.

that of other developed countries. In 1982, most Haitians with HIV were men, who outnumbered infected women five to one. Testing of more than five hundred Haitian mothers in 1982 showed a staggering 7.8 percent of them were HIV positive; in poor urban areas, the rate peaked at 10.5 percent in 1991.

The risk factors in Haiti were similar to those in the US: 50 percent of those infected engaged in high-risk sexual "behaviors," including bisexuality or homosexuality, and 22 percent received blood transfusions. The sex tourism trade was also thriving in Haiti during the seventies, and patients with AIDS were showing up in larger numbers in Carrefour, a suburb of Port-au-Prince well known for both men and women sex workers. But the takeaway from the data is that being Haitian wasn't a risk factor for HIV, yet another reason why the 4-H designation was so very misguided. (Of note, it was only in 2015 that the FDA removed the ban on donating blood by men who have sex with men, regardless of their recent sexual encounters or the near 100 percent success rate in identifying HIV-positive blood. More than twenty other countries still maintain similar lifetime bans for blood donation by gay and bisexual men.)

Blood transfusions were another source of spread. Since 1970, at the Hôpital de l'Université d'Etat d'Haiti, blood was sold to patients when needed, and those who gave blood were paid for their donations. As such, donors tended to hail from lower socioeconomic strata.

At the Red Cross's Centre de Transfusion Sanguine de la Crois-Rouge, blood was free, but in order to receive a donation when ill, a family member was obligated to donate blood in exchange. Given that patients with HIV often had partners who also had HIV, the obvious donor pool was coming from a high-risk population as well.

Public health billboard recommending the use of condoms, Port-au-Prince, Haiti

As if the initial silent spread of the disease in Haiti wasn't heinous enough, there was another terribly efficient way for the virus to spread in the 1970s: via the lucrative plasma trade. Donors could sell their plasma for $3 a liter (at a time when a daily wage was less than two dollars) and come back within days to repeat the process, unlike a regular whole blood donation where you need months to regrow red blood cells. Plasma, the honey-colored, protein-rich liquid fraction of your blood that has no cells in it, was used for several diseases, including hemophilia.

In 1972, one company, Hemo-Caribbean, was reportedly selling five to six thousand liters to the US from Haiti each month. The reused plasmapheresis machines could have been a source of HIV to unsuspecting donors, though no retrospective cases have been definitely linked to donations from Hemo-Caribbean.

So was Haiti truly the source for HIV in the US? The genetic work certainly points to it. But without more corroborating studies involving archival blood samples that may not exist, it will likely be our best guess, and it will be an incomplete one that tragically reinforces a decades-old association between HIV and people of Haitian descent.

This is particularly frustrating because we know that before HIV ever touched the island of Haiti, it lived elsewhere. Since HIV was known to be alive and well in Africa, the theory goes that when the Belgian government left Congo (now the Democratic Republic of the Congo, or DRC) in 1960, it meant a huge departure of skilled people, including teachers, technicians, and healthcare professionals. In their place, the World Health Organization (WHO) stepped in, bringing recruits from all over the world to fill the void, including a large number of Haitians. By some estimates, it was north of four thousand workers. People from Haiti were a good fit for working in DRC because they spoke French, as did the Congolese, and their home country was not particularly hospitable under the dictator at the time, François "Papa Doc" Duvalier.

With a high number of Haitians mixing with the Congolese, no doubt sexual contact and sex work happened between the two groups. Because HIV was quietly spreading in the country, it could have jumped to Haiti around 1966. Further, genomic work showed that the HIV-1 subtype B that spread in Haiti and the US most likely originated from central Africa.

The proof that it came from *that* area of Africa? One of the earliest known specimens came from a 1959 blood sample from a Bantu man who lived near Kinshasa. Another sample came from a woman in 1960, also from DRC. HIV RNA were extracted, sequenced, and compared. And here's the kicker: The samples were quite different from each other—12 percent different, to be exact. They were both HIV-1, group M (M for "major," as in 90 percent of worldwide HIV infections, the same group that ended up in Haiti and beyond). But they differed enough to be associated with two separate subtypes. Which meant a good deal of evolutionary divergence in the virus had already happened in Congo by the late 1950s.

◆ ◆ ◆

The genes of the HIV virus contain a kind of evolutionary clock in them, a telltale rate of mutation that ticks at a steady pace. This meant that when scientists looked at two samples from, say, 1959 and 1960, they were able to "turn back the clock" and determine exactly when their common ancestor emerged.

It was more or less around 1908.

But the HIV origin story doesn't end there. Researchers found a version of HIV common in monkeys and apes called simian immunodeficiency virus (SIV). Sometimes SIV made primates sick and sometimes it did not. But the key was that the SIVs were so common. As we've discussed (see page 23), the idea of disease "spillover" from animals to humans made sense in this case, since SIVs were

The sooty mangabey monkey is a known animal vector for HIV-2.

similar enough genetically to HIV. The connection seemed an obvious home run.

But finding the origin of the "jump" between simian and human wasn't easy. Researchers looked at all the known SIV types, in sooty mangabeys and African green monkeys, in chimpanzees and gorillas, macaques, in red-capped mangabeys and greater spot-nosed monkeys. A few captive chimps had an SIV that was a perfect match for a spillover virus to HIV-1, but it couldn't be found in the wild.

Researchers did find that the sooty mangabey's SIV was responsible for the jump to cause HIV-2. And not the only one, either. There was evidence for multiple mangabey-to-human jumps. But HIV-1's ancestor remained elusive.

Finally, after finding a better way to test wild chimps (fecal sampling), the researchers hit the medical jackpot—SIV in chimpanzees that was closely related to HIV-1. The crossover was likely from the bloody results of hunting a chimp and butchering it, with some chimp blood likely entering a cut on the hunter. In Congo. In 1908.

As for the chimps with the particular strain of SIV that likely spawned HIV-1, it turns out there was yet another jump. Genetic analysis of the chimp SIV in question showed that it too was a Frankenstein mash-up of two SIVs from greater spot-nosed monkeys and red-capped mangabeys. Chimps are omnivores known to partake in fellow primate flesh when given the chance to kill their smaller cousins. In West Africa less than a million years ago, hungry chimps consumed both species of infected monkey and their SIV genetic material did a tango inside a chimp cell at some point, glued themselves together, and voilà! The progenitor to HIV was born.

Not patient zero. Not Gaétan Dugas, or a person from Haiti, but chimp zero—the true beginning.

HIV Vaccine

As of 2019 there were 38 million people living with HIV, and almost two million more get newly infected each year. Of those, only 67 percent are able to receive medications. No matter how advanced our pharmacological weapons become, and despite knowing all too well how HIV is spread, the virus is still thriving.

So where is the vaccine? Not here yet.

This is because HIV is a particularly complex virus. First, natural resistance is exceedingly rare. HIV is not like the measles or smallpox, which are self-limited and confer lifelong immunity. At this moment, if you're infected with the virus you can never be "over" the infection (unless you're one of the very rare individuals who are homozygous, or have two copies, of the CCR5-delta 32 mutation that makes you resistant to HIV. One unusual case included Timothy Brown, aka the Berlin Patient, who had HIV, but received a bone marrow transplant with the homozygous CCR5-delta 32 mutation that gifted him with resistance to the infection). Further, we've always used that template—the body's innate ability to recognize and eliminate disease-causing microorganisms—to understand how a vaccine could work.

The second obstacle to a vaccine, also tied to why there's no natural immunity to HIV, is that the virus changes so frequently it's constantly fooling our immune system. Even if you made a good vaccine to one HIV virus, it would easily change and elude that vaccine's ability to protect someone.

If you look at the HIV family tree, with its myriad types and all the constant changes happening in the genome, it's hard to find one vaccine that will do the job.

Finally, there isn't a good animal model to work on that would help develop and test a vaccine, as has been done in the past with vaccines for tetanus, diphtheria, and typhoid. So far, vaccine research on animals doesn't seem to correlate with how they work in humans. Researchers are still working on SIV as a model, but it hasn't been able to predict how vaccines would work in humans.

But there is hope. In 2009, a two-part vaccine trial seemed to prevent HIV-1 in about 30 percent of those who got it. The "prime" vaccine was designed to stimulate cellular immune responses (T-cells) and the "booster" vaccine stimulates antibody responses (generated by the B-cells). Its protection rate wasn't good enough to bring it to a larger trial, but it was a start. There is also work being done with a possible cytomegalovirus vector vaccine, working with the CCR5-delta 32 gene therapy, and even using antibodies themselves to neutralize the virus.

In the meantime, education, availability of reproductive and medical care, use of prophylactic antiviral medication, and safe-sex practices all help. Furthermore, treatment of HIV is the most effective prevention, which is encouraged by the international push to achieve and understand the U=U concept (undetectable equals untransmittable). Unfortunately, many of these key pieces are sorely lacking in countries around the world.

The fight is far from over.

INDIGENOUS PEOPLES & THE COLUMBIAN EXCHANGE
The "Exchange" Was Not Equal

Impact: The utter, but not fully recorded, devastation of great numbers of North and South American Indigenous people

When: 1492 to today

Who: The Christopher Columbus and subsequent expeditions

What Happened Next: Indigenous people died by the millions

Little-Known Fact: Christopher Columbus himself may have been one of the patient zeroes who brought a new variant of syphilis to Europe.

The "Columbian Exchange."

That's the innocuous label we use today for the cross-continental transfer of technology, plants, animals, cultures, and—crucially—diseases that began when Italian explorer Christopher Columbus and his crew first arrived in the Caribbean in 1492. The new diseases quickly ravaged across the Americas following Indigenous trade routes, devastating and destroying whole communities well ahead of European exploration.

By the time Hernando de Soto and his expedition of European colonizers reached the Mississippi Valley in 1540, whole cities that were part of the once thriving Mississippian culture had been abandoned in the wake of epidemic disease.

Some eighty years later, a colonist recorded the eerie devastation in the Massachusetts forests, writing "And the bones and skulls upon the several places of their [sic: Indigenous] habitations made such a spectacle after my coming into these partes, that, as I travailed in the Forrest near the Massachusetts [Bay], it seemed to mee a new found Golgotha."

To put data to the carnage, it is believed that the pre-Columbian population of North and South America combined was anywhere between 40 million to 100 million people. Mortality rates from the newly introduced diseases are estimated to be between 60 and 95 percent. So it's likely that anywhere between 24 million to 95 million Indigenous inhabitants of the Americas died from European diseases, many of them before they had ever seen a white European in the flesh.

The primary killers were smallpox and measles, which would have been similarly decimating when first introduced to European populations centuries earlier. Both are zoonotic diseases (see page 12), meaning at some point in history the viruses jumped from animals to humans. And when that happened, as they spread along Eurasian trade routes, huge swaths of populations surely died. Compounding the destruction, both measles and smallpox are endemic, which means they stick around in pockets of a given population. So when they're introduced to a new, untouched population . . . well, from the perspective of the virus it's equivalent to winning the lottery.

American historian Alfred Crosby described the impact of a new disease like smallpox on a previously unexposed population as a "virgin-soil epidemic." Along with measles and smallpox, Europeans introduced a veritable smorgasbord

Depiction of an Indigenous North American shaman with a sick patient

of "virgin-soil" epidemics to Indigenous populations. The list, long and distinguished, includes bubonic plague, chicken pox, cholera, the common cold, diphtheria, influenza, malaria, scarlet fever, typhoid, typhus, and tuberculosis. (Meanwhile, the only major disease that may have traveled the other way across the Atlantic was a possible strain of syphilis.) First introduced to the Americas through European carriers with low-grade infections, or through colonial outbreaks that were not sufficiently quarantined, each of these diseases rapidly became epidemic, razing Native American populations with no built-in immunity against any of them.

But the worst European disease to spread across the continent by far was smallpox, which nearly wiped out entire Indigenous populations singlehandedly. In Hispaniola, where Columbus first landed, the estimated 1492 population of 1 million inhabitants collapsed to a scant few hundred within fifty years. The Hurons, a powerful tribe when first encountered by Europeans, responsible for controlling much of the fur trade in Canada, lost 50 percent of its population in a single smallpox outbreak in the winter of 1639.

During the Revolutionary War, a soon to be continent-wide smallpox epidemic erupted in Boston in 1775, sickening both Colonial and British soldiers, before spreading to the Indigenous allies of both sides, including the Iroquois, Delaware, Seneca, and Cayuga nations. When the epidemic finally subsided, it had spread along Indigenous trade routes as far south as Mexico, west to the Pacific Coast, and north all the way to Alaska. By the time the English sea captain George Vancouver sailed to the Pacific Northwest in 1792, he found

Members of Florida's Indigenous Timucua tend to the sick.

Smallpox Blankets

In the centuries of conflict between European colonizers and Native Americans, few atrocities strike terror into hearts quite like the story of colonial settlers purposefully giving blankets to Native Americans that were infected with smallpox. Cited frequently as an early example of biological warfare, the story goes that traders and soldiers would give smallpox-laced blankets to unsuspecting tribes, igniting outbreaks.

As is often the case, the truth is more complicated.

Only once in recorded history is there a documented incident where colonists knowingly gave out infected blankets. And while once is one time too many, the twist is that it probably didn't work.

In the spring of 1763, members of the Delaware, Shawnee, and Mingo tribes laid siege to British-occupied Fort Pitt (in present-day downtown Pittsburgh). After several long months of siege, two emissaries from the Delaware tribe visited the fort in July and attempted to persuade the British to abandon it. The British held fast, and when the emissaries asked for provisions before they returned to their camp, the soldiers gave them two blankets and a handkerchief from the hospital ward. Unbeknownst to the Delaware tribesmen, the fort's hospital ward was dealing with a smallpox outbreak.

This was the first and only incident of attempted biological warfare via smallpox blankets verified in the historical record. While brutal, and decidedly against the British ideals of civilized warfare in the eighteenth century, the tactic also almost certainly failed. As scientists have since pointed out, any virus particles on the blankets were likely too old by that point to be contagious.

The horrific nature of the attempt has nevertheless burnished the account into legend.

deserted village after deserted village, where the "skull, limbs, ribs, and back bones, or some other vestiges of the human body, were found in many places promiscuously scattered about the beach, in great numbers." Smallpox had beat him there.

More was yet to come. In April 1837, a steamboat departed St. Louis to make its way to Fort Union, located on the Montana–North Dakota line. By the time it reached Leavenworth, Kansas, one of the deckhands was showing signs of smallpox. However, the captain of the steamboat declined to quarantine him.

Three Arikara women joined the steamboat at Leavenworth as it continued north toward Fort Clark, a trading post near the mouth of the Knife River in South

A depiction of an Indigenous Midew, or medicine man, from the Maritimes, New England, and Great Lakes regions of North America

Dakota, close to several Arikara and Mandan villages. When the Arikara women disembarked at Fort Clark, they showed no signs of infection, but they were most likely carrying the disease. The steamboat docked at Fort Clark for twenty-four hours, during which workers unloaded a variety of goods and furs from the boat, resulting in a lot of human traffic between the fort and the steamboat.

Francis Chardon, a Fort Clark trader, recorded the first death of a Mandan tribesman from smallpox a month later, in July 1837. Within six scant months, the Mandan tribe was virtually extinct. On August 11 that year, Chardon wrote, "I Keep no a/c [sic, account] of the dead, as they die so fast it is impossible." From a population two thousand, only twenty-seven Mandan survived by October 1837.

The disease then spread to Fort Union on the Upper Missouri River, where, despite sincere efforts to contain the epidemic and vaccinate local tribes, it broke out among the Assiniboine, devastating them. Bodies piled high everywhere. A trader at Fort Union reported "such a stench in the fort that it could be smelt at a distance of 300 yards."

The epidemic then hitched a ride on a logboat from Fort Union to Fort McKenzie in Montana, where it infiltrated the Blackfoot Confederacy, and from there exploded into an outright epidemic among the Great Plains tribes. Two-thirds of the Blackfoot Confederacy died from smallpox, 50 percent of the Assiniboine and Arikara tribes, a third of both the Crow and Pawnee.

By the time the pandemic began to burn out in 1839, the Commissioner of Indian Affairs reported on the casualties: "No attempt has been made to count the victims, nor is it possible to reckon them in any of these tribes with accuracy; it is believed that if [17,200, the death count for the Upper Missouri River Indians] was doubled, the aggregate would not be too large for those who have fallen east of the Rocky Mountains."

◆ ◆ ◆

Three centuries earlier, a similar scene played out in the Mexican highlands.

"In the cities and large towns, big ditches were dug, and from morning to sunset the priests did nothing else but carry the dead bodies and throw them in ditches," wrote a Franciscan monk who witnessed a devastating outbreak of disease among the Aztec people in 1576.

Spanish colonizer Hernán Cortés, with a band of several hundred men, landed in Mexico in 1519 intent on conquest. They encountered the rich, vibrant, and highly developed Aztec culture, with a vast empire inhabited by approximately 25 million people. That population plummeted to a scant 1 million in the next one hundred years, as the Aztec were killed off in battle and absolutely destroyed by several waves of infectious diseases that arrived with the Europeans. When a smallpox-infected enslaved person arrived from Cuba into Mexico in 1520, an outbreak was right on his heels that killed nearly half the Aztec population, including their Emperor Cuitláhuac. And that's largely how Cortés managed to bring down the mighty Aztec Empire with only six hundred men.

More devastating outbreaks of disease were coming. The two largest of these outbreaks were called the cocoliztli, the Nahuatl word for "pestilence," in Aztec recorded histories. One occurred in 1545, the other in 1576. Between the two of them, an estimated seven to eighteen million people died.

What exactly the cocoliztli were is a question that medical historians have long struggled to answer. It could have been the return of smallpox, measles, or typhus, among other candidates. Researchers in 2017, however, discovered compelling DNA evidence that both outbreaks were actually caused by a particularly virulent and deadly strain of salmonella, brought over by the European colonizers.

Now very rare, this particular strain of the bacteria, *Salmonella paratyphi* C, was known to be circulating in Europe before the "conquistadors"

A painting of Spanish colonizer Hernán Cortés meeting Tenochtitlan (now part of Mexico) ruler Moctezuma II in 1519

A depiction of Incan ruler Atahualpa's funeral in 1533

arrived in the Americas. Some people infected with the bacteria can carry it without falling ill, which explains why some of the soldiers on the Cortés expedition were likely asymptomatic carriers. Without any previous outbreak of this deadly strain of salmonella, the bacteria would have run rampant among the Aztec people. It is transmitted through fecal matter, and the collapse of social order in the wake of Spanish colonization could have created favorably unsanitary conditions for the bacteria to thrive. Once infected, victims would have experienced an enteric fever very similar to typhoid, with symptoms of vomiting and body rashes—and a particularly high mortality rate.

Two empire-wide outbreaks of salmonella poisoning may have been a major reason the Aztec Empire collapsed.

It was smallpox, however, that brought down the Inca Empire, the largest of all the empires in pre-Columbian Americas. By the time Francisco Pizarro arrived on the scene in Peru in 1532, influenza, typhus, measles, and, crucially, smallpox had already traveled along trade routes from Central America into the Andean highlands where the Incas were based. The Inca population hovered around 10+ million before the arrival of smallpox. In addition to taking out between 60 to 90 percent of the Inca population, smallpox also unexpectedly killed the Inca emperor Huayna Capac, leaving the once-mighty empire embroiled in a civil war of succession. Similar to the situation in Mexico, the smallpox outbreak made it possible for Pizarro—with 168 men, 27 horses, and one cannon—to conquer an advanced civilization occupying a territory the size of Spain and Italy combined.

Maya civilization, however, experienced a rapid decline before the arrival of Europeans, which some scholars have attributed to disease. The classic Maya collapse, which resulted in a reduction in Maya population and the abandonment of elaborate Maya cities in the lowlands of Central America, may have actually been caused by an epidemic.

The Maya civilization had flourished in the region between the second and eighth centuries but then experienced a dramatic collapse between the eighth and ninth centuries that has long puzzled archaeologists and historians.

One theory is that the Mayans may have been hit by an outbreak of Chagas, a parasitic disease caused by *Trypanosoma cruzi* and spread by insect bites in tropical regions such as the Maya lowlands. Combined with a rapid increase in enteropathogens (organisms that cause disease of the intestinal tract, leading to acute diarrhea), the Mayans may have had a real mess on their hands.

Another theory holds that sustained periods of severe drought led to outbreaks of hemorrhagic fever that in turn contributed to steep population losses and the eventual abandonment of Mayan cities.

Regardless, the Mayans were also hit by the arrival of a Eurasian disease in 1520 or 1521. Contemporary accounts of disease symptoms point more toward influenza than smallpox, but the effect was similarly devastating. The Kakchiquel Mayas recorded the horrific impact, "Great was the stench of the dead. After our father and grandfathers succumbed, half of the people fled to the fields. The dogs and vultures devoured their bodies. The mortality was terrible."

The entire history of the Americas might have played out very differently if the disease transmission pattern had gone the other way. If European soldiers and settlers had been repeatedly wiped out by native epidemics at the same 60 to 90 percent mortality rate of smallpox among the Indigenous people, it's possible they would have even given up entirely on their early attempts at settlement. It could have meant another century or two passed before Europeans would've been able to arrive in the Americas with sufficient quantities of immune people to make any kind of permanent settlement possible. And maybe the history of the "New World" would have played out very differently as a result.

Triatomine bugs can carry Trypanosoma cruzi, *the parasite that causes Chagas disease.*

Uncontacted Peoples

Sentinelese youth looking to ward off outside contact

Fear of introducing a variety of virgin-soil epidemics into defenseless populations has mitigated attempts by the modern world to engage with the few remaining "uncontacted tribes" still left in remote parts of the globe. No one really knows how many such tribes remain, but guesses typically hover around one hundred scattered mostly in Amazonia, with several more in New Guinea and on distant islands in the Indian Ocean.

Historically, such contact has proven disastrous for the tribe, who often lose up to 50 percent of their members to the introduction of new pathogens and diseases, even with modern medical care available. As a result, more enlightened "hands off" policies have been implemented by the governments that control territories where uncontacted tribes live.

Some nations, such as Bolivia, Brazil, Columbia, and Peru, have created reserves or national parks around known tribal territories to help protect the uncontacted peoples who still live there. Ecuador has even gone as far as to pass governmental legislation that mandates self-determination, equality, and no contact to these tribes.

Still, contact threats abound, particularly from the intrusion of illegal mining and lumbering operations in the Amazon rain forest. Another contact risk comes in the form of Christian missionaries, who sometimes subvert government policy in their attempts to introduce and convert tribes to their religion.

The risks, however, go both ways. In addition to protecting uncontacted peoples from disease, governments also are trying to protect adventure seekers from them. Some tribes are less than welcoming toward outsiders, and speculation as to why often involves previous contact that led to widespread illness and death among them. As recently as 2018, a twenty-seven-year-old Christian missionary from the United States landed on North Sentinel Island in the Bay of Bengal, home to the fiercely protective Sentinelese people. He was promptly shot with an arrow and buried in sight of the charter boat that brought him to the island. The message was clear: Leave us alone.

A modern-day twist on the "Columbian Exchange."

TYPHUS
Patient ZERO
Roland Jenks

Cause: *Rickettsia prowazekii* (bacteria)

Symptoms: Sudden onset chest rash, fever, aching joints, delirium, malaise, intense stink

Where: Any place with cold weather where people lived in deeply unsanitary conditions

When: Throughout known human history, including modern outbreaks in selected areas

Transmission: The rubbing of infected flea, lice, or tick feces into an open wound, generally a bite made by the same, or breathing in dried fecal matter from an infected arthropod

Little-Known Fact: Anne Frank likely died from a typhus infection in Bergen-Belsen, a World War II German concentration camp.

I n the summer of 1577 Roland Jenks, an Oxford bookseller, made the mistake of selling "popish" books (or those with Roman Catholic themes), some of which ended up in the wrong hands. As England had broken with the Catholic Church forty-three years earlier in 1534 and banned its teachings, Jenks ended up in jail.

In Elizabethan England, jail was possibly the worst place to experience the so-called Golden Era. Imprisoned in Oxford Castle, the "foul-mouthed and saucy" Jenks languished in the stench and squalor of a prison cell for two days, waiting for his trial to begin. When it did, on July 4, it seemed most of Oxford turned up for the event. For centuries in England, superior courts held occasional sessions in English counties for the trial of civil and criminal cases. These sessions were called "assizes." When Jenks's case came up for trial at the Oxfordshire assize, the courtroom was packed to the brim. Unsurprisingly to those in attendance, Jenks was found guilty. His sentence was three days in the pillory (read: locked in wooden stocks on public display).

Jenks responded by loudly laying a curse on the courtroom and the city, a fact many would remember in the dark days that followed.

Jenks can hardly be blamed for cursing the city, as pilloried victims were subject to all manner of disgusting things thrown at them while on public display,

Near this Spot stood the ancient
Shire Hall.
unhappily famous in History as the Scene in
July 1577.
of the BLACK ASSIZE.
when a malignant disease, known as the Gaol fever, caused the death, within forty days, of
THE LORD CHIEF BARON (SIR ROBERT BELL)
THE HIGH SHERIFF (SIR ROBERT D'OYLY of Merton),
and about three hundred more.

The Malady from the stench of the Prisoners developed itself during the Trial of one Rowland Jenkes, a saucy foul mouthed Bookseller, for scandalous words uttered against the Queen.

Anno 1875
J.M.D.
pie posuit

A plaque that still hangs on the wall in the Old County Hall of Oxfordshire, England.

including feces and dead animals. Furthermore, you were likely to lose portions of your ears in the messy business of having them nailed to the crossbeam so that you were forced to face your tormentors. Despite the harsh realities of his pillory sentence, Jenks got off lightly, it turned out.

Within forty days of the trial, accounts reported that every single attendee of the trial fell ill from "the spotted fever," which killed all the men in the courtroom, except Jenks himself. And while Jenks was almost certainly infected with the same thing the other men at the trial died from, he somehow managed to survive the outbreak. Curiously, historical records indicate that women and children were spared from the infection. (One theory is they were sitting in a different part of the courtroom.) Two judges were also victims, as were the lord high sheriff, the lord chief baron, the court clerk, the coroner, every witness, and all members of the jury. The entire event became known as the "Black Assize."

Oxford residents were terrified of what they quickly dubbed the Jenks Curse, as the disease ravaged the town for just over a month. Those who caught it were feverish and delirious, with intense headaches, pains in their muscles and joints, and rashes of bright red spots covering their skin. They also emitted a terrible stench. Many of them, once afflicted, died. All told, over the course of that summer in 1577, five hundred people were killed by the fever, including one hundred members of Oxford University, before the disease disappeared just as abruptly as it arrived.

While the sixteenth-century residents of Oxford didn't yet have the medical understanding to describe it, they were suffering from an outbreak of typhus, an acute infectious bacterial disease transmitted most often by body lice, but that can also be carried by a host of other arthropods, including mites (chiggers), fleas, or ticks. There are different types of typhus—murine (flea-borne), scrub (chiggers-borne), and epidemic (louse-borne). But it's epidemic typhus that has been a scourge of humankind for centuries, with its 10 to 40 percent mortality rate. The disease has laid low entire armies and demonstrably changed the course of human history on multiple occasions.

Body lice thrive in overcrowded and unsanitary conditions. They particularly love layers of warm, dark clothing, and as a result are a well-known pestilence in cold parts of the world, like Northern Europe. By contrast, the body louse is virtually unknown to Indigenous peoples who live in warmer parts of the world where protective layers of clothing are unnecessary. For centuries, body lice were an accepted part of European living, as unavoidable as the setting sun. However, in the filth and squalor of the Middle Ages, where people lived in close quarters and rarely if ever bathed, typhus was endemic.

When hungry, body lice living on someone crawl out of their host's clothes and onto the skin in search of food. And for a louse, food means human blood. If the louse sucks blood from someone suffering from typhus, the louse in turn

A color-enhanced scanning electron micrograph image of a human body louse (Pediculus humanus corporis)

becomes infected. The louse will soon die from typhus, but it's a Pyrrhic victory for humanity. Because if the dying louse moves to another person—and then defecates on them—its infected droppings can easily be rubbed by accident into a scratch or wound, including, ironically, a wound from a louse bite. This in turn allows the bacteria to infect the next person. Breathing in the dried louse feces found in clothing or bedding is another equally stomach-churning method of acquiring typhus. In this method of transmission, the infection enters the human host through mucous membranes in the nose or mouth.

Historically, once infected with typhus a person had up to a two in five chance of dying. In addition to the symptoms already mentioned, typhus sufferers also experience a general malaise. Many victims huddle in a sort of listless stupor while in the thrall of typhus. If they are lucky enough to survive, they remain infected by the bacteria for life. Under periods of stress, or with a compromised immune system, they can experience a milder form of typhus known as Brill-Zinsser disease. What's more, they can then become a source for another typhus epidemic—a brand-new patient zero. And the disease's ability to recur goes a long way toward explaining the ebb and flow of typhus epidemics throughout history.

While the 1577 outbreak in Oxford was particularly lethal, incidents of "gaol fever" were, if not exactly common, occurring with enough regularity that courtrooms attempted preventative measures. A misunderstanding of the disease as being airborne, caused by the "foul air" of a prison, led to creative solutions such as hanging a variety of herbs from the ceiling of assizes and distributing sweet-smelling scents. Refreshing vapors of garlic and vinegar were enthusiastically inhaled.

In the Black Assize, the Jenks Curse was likely the result of a typhus outbreak in the jail itself, infecting prisoners and guards, who marched back and forth from the jail to the crowded courthouse all day for back-to-back trials. Jenks, as a curious survivor of the affair, simply got the blame.

After losing his ears to the pillory, Jenks quickly left Oxford for France, a Catholic country better suited to his religious and political sympathies. The unlikely patient zero of the Oxford typhus outbreak of 1577 lived another thirty years. Five hundred other residents who fell victim to his "curse" weren't so lucky.

◆ ◆ ◆

The Plague of Athens

During the second year of the Peloponnesian War (430 BCE), a mysterious epidemic broke out in the city-state of Athens. Dubbed the "Plague of Athens," the disease rocketed through the city, eventually leaving between 75,000 and 100,000 people dead. Having lost a devastating 25 percent of their population, including Pericles, its general, the city—and its military forces—never recovered, ultimately losing the war to Sparta and its allies. The plague also significantly altered Athenian society, causing a massive redistribution of wealth, as many of the richest citizens died in the epidemic. Religious beliefs and adherence to law also broke down. By the end of the epidemic, Athens had lost its status as a major player in ancient Greece.

Historians have long puzzled over the nature of this epidemic. For many years, it was thought to have been an outbreak of the bubonic plague, but more recently medical investigators have advanced other theories.

"Epidemic typhus fever is the best explanation," said Dr. David Durack, consulting professor of medicine at Duke University, in a 1999 press release. "It hits hardest in times of war and privation, it has about 20 percent mortality, it kills the victim after about seven days, and it sometimes causes a striking complication: gangrene of the tips of the fingers and toes. The Plague of Athens had all these features."

If it was indeed typhus, then the Athenian defeat in the Peloponnesian War can be added to the list of times the disease has reared its ugly head and changed the course of history.

Plague in an Ancient City, *a painting by Michiel Sweerts, ca. 1652*

One of several infectious diseases caused by pestilence, epidemic typhus devastated armies, slums, and prisons long before it was first described by Girolamo Fracastoro, a Florentine poet and physician in 1546, just a few decades ahead of the Jenks epidemic. Effective preventative treatments for the disease only arrived in the mid-twentieth century. Fracastoro described typhus in his treatise on infectious diseases, *De contagione et contagiosis morbis*, after observing Italian outbreaks in 1505 and 1528. The physician was able to separate typhus symptoms from those of other pestilential diseases that thrive in similarly overcrowded, unsanitary conditions (think dysentery, scurvy, and typhoid) because typhus is unique in its sudden onset and characteristic chest rash.

While the disease is mostly spread by body lice, epidemic typhus itself is a bacterial infection, specifically from the species *Rickettsia prowazekii*, which is transmitted through the lice carriers. (Head lice is a potential or theoretical vector, and it's occasionally spread via bites of a flea found on flying squirrels in the southeastern US.) It would be several hundred more years, however, before the bacterial origin of the disease was identified. (Fracastoro wasn't able to specifically identify lice as carriers of the disease.)

Although no one had yet identified the "how" of typhus transmission, it became associated with the colder months and was particularly prevalent during Northern Europe's harsh winters. When multiple layers of clothing combine with unsanitary conditions, it produces the perfect environment for a body louse. And when a body louse thrives it truly thrives, growing its population at an astonishing 10 percent a day.

For readers in the twenty-first century, it's almost impossible to imagine how unsanitary Northern Europe was in the Middle Ages, even among the wealthy. These conditions made the body louse—and therefore typhus—an inevitable part of medieval life. Most poor people lived in airless, lightless hovels where family members would huddle together for warmth at night in threadbare clothes that were rarely, if ever, changed until they wore out. The wealthy, meanwhile, infrequently washed themselves or their many layers of clothes, which they wore for months at a stretch in the winter. A body louse could scarcely have been happier with the conditions.

A vivid example: When Thomas Becket, Archbishop of Canterbury, was murdered in December 1170, his body was prepared the next day for burial. He was discovered to be wearing a large brown mantle, under that a white surplice, under that three separate layers of wool coats, under those the robe of the

Thomas à Becket

Benedictine Order, under that a shirt, and finally next to his actual skin, a linen haircloth. As his body cooled after death, the multitudes of lice and other insects infecting these layers of clothes began to crawl through each successive layer in a frantic search for a new host. An observer wrote, "The vermin boiled over like water in a simmering cauldron and the onlookers burst into alternate weeping and laughter."

Under these "perfect" unsanitary conditions, it's little wonder that typhus spread around Europe for centuries with the advance and retreat of armies.

Numerous epidemics have brought down armies, from ancient times to more recently, and many of those campaigns could indeed have been derailed by typhus, though it's impossible to know for certain (see sidebar, page 159). The first confirmed typhus epidemic to break an army occurred in 1566. Maximilian II, the Habsburg Holy Roman emperor, was leading an army of 80,000 men to face a Turkish army led by Sultan Suleiman that was invading Hungary. While the Austrian army encamped at Komorn, a violent outbreak of typhus ravaged them to such a degree that they had to give up their campaign and were not able to aid their besieged allies at Szigetvár. The Turks eventually captured the city.

In the Thirty Years' War (1618–48), which spread infectious diseases up and down continental Europe, typhus raged so savagely that it brought down two opposing armies before they could even do battle. In 1632, the Swedish army, led by Gustavus Adolphus, retreated into the walled city of Nuremberg, Germany, where they were besieged by the Imperial Army forces commanded by Albrecht von Wallenstein. The siege led to a typhus epidemic on both sides of the wall that killed an estimated 18,000 soldiers and so weakened both armies that both were forced to withdraw. In 1663, during the English Civil War, Charles I and his army of 20,000 men were forced to abandon a plan to march on London as typhus ravaged the ranks. An opposing parliamentary army led by Robert Devereux, Earl of Essex, was similarly laid low the same year.

But typhus's greatest victory in the annals of military history came in the Napoleonic Wars. The disease famously contributed to the destruction of Napoleon's "Grande Armée" in its catastrophic invasion of Russia in 1812.

Over 600,000 men—the largest army ever assembled in the history of the world to that point—crossed the Neman River with Napoleon in June of 1812 intent on defeating the Russian Empire. As the Russian army retreated before Napoleon, they famously employed scorched-earth tactics, burning everything—villages, farms, crops—that may have been of possible use for the French army as food or supplies along the way. The Russian retreat, avoiding unnecessary battles as they pulled back, was undertaken because the Russian winter was coming—and would hopefully do the killing for them.

The French made it to Moscow, but found a smoldering ruin of a city courtesy of the Russians, who'd set fire to it rather than let the French capture it. After waiting for a peace offer that never arrived, the French began a gradual retreat back to Poland just as the Russian winter set in. By then typhus and dysentery, brought on by cold weather, dirty water, and lack of sanitation, had become endemic among the troops.

Sergeant Adrien Bourgogne wrote in his journal of the rapidity of the infection's spread: "I slept for an hour when I felt an unbearable tingling over the whole of my body . . . and to my horror discovered that I was covered with lice! I jumped up, and in less than two minutes was as naked as a newborn babe, having thrown my shirt and trousers into the fire. The crackling they made was like a brisk firing, and my mind was so full of what I was doing that I never noticed the large flakes of snow falling all over me."

A painting called Napoleon's withdrawal from Russia, *by Adolph Northen, 1851*

As the remnants of the army retreated, exhausted and ill, they were hammered repeatedly on all sides by guerrilla bands of Cossacks and Russian peasants. Then the Russian winter arrived in earnest, and typhus ripped its way through the ranks as soldiers huddled together in desperate attempts to ward off the piercing cold. Thousands of troops, too weak or sickened to keep up with the retreating army, simply froze to death.

As Napoleon's troops continued retreating, they came across the hastily built hospitals they had set up for their sick and wounded on their initial march to Moscow. The hospitals and their sick were in deplorable states. Overcrowded, unsanitary, packed to the brim with sick and dying men in the middle of a winter, each was also embroiled in its own typhus epidemic.

By late November, when the Grande Armée finally limped its way back to Poland, only 27,000 men of the original 600,000 were left. And most of the bedraggled band of survivors were infected with typhus. Over half a million soldiers perished in one of the greatest military disasters in history, many of them either directly or indirectly from endemic typhus. The tide of human history was unknowingly turned by the simple body louse.

♦ ♦ ♦

It was not until 1897, eighty-five years after Napoleon's disastrous retreat, that Charles Nicolle, director of the Institut Pasteur in Tunis, North Africa, correctly identified the body louse as the agent of typhus. Observing typhus patients at the hospital in Tunis, he noticed that those who were stripped of their clothing, shaved, and bathed before entering the hospital did not infect any other patients. He concluded, rightly, that the body louse was the culprit.

With the louse identified as the most frequent carrier of the disease, the patterns of typhus epidemics across history made sense. Nicolle's findings about the body louse were revealed just in time to make no small difference in World War I, where delousing troops on active duty, especially on the Western Front, became a regular preventative practice. The Eastern Front, however, did not fare as well. In 1914, Serbia suffered a major typhus epidemic around the start of the war, with an estimated 150,000 dying from the disease. A further jaw-dropping 3 million people were thought to have died in the collection of Eastern European countries that would become the Soviet Union.

In contrast, delousing became a common practice on the Western Front as enormous quantities of body lice spread rapidly among armies on both sides of

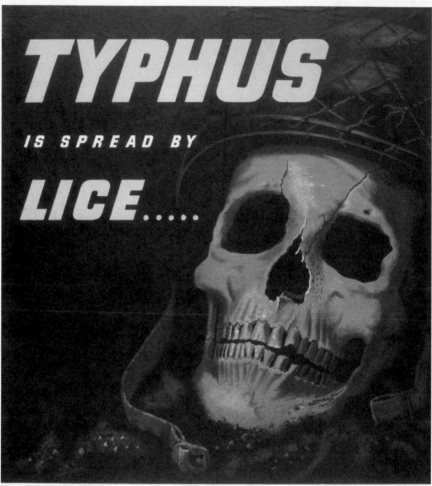

the trenches. The common wartime habit of picking off and then killing your body lice was memorialized in Erich Maria Remarque's classic novel of World War I, *All Quiet on the Western Front*:

"Killing each separate louse is a tedious business when a man has hundreds. The little beasts are hard and the everlasting cracking with one's fingernails very soon becomes wearisome. So Tjaden has rigged up the lid of a boot-polish tin with a piece of wire over the lighted stump of a candle. The lice are simply thrown into this little pan. Crack! and they're done for."

Soldiers, as well as prisoners, were bathed, shaved, and disinfected through steam treatments, fumigation, and anti-lice powders. While effective at preventing a typhus epidemic from breaking out on the Western Front, these measures were not enough to prevent another louse-carried disease from running rampant among the troops. Named "trench fever," the disease was significantly less lethal, but still a major irritant to soldiers who would often resort to burning lice off themselves with lit cigarettes. Trench fever also caused a substantial drain on available manpower during the war, second only to the spread of influenza in the trenches.

By the time World War II began, the first effective and readily available vaccine for typhus had been developed by Herald R. Cox of the US Public Health Service. Cox produced a vaccine by grinding up the guts and feces of infected lice and culturing the bacteria in chicken eggs to create vaccine components in large quantities. The resulting "Cox vaccine," while not fully effective at preventing typhus, did significantly reduce its severity and helped prevent epidemics from springing up in the conflict.

Typhus was rampant, however, in the atrocious conditions of the Nazi concentration camps. Thousands of prisoners died of typhus in Theresienstadt, Auschwitz, and Bergen-Belsen, where some 17,000 people are thought to have perished from the disease.

Fifteen-year-old Anne Frank and her sister Margot were transferred in late 1944 from Auschwitz to Bergen-Belsen, just before a major typhus epidemic swept its way through the camp. Both likely died in the outbreak. Gena Turgel, a survivor of Bergen-Belsen who worked in the camp hospital, revealed some details of Anne Frank's final days in an interview with the British newspaper *The Sun*: "Her bed was around the corner from me. She was delirious, terrible, burning up." Turgel continued, "The people were dying like flies—in the hundreds." Mass graves of typhus victims at Bergen-Belsen can be seen in footage taken by British troops who arrived to liberate the camp in 1945.

The troops who liberated Bergen-Belsen were also armed with a powerful new chemical to delouse the camp's victims: DDT. The insecticide turned the tide of a typhus epidemic in Naples in the winter of 1943–44 and was quickly put into use by the Allied armies to delouse troops, prisoners, and civilians. It was so effective—and so popular—that people enthusi-astically lined up for a thorough delousing administered via a DDT "blowing machine." Much decried now for its significant negative environmental and health impacts (DDT can poison the nervous sys-tem and possibly cause cancer), it was an enormously effective agent in stopping the spread of typhus toward the end of World War II.

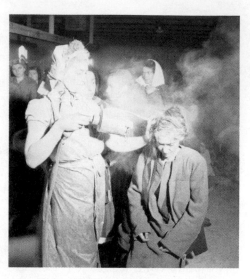

A recently freed World War II German concentration camp victim is sprayed with DDT to kill lice that could spread typhus.

Shortly after the war ended, antibiotics began to be used against the bac-teria that caused typhus. These antibiotics have proven so effective that typhus epidemics have essentially been eliminated, with some occasional exceptions in cold and impoverished regions of South America, Asia, and Africa during times of war and famine. The Cox vaccine and others are no longer in use. Antibiotic treatment almost invariably leads to a speedy recovery.

And so typhus, the great scourge of European armies for centuries, a disease that has killed millions of people and substantially altered the course of history on multiple occasions, was finally laid low. The body louse, however, is still very much with us. Having long ago tied its fate to that of humanity, the parasitic body louse will likely travel with us all the way to the end of our species.

MEASLES
Patient ZERO
The Faroe and Fiji Islands

Cause: *Measles morbillivirus* (virus)

Symptoms: Fever, cough, rash, inflamed eyes, runny nose

Where: Worldwide, but harder-hitting in areas where the virus was new

When: Faroe Islands (1846), Fiji (1875)

Transmission: Airborne, highly contagious

Little-Known Fact: During several nineteenth-century outbreaks, an estimated 100 percent of affected populations were infected by measles.

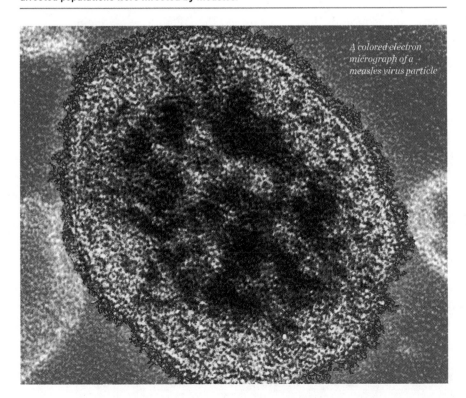

A colored electron micrograph of a measles virus particle

magine yourself on an island paradise. You glance around from your hammock, which is strung between two palm trees. There's not a cloud in the sky and the sun beats warmly down on you. The waves from the pleasantly calm ocean are lapping the sandy beach. You decide to take a quick swim, but just as you're about to enter the ocean, you look up and see a strange boat on the horizon, the likes of which you've never seen, heading toward your beloved little island. What's your first reaction?

In epidemiological thinking, the most appropriate reaction would be fear.

Isolated by geography, islands are uniquely prone to outbreaks of new diseases. This was acutely true before the advent of global travel, when island populations tended to remain largely homogenous. But the same isolation of gene pools that created unique ecosystems known for their biodiversity and natural beauty also contributed to their fragility. Almost half of global animal extinctions in the past four hundred years were island species. The introduction of new species—or new diseases—can wreak havoc on an ecosystem previously unexposed to them.

At first glance, the Faroe Islands and Fiji couldn't be more different. The Faroes lie in the far Northern Hemisphere, a group of rocky, forestless volcanic islands in the North Atlantic Ocean located between the Shetland Islands and Iceland. Seventeen of the islands are inhabited. Known for their rugged beauty, the mountainous landscapes and steep coastal cliffs of the Faroes attract outdoor recreationists and birders from all over the world.

Fiji, by contrast, is a tropical archipelago of more than three hundred islands (about a third are inhabited) in the South Pacific Ocean northeast of Australia. World renowned for their palm-lined beaches, coral reefs, and crystal-clear lagoons, the Fiji Islands attract travelers looking to scuba dive, surf, and beach-sit in a tropical paradise.

Despite their distance apart, climates, and cultures, both island populations share something in common: They were both absolutely devastated by the arrival of measles in the nineteenth century.

It happened first in the Faroe Islands. On March 28, 1846, a carpenter and cabinetmaker returned home to the islands after visiting friends in Copenhagen, Denmark. Unfortunately, the Faroese cabinetmaker had a Danish friend who was sick with measles—but the carpenter stopped by to see him anyway. Measles is highly contagious, so it was perhaps unsurprising that the carpenter caught it during the visit. Because of the latent period before measles symptoms emerge,

The Klaksvík fjord harbor, Faroe Islands, 1924

however, he made it all the way home before he realized what he had. And by then it was too late—he'd just reintroduced measles to the Faroe Islands.

Although the islands had experienced a previous measles outbreak, it was sixty-five years prior. Almost everyone who survived the previous epidemic was now dead. By the time of the 1846 outbreak, the virus was let loose in a largely non-immune population—where it thrived.

Among a population of 7,782 people, a staggering 6,000 came down with measles by the end of 1846.

♦ ♦ ♦

Measles, also known as rubeola, is a highly contagious disease caused by the *Measles morbillivirus*. It spreads easily between people when someone infected sneezes or coughs in the presence of someone uninfected. In fact, it's so deeply contagious that, unless you've been vaccinated or have immunity from a previous exposure, 90 percent of those living with an infected person will get the disease themselves. Its trademark symptoms are a high fever and reddish rash that covers most of the body. And while measles isn't often fatal in the US, it can be crushing when it breaks out in a population previously unexposed to it.

The tricky thing about measles is that, while it has an extremely high contagious rate, it also takes a while for a newly infected person to display symptoms. After initial exposure, it can be anywhere from ten to twelve days before someone shows signs of the disease. In other words, it's well designed to launch an epidemic.

Measles symptoms begin with a high fever, cough, runny nose, and red, watery eyes. A couple of days after the first symptoms arrive, small white dots called Koplik spots may appear in the mouth. Another day or two and the trademark rash breaks out, often beginning as flat red spots on the face near the hairline, which then spread down and cover the whole body, with some of the spots merging together. As the rash develops, a high fever typically returns. Most people are able to recover from measles on their own, but even so about one in five will need to be hospitalized. Some victims go on to develop severe complications, such as pneumonia or encephalitis. Today, the fatality rate in the US is about 0.2 percent, but in developing countries it can be as high as 10 percent.

Measles was first described by the Persian philosopher and physician al-Razi (also known as Rhazes), who wrote *A Treatise on the Smallpox and Measles* in 910. It was the first time a distinction was made between those two diseases, which are both characterized by skin rashes. Al-Razi considered measles the worse of the two, writing that the disease was "more to be dreaded than smallpox except in the eye." (Today the reverse is true, but it's possible that historical strains of measles, prevalent 1,000 years ago in Persia, were more virulent than those circulating today.) Al-Razi's treatise was translated into Latin and made its way to Western societies as well.

The origins of measles are murky. Some medical historians believe the disease evolved in Mesopotamia around 3000 BCE, right around the time we first

Geisha Accompanying Dancing Measles with Samisen, *by Tsukioka Yoshitoshi, 1862*

domesticated animals and built small-scale permanent settlements. Measles needs a certain density of human population to spread and, once that happened, the virus may have evolved out of canine distemper in domesticated dogs or bovine rinderpest in cattle. If it's the case that measles is a zoonotic transfer from either domesticated dogs or cattle, the timeline makes sense. Canine distemper is a highly contagious, airborne viral infection that shares the same incubation period as the measles virus. It has similar symptoms, with respiratory problems and skin rashes. Rinderpest, meanwhile, is also a highly contagious, airborne viral disease in cattle. It does not, however, create skin rashes. All three viruses (measles, canine distemper, and rinderpest) are part of the same *Morbillivirus* family.

Regardless of its origin, measles has been with us for a very long time. It was endemic in Europe, the Middle East, and Asia for much of human history, but was totally absent from North or South America until the arrival of Europeans at the end of the fifteenth century. Along with smallpox, measles ripped through Indigenous populations with a devastating mortality rate (see Indigenous Peoples & the Columbian Exchange, page 146). The Indigenous population of the Americas is thought to have been knocked down by 60 to 95 percent from the introduction of new diseases alone.

◆ ◆ ◆

The Faroe Islands outbreak became the subject of a classic epidemiological study by Peter Ludvig Panum, a Danish physician who arrived on the islands shortly after the outbreak erupted. Panum made numerous early discoveries about the behavior of measles, specifically how it moves through a population. He visited fifty-two villages around the islands and treated about one thousand people. He would later write a treatise about his experiences, the creatively titled *Observations made during the epidemic of measles on the Faroe Islands in the year 1846.*

The Faroese of the mid-nineteenth century were largely free from infectious diseases. Panum found little to no evidence of tuberculosis or syphilis, both of which were raging across Europe at the time. But the islands were also deeply impoverished, and famine was common. Other diseases such as rheumatism and bronchitis, as well as skin problems, were pervasive.

After the Faroese cabinetmaker with measles returned from Copenhagen, the disease sprinted across the islands despite their many remotely inhabited pockets. Panum identified one of the reasons: fishing. The Faroese fished communally, where men from a variety of villages would gather together to hunt whales and other aquatic creatures. While at sea, they were in close quarters with each other, then returned home to their respective villages. If anyone was on the boat with measles, almost certainly the majority of the others on board would catch the disease and bring it back to their home villages. And so an epidemic that would infect more than 75 percent of the population unfolded, killing just over one hundred people.

Panum continued to help out as the epidemic raged, putting together the epidemiological puzzle pieces as he worked. He was the first to observe that measles spread through direct human contact. He also noted that, while measles elsewhere was typically a childhood disease, on the Faroes it was striking everyone regardless of age. He also correctly observed that the isolation of the islands had protected the residents from exposure to illness, but when a new disease arrived, it struck with a vengeance. He concluded, rightly, that the survivors of measles are conferred lifelong immunity. This discovery came by virtue of interviewing ninety-eight elderly inhabitants who caught measles in 1781, the last time it visited the islands. Not one of those ninety-eight fell sick during the 1846 outbreak, while nearly everyone else on the island who was born after that outbreak came down with measles.

Panum was able to use the isolated geography of the islands to pinpoint infection between people and also incubation time periods. In fact, he got so good at accurately predicting the average fourteen-day incubation period before someone broke out in a rash that his powers of scientific observation were considered magical by some.

By the time the Faroe Islands outbreak ran its course, the death toll was quite low considering the huge percentage of the population who caught the disease. It would not, however, become the precedent when the disease arrived at other remote island chains.

◆ ◆ ◆

About three decades later in the Fiji Islands, a far worse outbreak would start with geopolitics. Located about halfway between Hawai'i and Australia in the South Pacific Ocean, Fiji is a collective of more than three hundred islands inhabited since at least 2000 BCE. In 1874, the island chain was loosely ruled by a monarchy.

Portrait of Cakobau, *self-proclaimed king of Fiji, ca. 1858*

In a calculated strategic move, the self-proclaimed king of Fiji, Cakobau (his rule was never universally accepted among the various island chiefs), decided to cede the islands to Great Britain. Cakobau was having difficulty governing a huge island chain that not only disputed his rule but was also in regular contact with a variety of European traders and would-be plantation owners. Without a strong central government, all the local competing interests and claims were boiling over into social unrest. Cakobau thought it best if they gained a European protector who might demand some stability, and eventually proposed formal annexation to the British.

Despite initially turning down a similar offer made by the king twenty years earlier, the British government agreed this time around. To consummate the deal, the king of Fiji would have to travel to a seat of British government, the closest being Sydney, Australia. So the king, his wife, their son, and a retinue of the Fijian court numbering between fifty and a hundred people all boarded the HMS *Dido*, bound for Sydney in early 1875.

Their ship docked in Sydney harbor and the king, his family, and companions were subsequently given tours of the city and its various industrial accomplishments of the late Victorian era. The Fijian royals were duly impressed; the scientific wonder of magnets and elevators in particular caught King Cakobau's attention.

What occurred to no one, somehow, was that Sydney, a bustling city in the far corner of the British Empire, was also in the midst of a mild measles epidemic. Relatively common in European nations and their colonies, these epidemics rarely advanced to anything more serious because enough of the population was immune to measles at any one time to provide a level of herd immunity. But not a single Fijian who lived on the islands had been exposed to the disease before.

King Cakobau's son, Timoci, caught it first. But again, because of the lengthy incubation period of measles, no one realized he had it until the ship carrying him and the entire royal entourage was halfway home again. The ship's doctor recognized the disease immediately and had Timoci quarantined in a small hut on the deck until they arrived at port in Fiji. And that was about the last medically sound action taken in the ensuing tragedy.

When the ship approached port in Fiji, by all rights it should have flown a quarantine flag, warning those on shore that an infectious disease was on board the ship. The captain, however, neglected to take this critical step. So as the ship anchored in the harbor, it was greeted by the island's police force, who came out en masse to welcome the returning royals. All 147 of them—the entire police force on the island—caught measles as a result.

A Fijian group dance, 1845

Shortly after King Cakobau returned ashore, a gathering was called among all the head chiefs from the collected islands. The chiefs all traveled by boat to meet with the royal family at Levuka, the ancient capital of the nation, for several days of festivities. It was, in practice, the worst thing they could have done under the circumstances, though no one knew better at the time. After the gathering, which served as a superspreader event, measles naturally traveled back with the chiefs to every inhabited island in the country.

Within four months, the nation was embroiled in a full-scale epidemic. Fiji's population at the time was estimated at 150,000, none of whom had ever been exposed to measles. And they *all* caught it—the contagious rate was just shy of 100 percent. It was utterly devastating.

The mortality rate of the outbreak was also extremely high for a measles epidemic: 20 to 25 percent of the population died, almost 40,000 people total. Entire communities came down with the disease at once, leaving no one to care for the sick or bury the dead. Or grow, harvest, or hunt for food. Some Fijians ran away from their houses, refusing to have anything to do with sick relatives. The overall atmosphere was one of panic and terror. Some reports stated that there were Fijians dying from fear alone.

In a paper given to the Epidemiological Society of London in 1877, William Squire described the outbreak: ". . . during the worst of the epidemic . . . people seized with fear had abandoned their sick . . . the people chose swampy sites for their dwellings, and whether they kept close shut up in huts without ventilation, or rushed into the water during the height of the illness, the consequences were equally fatal. Excessive mortality resulted from terror at the mysterious seizure, and [from] the want of the commonest aids during illness . . . Thousands were carried off by want of nourishment and care, as well as by dysentery and congestion of the lungs."

Another problem that escalated the outbreak's tragedy: There weren't any medical practitioners in Fiji with measles experience. As a result, a variety of unproductive remedies were tried, ranging from harmlessly ineffective to outright deadly. Many victims, burning with fever, resorted to lying down for long periods in cold streams or on wet ground. But the strategy only added to the death toll by causing outbreaks of pneumonia and dysentery along with measles.

The atmosphere of the islands was well-described by a magistrate in a letter quoted in a paper read before the Epidemiological Society of London by Bolton

G. Corney, Colonial Surgeon in Fiji, in 1884. The magistrate, stationed in Lau, happened to be away during the initial outbreak, only to return to the islands to find them embroiled in a seemingly post-apocalyptic epidemic: "On my return here, I found death, desolation, and starvation . . . Whole families have been carried off, and, but for the incessant beat of the death-drum, one might fancy the place deserted."

♦ ♦ ♦

Measles continued to spread globally until it infected every part of the inhabited world. The last "virgin-soil" epidemic of measles actually occurred within living memory.

Greenland managed to avoid measles as it spread to every corner of Earth until the mid-twentieth century. But in 1951, the disease finally found its way there when a sailor returned home from a voyage . . . and checked himself into a clinic with the telltale rash on his face and body. Within a few months, out of a population of 4,262 people in southern Greenland, 4,257 had measles.

Only five people avoided the disease, for an attack rate of 99.9 percent. Seventy-seven of them died.

In 1963, American biomedical scientist John Enders and his colleagues finally developed the first vaccine for measles. In short order, the vaccination became part of a mass immunization program that reduced measles cases in the United States and other developed countries by as much as 99 percent.

The measles vaccine today is typically combined with a mumps and rubella vaccine—known as the MMR—and is a regular part of the immunization package delivered to children in developed countries. In the developing world, however, measles still remains a major childhood disease. There are about 35 million cases each year that lead to 600,000 deaths, making it the leading vaccine-preventable killer of children worldwide. In response, the World Health Organization launched a number of programs around the globe to "make measles a memory" everywhere.

As with smallpox before it (see page 329), we have a fighting chance to completely eradicate measles. And we're making good progress. In 2000, about 72 percent of the world's children received the measles vaccination; by 2016 the rate had gone up to 85 percent.

Hawaiian Royal Tragedy

The Faroe and Fiji Islands weren't the only island groups to suffer notably from a first exposure to measles. In May 1824, when Hawai'i was still a sovereign nation, its monarchs King Kamehameha II and Queen Kamāmalu traveled to London to establish an alliance with King George IV. The young couple, both in their twenties, were hopeful that an alliance could be established between the Kingdom of Hawai'i and Great Britain. In a scene that would play out about a half century later with the Fijian delegation in Sydney, the Hawaiian royals were given tours of many of London's classic sights, their itinerary reading a lot like a modern-day tour: They visited Westminster Abbey and attended opera, theater, and ballet performances. Unfortunately, one of their tour guides thought the couple might be interested in observing Britain's first-rate (for the time) medical practices in operation at the Royal Military Asylum.

King Kamehameha II of Hawai'i, 1824

No one present at the time could have guessed the consequences.

Both King Kamehameha II and Queen Kamāmalu came down with measles after their visit to the asylum, as did most of their friends, relations, and servants who had accompanied them from Hawai'i. None of them had ever been exposed to the disease.

Three weeks later, King Kamehameha II was dead. Queen Kamāmalu followed him into the grave the following week. The royal couple was laid in state at the Caledonian Hotel in London before being shipped back to Hawai'i for burial.

Measles would finally make its way to the Hawaiian Islands twenty-four years later when the first outbreak struck in 1848.

HANSEN'S DISEASE (LEPROSY)
Patient ZERO
Unknown

Cause: *Mycobacterium leprae* (bacteria)

Symptoms: Ulcers; thick, dry, or stiff skin; loss of eyebrows and eyelashes; eye problems; numbness in the extremities; nerve, skin, and bone damage; severe disfigurement; asphyxiation

Where: Worldwide at its peak in the European Middle Ages

When: 1100s to 1800s

Transmission: Airborne transmission through aerosolized droplets

Little-Known Fact: Hansen's disease is very hard to get, as 95 percent of people are naturally immune.

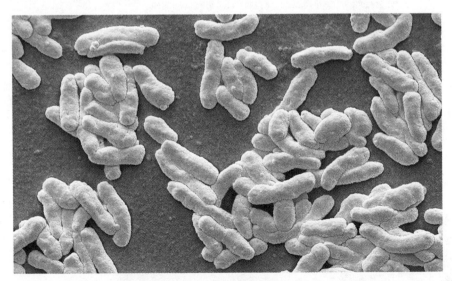

Scanning electron microscopy of Mycobacterium leprae, *the bacteria that causes Hansen's Disease*

B y 1866, the Kingdom of Hawai'i had been devastated by the arrival of numerous infectious diseases over the previous decades. Measles, smallpox, and syphilis had all become endemic on the islands, reducing a population thought to have once been over half a million strong to just 80,000.

And then leprosy arrived.

Now known as Hansen's disease, it's caused by a chronic bacterial infection from the bacillus *Mycobacterium leprae*. The disease has an extremely long incubation period, as the bacterium grows quite slowly. The bacterium also appears to take different forms in response to how a given host's body attempts to fight it. When the disease finally breaks out, it causes skin nodules to appear, as well as ulcers; thick, dry, or stiff skin; loss of eyebrows and eyelashes; muscle weakness; and eye problems.

Hansen's disease can also cause a dangerous numbness in the extremities, which in turn leads to infection-related injuries, particularly in the hands and feet. This is how leprosy victims in the past gained a reputation for having fingers and toes "fall off." Though they didn't fall off exactly; the body reabsorbed them little by little after repeated bouts of injuries.

In serious cases, Hansen's disease leads to significant nerve, skin, and bone damage and severe disfigurement. People in advanced stages of the disease run the risk of dying from suffocation, as Hansen's causes nodule growth that eventually blocks nasal and throat passages.

 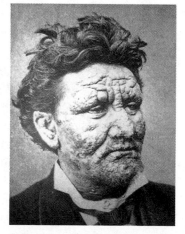

Historical photographs documenting the range of facial disfigurements in Hansen's disease.

However, despite what you may think from the way it has been portrayed in books and films, the formidable and long-stigmatized leprosy is easily the least contagious of all major communicable diseases. In fact, 95 percent of people today are naturally immune—their bodies can fight off the disease's bacterium without help.

But for hundreds of years, Hansen's was thought to be contracted through direct contact with patients. As a result, sufferers were forced to adapt to some of the most severe quarantine measures in history. The term "leper" has been widely adopted in English as a synonym for outcast and is partly why the infection is now called Hansen's disease. Recent evidence also suggests that, rather than by touch or passing of fluids, it is passed via the respiratory route, or by breathing in infected respiratory drops (much like COVID-19). Which means that to get infected with Hansen's disease you need prolonged, extensive contact with someone who has an untreated case of it, and *also* be one of the 5 percent of humans who are susceptible to it. And while there is still no vaccine for Hansen's, the disease can be successfully treated with antibiotics. It remains endemic in parts of the world, with around 250,000 people diagnosed with it each year, but the World Health Organization (WHO) is working hard to eradicate it for good.

◆ ◆ ◆

In 1866 when Hansen's first arrived in Hawai'i, native Hawaiians were not familiar with the disease and therefore didn't have the first idea how to treat it. Their only model was the prevalent Euro-American response to Hansen's in the nineteenth century: to criminalize the disease and quarantine its victims. Taking its cue from the colonial powers, the Hawaiian legislature passed a law in 1866 to do as the Europeans and Americans did—they criminalized Hansen's and quarantined its victims, with a twist. They banished them to an isolated stretch of the northern coast of the Hawaiian island of Moloka'i. There, two villages were constructed that were separated from the rest of the island by a mountain chain. These villages would be the epicenter of the leper colony of Moloka'i, where as many as one thousand Hansen's patients would live at any one time on a scattered collection of somewhat self-sustaining farms.

While the kingdom intended for the colony to be self-supporting, the reality of living with advanced cases of Hansen's disease made that largely impossible. Some half-hearted attempts were officially made to support the colony, but the residents were often close to starvation and living in abject poverty.

In response to the situation, the Catholic diocese in Honolulu decided the colony would benefit from the spiritual, medical, and practical support of a priest. The vicar apostolic asked for volunteers, knowing it was a high-risk assignment that could quite possibly result in the priest contracting Hansen's and dying from the disease. (Medical consensus at the time still considered the disease highly contagious.)

Four priests volunteered for the mission. Belgian-born Joseph de Veuster, known as Father Damien, was selected for the assignment in 1873. When he arrived at Moloka'i, he found a quarantine colony in near ruin. Damien recollected the condition of the colony upon his arrival in an official report addressed to the President of the Board of Health in 1886:

> The smell of their filth, mixed with exhalation of their sores, was simply disgusting and unbearable to a newcomer. Many a time in fulfilling my priestly duty at their domiciles, I have been compelled not only to close my nostrils, but to run outside to breathe the fresh air . . . At that time the progress of the disease was fearful, and the rate of mortality very high. In previous years, having nothing but small, damp huts, nearly the whole of the lepers were prostrated on their beds, covered with scabs and ugly sores, and had the appearance of very weak, broken-down constitutions.

The Kalawao Girls Choir with Father Damien, Moloka'i, Hawai'i, ca. 1870

Deeply sympathetic to the plight of those afflicted, Father Damien assured the residents of the colony that he was there to help and support them. When the Bishop Louis Maigret, who oversaw the Jesuit mission on Hawai'i, introduced Damien to the colony, he said "Here is Father Damien, who wishes to sacrifice himself for the salvation of your souls."

Prophetic words.

Father Damien kept his promise, living closely in and among his newly adopted flock for the next fifteen years. He joined forces with the native Hawaiian superintendent, William P. Ragsdale (himself a Hansen's sufferer), and set upon a vigorous mission to improve the living and medical conditions of the colony.

In addition to serving as the colony's priest, Father Damien helped provide medical treatment for the patients, including dressing their ulcers and tending their wounds. He built homes and furniture, made coffins and dug graves, organized the construction of civic buildings such as hospitals, orphanages, and churches, and taught in the local school.

Damien's arrival and his construction work with Ragsdale served as a catalyst for the colony, helping it to gradually create better living conditions for its residents. Father Damien shared food and smoked pipes with the residents, inspiring others with his complete lack of fear about contracting the disease.

Damien did eventually catch Hansen's eleven years into his time on Moloka'i. This development was unfortunate both for Father Damien and for the understanding of the relative contagiousness of the disease in the late 1800s. By contracting the dreaded leprosy himself, it reinforced the popular idea at the time that it was highly contagious.

Father Damien realized he had contracted the disease in 1884 when he accidentally put his foot into water that was scalding but didn't feel a thing. Incidents like this were a common way for people to discover they had leprosy. The nerve damage, rendering your digits unable to feel pain, is an early sign of the disease.

Father Damien lived for another five years on Moloka'i with progressively worsening Hansen's disease. But as the infection advanced, his energy increased as he sought to complete a variety of projects he'd begun in the colony. The disease ultimately killed him in 1889. By that time, the culmination of his good works among such a vulnerable population had put him directly on the path to beatification—one that concluded in 2009 when Father Damien officially became Saint Damien.

His patronage? Those with leprosy, of course.

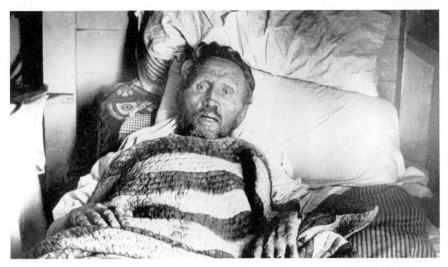

Father Damien nearing death from Hansen's disease, 1889.

However, shortly after Father Damien's death, his legacy came under fire. Dr. Hyde, a Presbyterian minister in Honolulu, wrote a letter in 1889, later leaked to the press, in which he criticized Father Damien for being a "coarse, dirty man, headstrong and bigoted," who had contracted leprosy as a result of his own "vices and carelessness." What's more, Hyde accused Father Damien of taking credit for reforms in the Moloka'i colony that were actually initiated by the board of health.

Renowned writer Robert Louis Stevenson leapt to Damien's defense. Stevenson visited Hawai'i in 1889, the same year Damien died, for an extended stay. Stevenson was himself suffering from tuberculosis, an infectious disease that was also incurable at the time. Stevenson traveled to the islands with his family in hopes of finding a climate amenable to his disease. Interested in Damien's work and legacy, Stevenson visited Moloka'i for a week's stay, during which he interviewed people from a variety of religious and cultural backgrounds about Father Damien's work.

The result was Stevenson's conviction of Father Damien's virtues. In an open letter written in response to Hyde, Stevenson declared that Father Damien was a man who, "with all his weaknesses" was still "essentially heroic, and alive with rugged honesty, generosity, and mirth."

As for Moloka'i itself, the colony remained in existence until 1969. While record keeping was far from perfect, evidence suggests that during the hundred

Carville

Carville resident Stanley Stein, founder and editor of The Star

The only leprosarium to ever exist in the United States was founded in 1894 at a former plantation house in Louisiana called Carville. The institution remained the country's one and only leprosarium for more than a hundred years.

Initially, Carville had all the amenities of what it was: an abandoned plantation. And the first few residents sent there suffered in what they described as a "swampy hell-hole." Conditions began to improve with the arrival of the Sisters of Charity, an organization of nursing nuns who would continue to staff Carville for the next century. They eventually transformed the plantation into a carefully maintained estate that became a "place of refuge, not reproach."

Nevertheless, patients were under strict isolation regulations that largely kept them away from the rest of society until the mid-twentieth century, when scientists began to better understand the low risk of infection from the disease. Residents were not allowed to vote or marry, or even live with their spouses if they weren't also patients.

One of the Carville residents, Stanley Stein, began a campaign in the 1930s to change public perception of Hansen's and advocate for reforms in how society approached those infected with it. He launched a magazine called *The Star*, in which he reported on conditions at Carville and any news on research into the disease.

The magazine did much to help change perception of the disease—and gave Stein a famous advocate and pen pal, actress and society page regular Tallulah Bankhead. The actress even convinced many of her friends to purchase multi-year subscriptions to the magazine. When Stein lost his sight, Bankhead had a bust of her face sent to him at Carville so he could appreciate her bone structure.

Advocacy from patients and nursing staff at Carville led directly to the replacement of leprosy with Hansen's disease as its official name. Over time, enforcement of the quarantine laws was allowed to lapse. (Staff ignored the fact that residents would occasionally "sneak out" on evenings to visit a local bar, for example.)

By 1999, there were so few residents left that the state of Louisiana made efforts to close down Carville. The few remaining residents could either leave and take a $46,000 annual stipend with them, or remain at Carville while the institution transitioned to other uses, or be transferred to a care facility in Baton Rouge. Some chose to stay, with the last residents not leaving until 2015.

years the colony existed, it hosted some 8,000 people, almost all of them Native Hawaiians. By law, they were declared legally dead when shipped to the colony, which goes a long way to demonstrating the general feeling toward leprosy sufferers throughout history—and why we now call it Hansen's disease.

◆ ◆ ◆

Hansen's disease in its contemporary form is quite rare and very difficult to spread. So why did it seemingly become endemic in the medieval era? The short answer is we still don't know. The long answer is a lot more interesting.

Leprosariums, asylums set aside to care for leper patients, started appearing all over Europe in the early Middle Ages. These special hospitals were built away from towns, but typically along a main road where there would be regular traffic. One of the (only) ways people with Hansen's were allowed to support themselves was by begging for alms in exchange for saying prayers for individuals. This translated as victims of the disease stationing themselves along a travel route and begging for money from passersby.

Spiritually, Hansen's sufferers were in a tight spot. In the view of the medieval Christian church, lepers were considered to be living through a purgatory on Earth. So, while they were purposefully excluded from the rest of medieval society, they were also viewed with deep spiritual curiosity by others. Church leaders declared that lepers could help others reduce their time in purgatory by saying prayers for them, so giving alms to those with Hansen's had a perceived benefit beyond simple charity. By carving out an odd little spiritual niche for them to occupy, medieval society found a way to support the infected and give them a religious context to understand their own suffering.

Their place on the spiritual stepladder aside, medieval society also condemned Hansen's sufferers (they sometimes even burned them alive), and set in place wildly restrictive rules governing their interactions with the rest of the world. In 1179, the Catholic Church declared that those infected with leprosy must not live with non-infected people. Once someone was confirmed (through dubious assertions from gate porters, policemen, priests, or monks) to have a case of Hansen's disease, they went through a ritual rite of passage called a "lepers mass." Kneeling before the altar, their face covered by a black veil, the leper endured the priest throwing dirt from the cemetery on their head three times to symbolize their death to the world. The new rules of their engagement with the rest of society were then declared to them:

> I forbid you to enter the church or monastery, fair, mill, market-place, or company of persons . . . ever to leave your house without your leper's costume . . . to wash your hands or anything about you in the stream or fountain. I forbid you to enter a tavern . . . I forbid you, if you go on the road and you meet some person who speaks to you, to fail to put yourself downwind before you answer . . . I forbid you to go into a narrow lane so that if you should meet anyone he might catch the affliction from you . . . I forbid you ever to touch children or give them anything. I forbid you to eat or drink from any dishes but your own. I forbid you to eat or drink in company, unless with lepers.

Having much of a life thus forbidden, the afflicted were forced to live in the leprosariums, an early form of quarantine. Intriguingly, however, they were not required to always *stay* on-site at the hospitals. As long as they followed the rules dictated by the "lepers mass," they were free to roam at will. An early and brutal form of social distancing was put in place.

Between 1100 and 1300, the infection rate for Hansen's seemed to grow exponentially across Europe, a curious development given what we now know about the disease being very difficult to catch. By 1300, a contemporary Benedictine monk estimated that there were an astonishing 19,000 active leprosariums across Europe, indicating just how widespread the disease had become.

Then suddenly by 1400, Hansen's cases became much less common—and medical historians are still puzzling over why. We know the bacterium itself hasn't changed or evolved much since the fourteenth century when Hansen's was endemic in medieval Europe.

One theory puts forth the idea that around the time Hansen's declined, tuberculosis was on the rise. The two diseases are closely related, and some think that one of the very few benefits inherent in catching tuberculosis was that it served as a natural vaccine against Hansen's. A competing theory is that quarantine procedures used to isolate the Hansen's population actually worked. A third theory holds that people with genes that made them susceptible to Hansen's caught, and then died from, the disease, while those with resistant genes survived. In other words, evolution may have once again altered history.

There's yet another theory, the most gruesome of the lot. If you think back to your history classes, what other major disease swept through Europe around the same time that Hansen's went into decline? Just about the same time leprosariums were in full swing across Europe, a black rat crawled off a ship in Venice and launched a little infectious disease called the bubonic plague (see page 50). The Black Death went on to annihilate between 30 and 60 percent of fourteenth-century Europe.

Perhaps by the time the plague had finally run its course in Europe, there just weren't very many people with Hansen's left alive.

The tide finally turned in the fight against Hansen's in the 1940s with the development of dapsone, an antibiotic proven effective in the treatment of the disease. However, the problem with dapsone is that a person needs to take it for a long time, often the rest of their life, in order to keep the Hansen's-causing bacteria at bay. This is difficult for medical practitioners to enforce and difficult for patients to keep up. Complicating matters, by the 1960s *M. leprae* began to develop a resistance to dapsone. Physicians added in rifampicin and clofazimine along with dapsone to create a potent multidrug treatment (MDT) for leprosy.

Beginning in 1981, the World Health Organization (WHO) began advocating this treatment for Hansen's. When followed, it kills the bacteria and cures the patient. Since 1996, WHO has been able to offer MDT free of cost as part of their Hansen's eradication efforts. They've since treated more than 20 million patients globally. WHO achieved elimination of Hansen's disease as a public health problem (defined as a prevalence of less than 1 case per 10,000 population) in 2000. While the disease is not eradicated yet (in 2018, there were 208,619 cases around the world), WHO has made significant strides in that direction.

A rattle (left) used by sufferers of leprosy to warn people of their approach; depictions (right) of leprosy in art from the Middle Ages tended to show it as facial spotting.

Armadillos

Armadillos have the ideal body temperature to host the Hansen's disease bascillus.

Armadillos bear a couple of distinguishing characteristics: They are the only mammal to wear a shell and the only non-human animal that can host the Hansen's disease bacillus. These residents of the southern United States, Latin America, and South America are closely related to anteaters and sloths, spending much of their day sleeping and the rest of it foraging for insects. The Hansen's pathogen needs a very particular temperature to stay alive—it can't be too hot or too cold—which is why the armadillo's average low body temperature of 90 degrees (32 degrees Celsius) makes it a perfect host for the bacillus. Transmission to humans, while low-risk, is possible when people handle, hunt, or consume armadillos. But how did armadillos get Hansen's disease in the first place? Scientists suspect we're to blame. We probably transmitted Hansen's to armadillos about 400 to 500 years ago, with the disease now affecting about 20 percent of armadillo populations. The lower lifespans of armadillos (twelve to fifteen years) compared to humans means they usually don't live long enough to be impacted much by the disease. But both species can help each other contain the disease by just keeping their respective distance.

SYPHILIS
Patient ZERO
Christopher Columbus*

Infectious agent: *Treponema pallidum* (bacteria)

Location: Europe, then spreading to the rest of the world

Symptoms: Genital and oral sores, reddish-brown rash, fevers, dementia, madness, abscesses, blindness, nasal collapse, aortic dissection

Transmission: Sexual

Little-Known Fact: Neurosyphilis, where syphilis attacks the brain, can actually cause bouts of euphoria.

"Through sexual contact, an ailment which is new, or at least unknown to previous doctors, the French sickness, has worked its way from the West to this spot as I write. The entire body is so repulsive to look at and the suffering is so great, especially at night, that this sickness is even more horrifying than incurable leprosy or elephantiasis, and it can be fatal."

Written by an Italian doctor present at the Battle of Fornovo in 1495, he was describing the first outbreak of epidemic syphilis in Europe. That outbreak, which observers unfailingly described as a new disease, struck hard among the soldiers and mercenaries of Charles VIII of France. At the time, the French were embroiled in the Italian Wars, a series of efforts by France to take the Kingdom of Naples on the grounds of a dynastic claim.

The first of the Italian Wars began in 1494 when Charles VIII assembled an army of 50,000 troops, including large numbers of mercenaries from all over Europe, and invaded Italy, bound for the Kingdom of Naples. Italy was then a collection of independent city-states through which the French quickly stormed, passing through Pisa, Florence, and Rome by the end of the year. Early in 1495, the army entered and occupied the Kingdom of Naples without encountering

*and/or those in his crew who returned to Europe after their first visit to the Americas

A colored electron micrograph of Treponema pallidum, *the bacteria that causes syphilis*

much resistance. The army was closely trailed by around eight hundred camp followers, including cooks, beggars, sex workers, and others. Stationed in Naples for several months while Charles set up a pro-French government, the troops were bored, and they soon found ways to occupy their time. It wasn't long before the sexually transmitted "Neapolitan sickness"—or the "French sickness," depending on which side you were on—was rampant among the French army, the trans-European mercenaries, the Neapolitan residents, and camp followers alike.

This "French sickness" was the first major European outbreak of syphilis—and it was met with shock and terror. When an army drawn from allied Northern Italian city-states seemed poised to cut off Charles's retreat route to France, the king led the French army back out of Naples, clashing with the newly formed League of Venice at the Battle of Fornovo in July 1495. It was here the first descriptions of a syphilitic outbreak appear in recorded history. Physicians described the symptoms with disgust, detailing the telltale eruption of pustules all over a patient's body, and later noting that the disease was sexually transmitted.

Beginning with genital ulcers, the disease progressed quickly to a fever, rash, and joint and muscle pains that were particularly bad at night. Within a few weeks, or sometimes months, large, painful abscesses and sores would break out all over the body that emitted a strong stench as well. These sores would soon ulcerate, destroying body parts such as the nose, lips, or eyes in the process. If the sores developed in the mouth and throat, the disease could quickly turn fatal.

"In the yere of Chryst 1493 or there abouts, this most foule and most grevous disease beganne to sprede amongst the people," wrote German scholar and poet Ulrich von Hutten in *Of the Wood Called Guaiacum*, his work on syphilis. And von Hutten was intimate with the subject: He spent the last fifteen years of his life slowly dying from syphilis.

The Battle of Fornovo ended in a draw, with Charles VIII making his way back to France, where his mercenary army dispersed to all corners of Europe, guaranteeing the spread of the seemingly new disease far and wide. An apocryphal quote attributed to the renowned French Enlightenment writer Voltaire summed up the genesis of the disease in Europe, "On their flippant way through Italy, the French carelessly picked up Genoa, Naples, and syphilis. Then they were thrown out and deprived of Naples and Genoa. But they did not lose everything—syphilis went with them."

When syphilis first broke out among Europeans, the disease moved through its stages quickly, killing many of its victims. Because this historical strain of the bacterium responsible for syphilis was particularly virulent, it spread wildly. Within a single generation, syphilis was endemic across Europe.

♦ ♦ ♦

A sexually transmitted disease, syphilis is caused by the bacterium *Treponema pallidum*. The bacterium actually looks cute—it's shaped like cavatappi ("corkscrew" in Italian) pasta—but its friendly appearance disguises a wicked intent. It spreads through contact with a syphilitic sore, called a chancre, on an infected person's genitals. The chancres are painless, one of the few mercies of this deceptive and complex disease. In addition, pregnant women with syphilis can transmit it to their unborn child.

The Battle of Fornovo, 1495

The syphilis strains we know today slowly work through four distinct stages. (Unlike the more virulent historical strain that swept through Europe in the late sixteenth century.) In the first stage, a chancre develops somewhere on your genitals, typically accompanied by swollen lymph nodes in the groin. Usually only one chancre appears, but sometimes there are several. The sore first appears about three weeks after initial contact and, being painless, often goes unnoticed. But even if you noticed the lesion, any alarm you felt would likely disappear when the sore did in a couple of weeks.

Things intensify in the second stage. After a reprieve of a few weeks, you have a one in four chance of suddenly breaking out in a spotty, reddish-brown rash. The rash will typically start on your trunk, but will proceed to cover most of your body, including your palms and the inside of your mouth. It won't itch, but it will be very noticeable. You may also get some wart-like lesions in your mouth or on your genitals in the second stage. The rash usually vanishes on its own within a couple of weeks but may come back repeatedly over the next year. People also often experience malaise in the second stage of the disease, as well as loss of appetite, sore throat, fevers, and body aches.

Ideally by this point you'll have sought out treatment for it, because you definitely don't want syphilis to advance any further. The disease is extremely easy to cure as long as it's in its primary, secondary, or very early tertiary stage. A single intramuscular injection of Benzathine penicillin will cure someone infected in these early stages. Literally just a single injection. (Credit goes to John Mahoney, Richard Arnold, and a serologist named AD Harris for their successfully experimentation with penicillin on syphilitic Marines in 1943.)

If you don't get treated, the disease will progress to its third stage, called latent syphilis. Here is where syphilis earns its reputation as a trickster. Because in the third stage the disease hides, or remains latent, and you have no symptoms. Even though the syphilis bacteria remains alive in your system, it might only be contagious in the early part of the third stage . . . and then never again. Syphilis is only spread via exposure to the lesions in the first and second stages. In the early part of the latent (third) stage, people are considered contagious but usually just in case lesions from the second stage are still lingering or are unseen. In the latter part of the latent phase, and in all of the tertiary phase, syphilis is no longer contagious. Latent syphilis can last for years. If you're one of the few lucky ones, you may never again develop symptoms.

Between ten and thirty years later, those with syphilis can enter the fourth and final stage, or tertiary syphilis, which begins by attacking your insides. Your brain, eyes, heart, liver, bones, joints, and nerves can sustain significant damage at this point. The disease can also start to disfigure you—abscesses and mounds of tissue called gummas may erupt all over the body while your face begins to be eaten away. Many late-stage syphilitic sufferers experience something called nasal collapse, which is as bad as it sounds. Inflammation of your aorta, the large artery that leads away from your heart, is also possible. Weakening and widening into an aneurysm, an inflamed aorta can even spontaneously split open

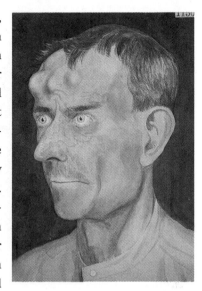

Gummas, or mounds of tissue, mark the fourth and final stage of syphilis.

(called an aortic dissection), killing you instantly. This can happen as much as thirty years after you were first infected.

Syphilis can also suddenly attack your brain at any stage during the infection. Called neurosyphilis, its symptoms include altered behavior, paralysis, and dementia. During the nineteenth century, when approximately 15 percent of the population in Europe and America may have been suffering from syphilis, victims reported a wide array of mental states, including euphoria, suicidal depression, megalomania, and paranoia. For a long time, syphilis was linked with cultural conceptions of "madness." As if that weren't enough, syphilis can also cause blindness. Ocular syphilis, like neurosyphilis, can set in at any time in the disease's progression.

After experiencing any and all of these symptoms of the disease, of course you could also die. When the syphilis outbreak that began in 1495 first spread through Europe, the death toll from it was astonishing. Then, for still-debated reasons, the disease settled down a bit over the next century, becoming less lethal and less virulent. Despite syphilis being readily treatable today, it remains the second leading cause of stillbirths globally, with some 300,000 deaths per year. Another 200,000 infants born from mothers with syphilis carry the risk of an early death.

◆ ◆ ◆

Depiction of Columbus first landing in the Americas

When Christopher Columbus returned from the Americas to Spain in 1493 with a crew of forty-four Europeans and ten Indigenous Americans, a virulent strain of syphilis was possibly present among them. Some historians believe Martín Alonso Pinzón, Columbus's pilot, was syphilis's patient zero in the subsequent European outbreak. (Several other crew members likely contracted syphilis as well in the Americas.) This is known as the Columbian hypothesis.

Soon after Columbus's return, some of his crew members joined the large mercenary army that Charles VIII raised for the Italian Wars. With the first major outbreak of syphilis right on their heels, for both contemporary chroniclers and later historians it was a natural assumption that the disease was previously unknown in Europe and was introduced there from the Americas. This assumption was backed up by later scientific evidence that syphilis was endemic in the Americas before the arrival of Columbus (confirmed via skeletal lesions characteristic of syphilis found throughout the Americas dating back several thousand years). The "newness" of the disease was also commented on by two late-fifteenth-century Spanish physicians, Gonzalo Fernández de Oviedo and Ruy Díaz de Isla, who were present in Barcelona when Columbus returned.

For his part, Isla wrote a treatise on syphilis, published in 1539, in which he reflected on the arrival of the disease in Barcelona: "The sickness owes its origin and birth from time immemorial to the island which is now called Hispaniola, as is widely and clearly understood. And since this island was discovered by admiral Don Christopher Columbus, who had relations and congress with the inhabitants of this island during his stay, and since this sickness is naturally contagious, it spread with ease, and soon appeared in the fleet itself."

Syphilis was thus thought to have been part of the Columbian Exchange (see page 146), the name for the swapping of plants, technology, and animals between Europeans and Indigenous Americans that also included new diseases.

Like many colonial narratives, however, the Columbian hypothesis of the origins of the 1493 outbreak of syphilis—and its progression to becoming endemic in Europe—is now up for debate. Recent scientific inquiry has revealed that the truth is a lot more complicated.

While we've known for many years now that syphilis was endemic in the Americas, genetic researchers recently investigating human bones in graveyards in Finland, Estonia, and the Netherlands have found syphilis DNA from long before Columbus returned from the Americas. They confirmed the bacteria was circulating in Europe before 1493. However, the genome of the syphilis strain found on the European bones differs significantly from the syphilis strains brought back from the Americas. It appears that the strains had a common ancestor that diverged sometime between the fourth century BCE and the twelfth century CE. Historians have theorized that the European strain of syphilis may have been a milder form of the bacteria that caused symptoms contemporary observers misidentified as leprosy, or Hansen's disease.

So we know now that syphilis was present in both Europe and the Americas. But what explains such a virulent European outbreak in 1493? One theory is that European syphilis bacteria recombined with American syphilis bacteria in the wake of the Columbian Exchange, uniting them to make an exceptionally strong strain of the bacteria. Another theory is that the European syphilis strain went through a mutation around the late fifteenth century that abruptly made it much more virulent. But we still don't know for sure. More sequencing and comparisons of modern, old, and ancient syphilis genomes need to be done in order to fully understand what happened.

Not that any of that particularly mattered to the thousands of people suffering and dying from syphilis in fifteenth- and sixteenth-century Europe.

Culturally, the disease was viewed with open disgust—its symptoms were vile and the sexual nature of its transmission gave it a lurid and morally galling reputation. Every country blamed its arrival on their enemies, so it was known by many names. The French called it the "disease of Naples" or "the Spanish pox." The Polish dubbed it "the Russian disease," and the Turks called it a "Christian disease." The English, Germans, and Italians all united in calling it "the French Disease." And the French coined the term "La Grande Verole" (The Great Pox) to distinguish it from smallpox because the lesions it caused were larger. The name "syphilis" didn't arrive until the publication of a 1530 poem by Italian physician-poet Girolamo Fracastoro. In the poem, a shepherd named Syphilus makes the mistake of blaming the sun god for a drought. The sun god, in his wrath, strikes Syphilus down with a hideous new disease, the symptoms of which were now all too familiar to European readers..

The German writer Ulrich von Hutten was one of the first syphilis victims to describe the disease, in his 1519 work *De Morbo Gallico (A Treatise of the French Disease)*: "Boils that stood out like Acorns, from whence issued such filthy stinking Matter, that whosoever came within the Scent, believed himself infected. The Colour of these was a dark Green, and the very Aspect as shocking as the pain itself, which yet was as if the Sick had lain upon a fire."

If physicians weren't somehow repulsed by syphilis, they certainly had no effective remedies to treat it. Initial attempts focused on trying to expel the disease from the body. Bloodletting techniques and laxatives were also tried to no avail. In the early sixteenth century, guaiacum, or holywood, was commonly employed as a treatment. This Caribbean tree was one of the early Spanish imports from the Americas and was enthusiastically embraced for purported miraculous effects. When infused in boiling water and ingested, the wood made patients sweat, which was thought to rebalance their humors. Physicians next landed on mercury as a primary treatment. If you've ever heard the phrase, "A night with Venus and a lifetime with Mercury," you can credit syphilis for it. Mercury is a very effective cathartic, so it did a vigorous job of helping patients empty their bowels. But whatever they expelled in their chamber pots wasn't syphilis.

The unfortunate truth about mercury is that it's highly toxic and caused about as many complications to those who ingested it as the disease itself. Too much mercury in the system leads to mercurial erethism, a neurological disorder characterized by depression, anxiety, pathological shyness, and frequent sighing. It also leads to uncontrollable body tremors. The general medical consensus at the time also recommended repeated mercury treatments, hence the "lifetime with Mercury." If the mercury level continues to build it can lead to tooth loss,

Among many notable syphilis sufferers were Karen von Blixen, who wrote Out of Africa *and acquired the disease unknowingly from her husband, and French novelist Gustave Flaubert.*

rotting jawbones, and gangrenous cheeks that produce facial holes. Many victims of syphilis in Europe suffered their way to an early grave, burdened with both foul-smelling abscesses and mercury toxicity. It's little wonder the disease also earned a reputation for causing insanity.

◆ ◆ ◆

Ironically, syphilis was also equated with artistic genius. As sexual promiscuity went hand in hand with the bohemian culture of nineteenth-century Europe, a great many artists, writers, musicians, and poets contracted syphilis. Indeed, a list of these afflicted artists reads like a who's who of the nineteenth-century European cultural scene: Beethoven, Schubert, Schumann, Baudelaire, Dostoevsky, Flaubert, Maupassant, van Gogh, Manet, Goya, Gauguin, Nietzsche, Schopenhauer, and Wilde were all either confirmed or likely victims of syphilis.

The French novelist Gustave Flaubert described the graphic impacts of syphilitic treatments in 1854 when he received mercury and iodide for a syphilitic tumor, or gumma: "For a week I was hideously sick," he wrote. "Terrific mercurial salivation, mon cher monsieur, it was impossible for me to talk or eat—atrocious fever, etc. Finally I am rid of it, thanks to purges, leeches, enemas (!!!), and my 'strong constitution.' I wouldn't be surprised if my tumor were to disappear, following this inflammation; it has already diminished by half... Meanwhile I'll keep stuffing myself with iodide."

Syphilis continued to haunt artists and writers well into the twentieth century. In 1906, James Joyce wrote, "I presume there are very few mortals in Europe who are not in danger of waking some morning and finding themselves syphilitic." He was probably one of them. Two years earlier, a young Joyce visited a brothel in Dublin and returned home infected with a venereal disease thought by some to be syphilis. He bore many of the symptoms of the disease as his life progressed—particularly continual problems with his eyes.

Karen Blixen, better known as Isak Dinesen, author of *Out of Africa*, had a confirmed case of syphilis that she contracted from her husband, Baron Bror Blixen, in the first year of their marriage in 1914. Bror caught the disease during one of his many affairs undertaken near their East African plantation, then promptly passed it to his wife. Bror, who had a mild case, managed to make it through his life hardly ever showing symptoms. Dinesen was not so lucky—she suffered horribly from the disease from 1914 until her death in 1962. She kept her diagnosis from all but her closest friends, referring to the disease as "life's bitter secret." Her syphilis affliction only became public knowledge sixteen years after she died, completely emaciated, following a lifetime of severe gastrointestinal illness.

The disease also found its way into the royal palaces of Europe. Henry III and Charles V of France both had syphilis, as did Henry VII and George IV of England, Maximilian I of the Holy Roman Empire, and Paul I and Ivan IV of Russia. The mental disturbances brought on by the disease are what some historians believe contributed to Ivan IV's more infamous moniker: Ivan the Terrible. In addition to executing great swaths of Russian nobility, Ivan killed his own son in a fit of rage. The once-great Russian empire soon descended into its catastrophic Time of Troubles, the blame for which can be laid in part at least on syphilis.

The irrational anger and wild mood swings associated with syphilis helped define another famous sufferer: Al Capone. In 1920, while on his way up in the Chicago mob, Capone was assigned to be a bouncer at a bordello owned by his boss "Big Jim" Colosimo. While working there, Capone frequently slept with sex workers and contracted syphilis. Ashamed of the disease, he refused to see a doctor. As a result, the disease was allowed to progress unchecked through his system, haunting the gangster for the rest of his life.

Capone was arrested in 1931 and imprisoned first in Atlanta, and then at Alcatraz when it opened in 1934. By that time, his brain was so adversely affected by syphilis that he couldn't follow the orders of his guards, not out of defiance but

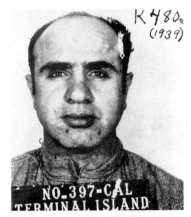

K480
(1939)

Mobster Al Capone in 1939, about seven years before he died from syphilis.

because he couldn't understand them. He was noted for wearing a "strange grin" on his face at unusual times and dressing up in his winter coat, hat, and gloves in his heated cell. His wife, Mae, used Capone's deteriorating mental condition to petition for his early release from his eleven-year sentence. The appeal was successful—Capone was officially diagnosed with neurosyphilis in 1938 and released to the care of his wife in 1939, eight years into his sentence.

By 1942, Capone was treated by Dr. Joseph Moore, a leading expert on syphilis who also managed to acquire a brand-new wonder drug for its treatment: penicillin. The drug, a game-changer in the fight against bacterial disease, finally gave physicians a safe and effective way to treat venereal diseases such as gonorrhea and syphilis. (Today, gonorrhea is no longer treated with penicillin as the disease has built up an alarming resistance to it and several other antibiotics, but penicillin remains incredibly effective against syphilis.) But it was too late for Capone. Too much damage had already been done. His condition continued to worsen each year for the rest of his life, contributing to his official cause of death, cardiac arrest, in 1947. By the time he died, Capone was said to have the mental capacities of a twelve-year-old, his brain eaten away.

◆ ◆ ◆

The realization that syphilis could be treated effectively with penicillin was akin to a divine reprieve to sufferers around the world. A single injection will cure a person who has primary, secondary, or early latent syphilis. So you would think that the Great Pox of Old Europe, with its cultural connections to artists and writers of a bygone era, had been eradicated by now. But syphilis today is actually on the rise, even in developed countries with easy access to health care.

The CDC attributes the rise in syphilis to decreases in sexually transmitted disease programs, decreased condom use among vulnerable groups, and a rise in poverty, drug misuse, and unstable housing that in turn reduce access to STI prevention. Until we can address underlying issues of poverty and access to STI prevention and treatment, the Great Pox is likely to stay with us.

Tuskegee

Blood being drawn from a Tuskegee study test subject

One of the most shameful episodes of racist medical malpractice in American history, the Tuskegee Study of Untreated Syphilis in the Negro Male, began in 1932. It took place in Alabama and followed 399 impoverished African American sharecropping men infected with latent syphilis. The "participants" were tracked by the US Public Health Service for the next forty years to monitor the progression of the disease. The problem was, the men didn't know they had syphilis—because no one told them. Instead, the doctors informed them they had "bad blood," offering free medical care in exchange for being monitored throughout their lives.

The particularly insidious part of the Tuskegee study is that it happened with the full understanding that syphilis is contagious. The victims weren't informed of their condition, so they surely spread it to other people as well. Worse, after penicillin was developed, the men weren't offered the drug, though it would have quickly cured them of the disease.

The so-called study, one of the worst medical scandals in American history, didn't conclude until 1972 when a whistleblower finally leaked it to the press. By then, many of the patients had died of syphilis, as had forty of their wives and nineteen of their children. It remains a despicable and unforgivable act of racism on top of a massive ethical violation. Public outrage over it led to some major and long-overdue reforms in medical experimentation laws, including the requirement of informed consent.

QUACKERY
From Mercury and Bloodletting to Hydroxychloroquine

Impact: False hope, frequent unnecessary pain and suffering, and often accelerated death when used in lieu of legitimate treatments

When: Since there have been disease and "healers" willing to offer treatments for profit or otherwise

Who: All manner of healers, physicians, politicians, and hucksters

What Happened Next: If you were lucky, nothing. If you were unlucky, pain and suffering and sometimes death.

Little-Known Fact: Quinine, a chemical relative of hydroxychloroquine, was used ineffectively to treat yellow fever and polio. It glows under black light in your gin and tonic.

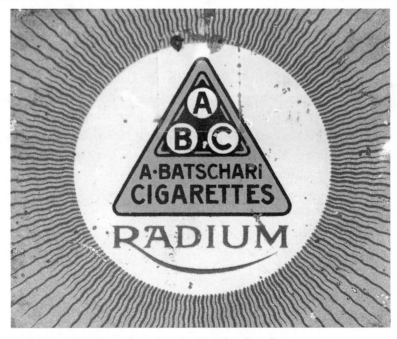

A one-two cancer punch: carcinogenic smoke and radioactive radium

T hroughout human history we've been hit with plague after plague, and our response is pretty much always the same: In desperation, we resort to a wide variety of unusual, ineffective, or outright dangerous remedies while hoping that science (or pre-science/understanding in general) catches up to the disease. The impulse is certainly understandable. Long before we even knew what they were, we longed for a measure of control over the mysterious microbes that occasionally wreaked havoc on our health. Even today, with all the astounding benefits of twenty-first-century medicine, we still get "creative" (some would say zealous) with remedies we hope might help reduce our suffering or cure a disease. This quackery has always been with us and likely always will be. As long as there is disease, there will be a quack willing to peddle snake oil. Here are some of the more unusual "remedies" once offered for the diseases in this book.

THE RELICS OF SAINTS

Disease: *Ergotism*
When: *Medieval period*
Where: *Europe*

In the eleventh century a French nobleman, Guérin la Valloire, declared himself cured of ergotism thanks to the relics of St. Anthony the Great. While suffering from the disease, Valloire seems to have stationed himself at the St. Anthony shrine at Saint-Antoine-l'Abbaye, Isère, which housed the saint's relics. When the disease faded away on its own, Valloire declared the relics as the cause of his sudden recovery. While he was wrong about the solution, the grateful Valloire and his father generously redirected some of their wealth to fund the beginnings of the Hospital Brothers of Saint Anthony. These Antonite monks would soon establish monastic hospitals all around Europe to help people suffering from St. Anthony's Fire (ergotism).

Side effects: *None.*

COLLOIDAL SILVER

Disease: *Lyme disease*
When: *Twenty-first century*
Where: *United States*

With a rich history in medical quackery, colloidal silver has also been marketed to people suffering from
Lyme disease. Proponents claim that silver supports
the immune system, while simultaneously killing
Lyme and other infections. It also is supposed to offer
general support for the immune system while it fights
off the disease.

Stan Jones: Libertarian, not afraid to ingest colloidal silver

Side effects: *If you ingest too much silver, your skin will turn blue. Permanently. Meanwhile your Lyme disease symptoms will continue unabated.*

HOMEOPATHY

Disease: *Ebola*
When: *Twenty-first century*
Where: *Africa*

Homeopathy, a medical system developed in Germany in the 1800s, is founded
on two unconventional medical theories: 1) "Like cures like," based on the idea
that a disease can be cured by a substance that produces similar symptoms as the
disease itself, and 2) "Law of minimum dose," which says the lower the dosage of
the medication, the higher the effectiveness. Many oral homeopathic remedies
are so diluted in water that there is a negligible amount or no trace of the original
substance. To date, despite rigorous attempts at testing, there is still no evidence
to suggest that homeopathy is effective for any condition.

After the 2014 Ebola outbreak in Liberia, homeopaths flew to the location
to ply their trade. Some examples of homeopathic remedies claimed to be useful
to treat Ebola include viper venom, rattlesnake venom, and six-eyed crab spider.
The Liberian government had approved the medical team and issued them visas
under the assumption that they were medical doctors. When the Ganta Hospital
in Liberia realized they had homeopaths instead of MDs on their hands the entire
group was banned from the Ebola Treatment Unit.

Side effects: *At worst, death from delaying treatment.*

Vinegar versus plague?

FOUR THIEVES VINEGAR

Disease: *Bubonic plague*
When: *Seventeenth century*
Where: *Marseilles, France*

According to legend, when an outbreak of the plague struck Marseilles in the seventeenth century, a band of thieves took advantage of the sudden lack of authority to repeatedly break into the homes of the plague-stricken, stealing anything of value. At first the authorities took little notice of them, assuming anyone brash enough to come into close contact with plague victims was bound to die soon from plague themselves. But the thieves continued their operation, seemingly unaffected. When the authorities finally captured the thieves, they were found to be dousing their clothes and face masks with a special herbal vinegar that "protected" them from the disease. A judge offered the thieves a bargain: They could go free if the thieves would relinquish their special vinegar recipe. They did—and Four Thieves Vinegar was born. While the exact original recipe has been lost to history (if it ever existed), a Four Thieves Vinegar recipe from early twentieth-century herbalist Jean Valnet is thought to be the most accurate: vinegar infused with wormwood, meadowsweet, juniper, marjoram, sage, cloves, horseheal, angelica, rosemary, horehound, and camphor.

Side effects: *Vinegar mixed with these herbs would indeed be mildly antibacterial. As for an impact on the bubonic plague? Not so much.*

HAND EXERCISER ATTACHED TO ANTENNAE

Disease: *Hepatitis C*
When: *Twenty-first century*
Where: *Egypt*

In 2014, an Egyptian military doctor announced that he'd found a cure for both HIV and hepatitis C (the latter is endemic in Egypt, affecting approximately 15 percent of the population). "Defeating the virus is very easy, but God grants wisdom to whoever he wants," said the doctor in a televised news conference that included a demonstration of the special antiviral devices used in treatment. The video featured patients hooked up to mysterious boxlike machines while doctors held on to devices that "looked like a hand exerciser attached to an antenna that swiveled, following the patients as they walked," according to the *New York Times*.
Side effects: *International embarrassment.*

CHLOROQUINE AND HYDROXYCHLOROQUINE

Disease: *COVID-19*
When: *2020*
Where: *United States*

In the early days of the COVID-19 pandemic, American president Donald Trump made inflated claims about hydroxychloroquine as a treatment for the disease.

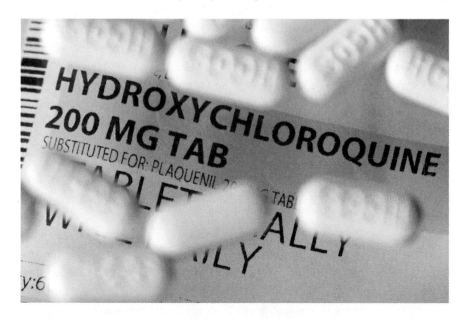

The anti-malarial drug, now most often used for rheumatological diseases, was the subject of early experimentation as a COVID cure. Some people took these statements to mean chloroquine in any form. One Arizona couple went rummaging in their closet and found chloroquine phosphate, a fish tank cleaner, which they promptly ingested. The husband died and the wife became seriously ill. Hydroxychloroquine, meanwhile, has not panned out as an effective COVID-19 treatment.

Side effects: *Vomiting. Arrhythmias. Death.*

HEROIC DEPLETION THERAPY: MERCURY AND BLOODLETTING

Disease: *Yellow fever*

When: *1793*

Where: *Philadelphia*

A powdered mercury salt called "calomel" is a white, odorless powder that was widely distributed as a medicinal aid between the sixteenth and twentieth centuries. When taken orally, calomel operates as a potent cathartic, causing a violent rush of diarrhea.

Such a "purging" was often recommended for a wide variety of ailments. During the 1793 yellow fever outbreak in Philadelphia, the city's primary physician, Dr. Benjamin Rush, recommended "heroic depletion therapy," consisting of repeated doses of calomel and copious bloodletting sessions. Some members of the College of Physicians of Philadelphia were horrified at these recommendations, which they labeled as "murderous" and "fit for a horse."

Side effects: *Heavy metal toxicity. Mercurial erethism. Tooth loss. Rotting jaw. Gangrenous cheeks. Ulcerated tongues and gums. Death.*

URINE

Disease: *Mad cow disease (bovine spongiform encephalopathy)*

When: *Twenty-first century*

Where: *India*

The head of India's Water of Life Foundation declared that cows suffering from mad cow disease could be cured if they were forced to drink their own urine. And yes, "Water of Life" in the view of the foundation is actually . . . urine. The foundation also espouses drinking your own urine for a variety of personal ailments as well. To date, there is no scientific evidence for the therapeutic use for urine.

Side effects: *While revolting to most people, drinking small quantities of your own urine is likely to be harmless. Large amounts? That's another story.*

HERBAL CONCOCTION MADE BY YOUR DICTATOR
Disease: *HIV/AIDS*
When: *Twenty-first century*
Where: *The Gambia*

Yahya Jammeh, authoritarian leader of the Gambia in West Africa from 1996 to 2017, declared early in 2007 that he could cure AIDS. His technique? A combination of a mysterious herbal concoction and Islamic prayers that would only work on Mondays and Thursdays. "Volunteers" for Jammeh's HIV cure were stationed in a special charity clinic run by the president for a six-week period. During that time, the president would rub an herbal paste on their chests in the mornings (on Mondays and Thursdays only, of course) while chanting prayers from the Quran. Patients were also forced to drink twice daily from a bottle filled with a yellow herbal mixture. The ingredients were never revealed. Participants had to forego antiretroviral drugs. Some nine thousand people are thought to have participated in the program. However, since the clinic's records were considered state secrets, no one knows for sure how many of them died.

Side effects: *Worsening of HIV infection from lack of conventional treatment. Death.*

BLOOD BATH
Disease: *Hansen's disease (leprosy)*
When: *Medieval period*
Where: *Europe*

In the Middle Ages, Hansen's disease (and disease in general) was explained by the humoral model. This model theorized that good health depended on the balance of four humors: blood, yellow bile, black bile, and phlegm. Hansen's disease, formerly known as leprosy, was thought to result from an excess of black bile that contaminated the blood. As a result, bloodletting to remove "corrupted" blood was advised, followed by a blood-spiked bath. The special bathwater was infused with the blood of an infant or virgin (both were thought to contain "pure" or "clean" blood). By sitting in this literal bloodbath, the patient would undergo the medieval magical thinking equivalent of a transfusion.

Side effects: *Risk of bloodborne disease transmission if the blood of your virgin isn't quite as pure as your doctor hoped. The virgins and the babies were the unfortunate losers in this scenario.*

LIGHT BATH

Disease: *Typhoid*
When: *1890s*
Where: *Battle Creek, Michigan*

For treatment of typhoid, John Harvey Kellogg (later of cereal fame) recommended sitting in a cabinet with fifty 110V electric lamps surrounding the body. The lamps were arranged in groups that could each be activated with a switch. Kellogg wrote about the treatment, "The electric-light bath

Fig. 21.   Arc Light to the chest.   See page 94.

Not very enlightened

prolonged to the extent of vigorous perspiration should be employed two or three times a week . . . Tanning the whole surface of the body by means of the arc light will be an excellent means of improving the patient's general vital condition." Scarlet fever, diabetes, and typhoid were all thought to scurry into the darkness in the harsh glow of all that light.

Side effects: *Profuse sweating. Typhoid wouldn't be impacted at all.*

ANCIENT EGYPTIAN MUMMIES

Disease: *Syphilis*
When: *Fifteenth and sixteenth centuries*
Where: *Europe*

"Mumia," a standard part of the European pharmacopoeia for some two centuries, was manufactured from stolen Egyptian mummies. The corpses were ground up, or boiled to retrieve a black oily substance that bubbled up to the surface. The medical thinking, to be generous, was that the body's spirit could be distilled via the corpse, which could then be used to cure almost anything. Mummy-infused poultices were commonly used to treat syphilitic sores.

Side effects: *Probably harmless to the consumer, but now you've contributed to an ethically bankrupt business that arose from the desecration of tombs and cultural sites.*

LEECHES

Disease: *Cholera*
Where: *England*
When: *Nineteenth century*

Owing to extreme dehydration, the blood of cholera victims is abnormally thick. In the humoral theory of disease, a doctor would attempt to remove whatever seemed wrong with the patient. So doctors employed leeches to help remove the "bad blood" from cholera patients. The leeches were also thought to relieve stress on the heart and lungs.

Side effects: *Cholera victims didn't need any help shedding additional liquids, so leeches would have likely made the condition worse.*

GARLIC AND OREGANO OIL

Disease: *Anthrax*
When: *Twenty-first century*
Where: *United States*

In the wake of the 2001 anthrax attacks in the United States, a variety of products were touted for treatment following exposure to anthrax spores. Garlic and oregano oil, both of which have some antimicrobial properties, topped the list. After a 2001 experiment on mice conducted by Dr. Harry Preuss of Georgetown University proved that oil of oregano was effective at reducing infection, quacks ran with the claim, cashing in on sales of oil of oregano. But it was a big leap from a supplement helpful at fighting off small-scale infections to one capable of fighting off a disease as devastating as anthrax. Similar preventative claims were made about garlic supplements as well, though neither it nor oregano oil will do much to help you fight off anthrax.

Side effects: *Pungent breath.*

HAIR OF THE DOG

Disease: *Rabies*
When: *First century CE*
Where: *Italy*

The Roman writer and naturalist Pliny the Elder describes rabies treatments in detail in his *Natural History* (77 CE). His most famous "cure" remains with us today as an expression to describe a hangover treatment: the hair of the dog. The origins of the phrase come from a rabies treatment where Pliny suggested that one should "insert in the wound ashes of hairs from the tail of the dog that inflicted the bite." He had several other suggestions as well, including inserting a maggot from the dead dog's corpse into the bite wound, or wrapping a wound with a linen cloth soaked with the menstrual blood of a female dog. Once someone decapitated the rabid dog, other options presented themselves: You could burn the head and put the ashes in your wound, or, if you were feeling peckish, you could just eat the dog's head.

Side effects: *The concept of "hair of the dog" may work for a hangover cure, but nothing was getting in the way of rabies's near 100 percent mortality rate.*

CORNUCOPIA OF QUACK REMEDIES

Disease: *Polio*
When: *Twentieth century*
Where: *United States*

One recommendation, recorded in a 1916 publication about the New York City polio epidemic, recorded this enormously involved period polio remedy:

> Place hydrogen conductors at soles of feet and hands, and cause attraction for this fine hydrogen by neg. electricity or neg. applications. Apply cantharides [Spanish fly] and mustard plasters. Diet must be high in fine oxygen, such as rice, bread and oxygen waters. Give oxygen through the lower extremities, by positive electricity. Frequent baths using almond meal, or oxidising the water. Applications of poultices of Roman chamomile, slippery elm, arnica, mustard, cantharis, amygdalae dulcis oil, and of special merit, spikenard oil and Xanthoxolinum. Internally use caffeine, Fl. Kola, dry muriate of quinine, elixir of cinchone, radium water, chloride of gold, liquor calcis and wine of pepsin.

Side effects: *So many.*

THE CARBOLIC SMOKE BALL

Disease: *Flu*
When: *1889 to 1892*
Where: *London*

Rubber ball filled with powdered carbolic acid (a sweet-smelling, if poisonous chemical substance derived from tar).

Instructions: You squeeze the rubber ball, sending carbolic acid smoke up through tubes connected to the rubber ball . . . and into your nostrils. Repeat three times per day for two months.

Side effects: *Over time, corrosive lung damage from repeated inhalation of carbolic acid. On the positive side, consumer protection laws. (See the court case* Carlill v. Carbolic Smoke Ball Company, *which is still studied today by law students around the world.)*

THE TOUCH OF A KING

Disease: *Tuberculosis*

When: *Eleventh century*

Where: *France and England*

Medieval kings held royal touching ceremonies, where anyone afflicted with tuberculosis could visit with the hopes of being touched by the monarch. As the king was a divine appointee of God himself and obviously imbued with godly healing powers, the mere touch of a king's hands was supposed to cure your disease.

Side effects: *Zero improvement to tuberculosis aside from the placebo effect, but potential for monetary gain (any peasant lucky enough to be touched by a king was given a special gold coin dubbed an "angel").*

ANTIMONY: THE EVERLASTING PILL

Disease: *Smallpox*

When: *Eighteenth century*

Where: *America and Britain*

A common method of treating smallpox in the eighteenth century was inducing "purges" to help the patient rid the disease through copious bowel movements. One method of purging was ingesting a pill made from the metal antimony. A trace amount of the metal would be absorbed by your intestines, and after you'd successfully evacuated your bowels, you could retrieve the antimony pill for reuse. This common practice led to its nickname the "everlasting pill."

Enhancing the urge to purge.

Side effects: *Possible bowel movement, vomiting, and a cost-effective strategy to keep generations of family members purging through the years. Not much help, though, with recovering from smallpox.*

TYPHOID FEVER
Patient ZERO
"Typhoid Mary" Mallon (1869–1938)

Cause: *Salmonella typhi* (bacteria)

Location: New York City, United States

Symptoms: Sustained fever, red spotting on neck and chest, weakness, stomach pain, headache, digestive problems, cough, loss of appetite

Transmission: Oral ingestion of fecal matter via unwashed hands, contaminated food, or drinking water

Little-Known Fact: It was never definitively established if Mary Mallon had a mental illness, but she believed almost to the end of her life that she was *not* contagious.

Mary Mallon became the poster child for asymptomatic disease spread.

n the summer of 1906, wealthy New York banker Charles Henry Warren rented a summer home for his family in Oyster Bay, Long Island. Warren's family of four took up residence at the house along with seven servants. One of them was a thirty-seven-year-old Irish cook named Mary Mallon. The Warren family considered Mary a fortunate hire. Though she was largely uncommunicative, she made excellent food—in particular her "peaches on ice" dish.

The quality of the food on the Warren family table surely suffered when, abruptly and for no apparent reason, Mary left their employment and went in search of work elsewhere. While her "peaches on ice" were likely missed for the rest of the summer, the culinary downgrade was about to be the least of the Warren family's problems.

Within a few weeks of Mallon's departure, six of the ten remaining people in the Warren house came down with one of the most dreaded diseases of the time: typhoid fever.

Typhoid isn't your usual stomach flu. The disease produces a variety of symptoms, including abdominal pain, headaches, rashes, and high fever. Without treatment, it also has a mortality rate of 10 to 30 percent. It is caused by the bacterium *Salmonella typhi*, which tends to pop up when food or water is contaminated with human feces.

In reality, the sanitation practices and reliable water supplies we take for granted are a relatively recent development in world history. Because typhoid

Early typhoid outbreaks were often the result of infected human waste seeping into drinking water.

Asymptomatic Havoc

While Mary Mallon was the first identified, she was by no means the only healthy carrier of typhoid fever. Concurrent with Mallon's first forced quarantine in 1906, the New York City health department made it a priority to start locating other healthy carriers. And they found a lot of them—alarmingly, many of them were also working in the food industry. By 1908, five carriers, including Mallon, were identified. Twelve years later, in 1920, eighty-five healthy carriers were under observation, and by the time Mallon died in 1938, there were almost four hundred healthy typhoid fever carriers registered in the city.

Aerosolized sneeze

The health department identified healthy carriers by tracking people who got sick with typhoid as they recovered. If you still tested positive for typhoid for three months or more after recovery, you were entered on the list of healthy carriers and forbidden from working in food handling. Like Mallon, not everyone took kindly to the suggestion. In 1922, a carrier who had reportedly caused an outbreak resulting in eighty-seven cases and three deaths in New York City was found to be working in the food industry again in New Jersey. He was required to change jobs and report weekly to the health department, but he was not forced into isolated quarantine like Mallon.

Mallon's infamous moniker, "Typhoid Mary," also entered the general lexicon to mean any person immune to a disease that they subsequently spread. The press first saddled Mallon with that nickname after her initial brush with the law in 1906. Then they began to use the phrase to describe other healthy carriers in similar situations. In 1910, after an outbreak of typhoid at a retreat in the Adirondack Mountains that sickened thirty-six people and killed two was sourced to the healthy camp guide, the *New York Times* subsequently dubbed him "Typhoid John." In Alsace, France, in 1913, another "Typhoid Mary" made headlines when an outbreak of typhoid that infected fourteen people was traced to Mlle Jansen, the healthy secretary of a businessman in Colmar. Jansen was subsequently brought to the Institut Pasteur in Paris for further study, where she was positively identified as a healthy carrier . . . and then subsequently offered permanent employment as an assistant librarian.

predates most of these practices, it was a plague of humankind for centuries. And it continues to haunt developing areas of the world with poor sanitation, including parts of Africa, Asia, the Middle East, and Central and South America. Even today, it is *still* responsible for upward of 200,000 deaths each year, infecting between 11 to 21 million people annually.

Typhoid fever typically breaks out in poor areas where clean water is in short supply and, historically, where sanitation is either nonexistent or not understood. And while major outbreaks tended to be concentrated in impoverished, urban areas, sometimes anomalous outbreaks, like the one at the Warren family summer home on Long Island, still occur.

After the Warren family returned to New York City, the landlord for their summer rental had a problem on her hands. How was she going to rent out the house again next summer if it was associated with a typhoid fever outbreak? Gossip among the monied class in New York spread fast, far, and wide. She needed to prove the house was no longer contaminated, so she hired a sanitary engineer named George Soper to investigate.

Soper's initial guess at what caused the outbreak was a bad batch of clams. However, a cursory investigation revealed that not every member of the household had eaten the clam meal. Soper then made a thorough investigation of the

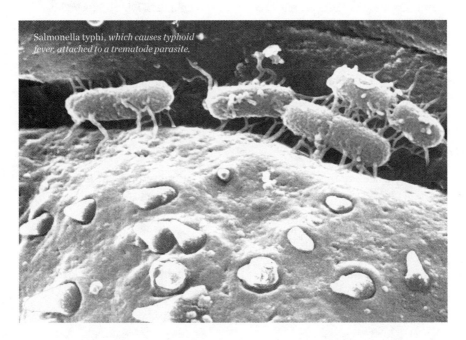

Salmonella typhi, *which causes typhoid fever, attached to a trematode parasite.*

property, but everything checked out. He knew that typhoid fever spread through contaminated water, so he took a close look at every possible point of human contact with water in and outside the house—the well, the overhead water tank, the cesspool, the privy, food supplies in the pantry, manure used to fertilize the lawn. He even investigated the neighbor's sanitation arrangements. Again, nothing looked amiss.

Soper concluded that a human carrier must have entered the house. By 1906, researchers in Germany had proven that people could be healthy carriers of the *Salmonella typhi* bacteria without showing a single symptom of the disease. The bacteria would show up in their urine and feces, however, so depending on their commitment to personal hygiene, healthy carriers could unintentionally spread the disease. While Soper was aware of this research, a healthy carrier had yet to be identified in the United States.

It wasn't long until he zeroed his suspicions on the chef who abruptly departed in the middle of the summer. While the level of heat typically generated in cooking is a reliable killer of *S. typhi* bacteria, Soper learned there was a particular dish the chef made that was beloved in the household, but didn't require heating: peaches on ice.

He wrote about his discovery later, "I found, however, that on a certain Sunday there was a dessert which Mary prepared and of which everybody present was extremely fond. This was ice-cream with fresh peaches cut up and frozen in it. I suppose no better way could be found for a cook to cleanse her hands of microbes and infect a family."

The case was seemingly cracked: The Warren family came down with typhoid fever because their chef defecated, didn't wash her hands, and then made them peaches on ice.

◆ ◆ ◆

On the heels of this first realization quickly came another: If Mallon spread typhoid fever at the Warren family household, she might be a rare healthy carrier of the disease. Suddenly, it was a race against time for Soper to track her down to stop her from further spreading typhoid.

But Mallon was not—nor had ever been—an easy person to find. She changed jobs frequently, moving from place to place, only maintaining an address for a few months at a time. By most accounts she was also a difficult person to talk to, both a bit mysterious and a touch combative. On investigating her employment

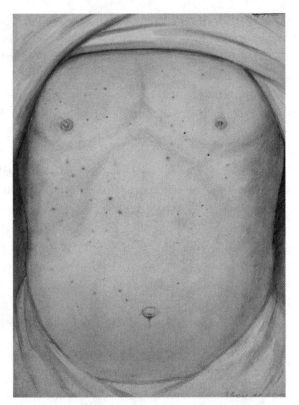

Telltale red spotting can indicate a typhoid infection.

history, Soper realized that Mallon had been moving between cooking jobs for a variety of families, often leaving typhoid fever in her wake. She was the common link between the outbreaks.

Soper eventually identified eight families for whom Mallon had previously worked.

Of those eight families, seven of them had experienced an outbreak of typhoid fever. "In each household," wrote Soper, "there have been four or five in the family and from five to seven servants. Four of the persons attacked have been laundresses. Two have been gardeners, permanently attached to the country places where the typhoid has broken out." In total, Soper counted twenty-two victims of typhoid fever.

It took Soper four months, but he finally found Mallon in March 1907. To his horror, he learned that "Mary was working as a cook in an old-fashioned, high-stoop house on Park Avenue [in Manhattan] on the west side, two doors above the church at Sixtieth Street. The laundress had recently been taken to the Presbyterian Hospital with typhoid fever and the only child of the family, a lovely daughter, was dying of it."

Soper confronted Mallon, explaining that he was quite certain she was a healthy carrier of typhoid fever. The conversation ground to a halt, however, when Soper asked for samples of her blood, urine, and feces. Mallon didn't take kindly to the request.

According to Soper, "She seized a carving fork and advanced in my direction. I passed rapidly down the long narrow hall, through the tall iron gate, out through the area and so to the sidewalk. I felt rather lucky to escape."

But Soper was persistent. He tracked Mallon down to her residence, a place of "dirt and disorder," where he again confronted her. He explained that while she wasn't herself ill, he had good reason to believe she was carrying the bacteria that caused typhoid fever. This meant she was spreading the disease among the people for whom she worked. Mallon angrily denied the accusation, insisting that she didn't have, nor ever had, typhoid fever. And for the life of her, she couldn't see why Soper was bothering her. After all, typhoid was everywhere—why was he picking on her over it?

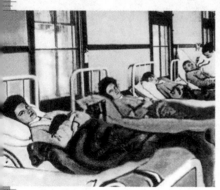

Mallon (foreground) hospitalized, ca. 1909

Soper knew Mallon was about to leave her Park Avenue job. He worried that she would disappear into the great maw of New York City only to spread the disease further. Soper went to the authorities, approaching the New York City Health Department and convincing them to investigate. On March 19, an investigator from the Health Department visited Mallon at her home, but she slammed the door in the investigator's face.

The next day the investigator returned, this time with the members of the New York City Police Department. Mallon fled, racing out her back door and scrambling over a fence. When the police officers finally caught up with her three hours later, they found her hidden in a closet under the front steps of a neighboring residence. She put up fierce resistance, kicking and swearing at the officers before ultimately being taken into custody.

George Soper, the sanitation engineer who identified Mallon as an asymptomatic spreader of typhoid.

It remains an open question if Mallon had an undiagnosed mental illness. She displayed symptoms of mental instability, often acting out angrily and irrationally. What we know for sure is that she was also an immigrant in a foreign country, alone, and fearful of the authorities.

Mallon's capture by the police prevented her from causing any further outbreaks in the short term. That year, 1907, approximately three thousand New York City residents were stricken with typhoid fever. Some medical historians have attributed most of those cases to outbreaks that began with Mary Mallon. Regardless, it was a great public relief when she was finally captured. It was also the start of a long and tragic isolation for Mallon that wouldn't end until her death thirty years later, in part at least because of her fiercely uncooperative nature.

◆ ◆ ◆

Once she was in custody, Mallon's stool was collected and tested for *S. typhi* bacteria. Sure enough, they were present. Soper's relentless investigation was a success, one that had undoubtedly saved many lives.

Soper visited Mallon in her forced quarantine at the Willard Parker Hospital. To his credit, the sanitation engineer tried his best to explain to her that he didn't fault her for the outbreaks. How could she have known she had typhoid? He said he would help free her if she would just improve her personal hygiene (i.e., wash her hands after using the bathroom) and answer a few questions before he concluded his investigation.

Mallon didn't respond. Soper wrote, "She pulled her bathrobe about her and, not taking her eyes off of mine, slowly opened the door of her toilet and vanished within. The door slammed. There was no need of my waiting. It was apparent that Mary did not intend to speak to me. So I left the place."

It's easy to view Mallon in an unfavorable light in the wake of these events. Her dismissal of scientific evidence, refusal to work with the authorities, and seemingly blatant disregard for public health all cast her as a . . . well, if not a villain, at least a person difficult to cheer for. Mallon, however, had not techni-

Mallon's cottage on North Brother Island in New York City, ca. early 1900s

cally committed any crimes and was being held against her will. She was a fiercely independent person, an immigrant, Irish-born in a time of heavy anti-Irish sentiment in America, and from a working-class background. It's hard to imagine a wealthy man from the Upper East Side, born and bred in New York City, having the same experience, even if he was identified as the source of typhoid fever outbreaks. Mallon's gender and ethnic and socioeconomic backgrounds certainly played a role in the way she was treated by authorities in New York.

Mallon was soon moved from the Willard Parker Hospital to quarantine at Riverside Hospital on North Brother Island, in New York's East River between the Bronx and Rikers Island. If there was an upside, her quarantine digs were better provisioned than what she had known before. Her cottage, which she had all to herself, included gas, electricity, and modern plumbing. This was likely an upgrade from what she could afford on a cook's salary. But it was also deeply lonely. Hospital staff delivered Mallon's food, but that was about it for human interaction.

She lingered in quarantine on the island for three years. During which hospital staff tried various methods to treat her (hexamethylenamin, laxatives, urotropin, and brewer's yeast), hoping to remove the typhoid-causing bacteria from her system so she could be released—to no avail. Mallon was destined to be a carrier for life. Immunization for typhoid fever wouldn't be developed until 1911, and effective antibiotic treatment wouldn't come along until almost forty years later, in 1948.

While Mallon settled into her lonely life on the island, she remained deeply unhappy with her situation and angry at authorities for imprisoning her there. As she wrote in a 1909 letter, "There was never any effort by the Board authority to do anything for me excepting to cast me on the island and keep me a prisoner without being sick nor needing medical treatment. . . . I have been in fact a peep show for everybody. Even the interns had to come to see me and ask about the facts already known to the whole wide world. The tuberculosis men would say, 'There she is, the kidnapped woman.'"

In 1909, Mallon sued for release. She claimed, correctly, that she had been denied due process under the law and had never been formally charged with a crime. But the courts had an overriding consideration: When someone is at risk for transferring a deadly infectious disease to the general public, a person could be forcibly held in quarantine without due process.

Nevertheless, the courts were sympathetic to Mallon's plea. In 1910, she was released from quarantine after she promised to never again work in food preparation and to regularly check in with the New York Health Department. The decision was likened to a quarantine probation.

Mulberry Street, New York City, ca. 1906

Mallon immediately broke her promise to abide by the ruling. Despite all the evidence presented to the contrary, she still didn't believe she carried typhoid fever. She remained convinced she was unfairly persecuted and had no intention of cooperating with the Health Department ever again. Instead, she proceeded to take cooking jobs to keep herself afloat; they paid more than the other wage-based work for which she was qualified, such as toiling as a laundress. So she was back in kitchens again, working short-term cooking jobs under false names, including Mary Brown and Marie Breshof. She also worked in restaurants, hotels, and even, ironically, hospital kitchens. That whole time she inevitably made people sick with typhoid fever.

Soper, meanwhile, lost track of Mallon while she hid behind her false names and short-term gigs in the big, bustling city. Five years went by this way until 1915, when Soper received a call from a doctor at the Sloane Hospital for Women. The doctor told Soper the hospital was in the middle of a typhoid outbreak, with twenty staff members sick with the disease. He further informed Soper that a woman named Mary Brown had been hired recently as the hospital cook and that "the other servants had jokingly nicknamed her Typhoid Mary."

Soper asked for a description of Mary and a copy of her handwriting, both of which identified her as the same Mary Mallon from the summer of 1906. He immediately contacted the New York Health Department, who again showed up to take Mallon into custody. The nickname Typhoid Mary stuck when the press got wind of the story.

And this time it was permanent. Mallon remained in quarantine for the rest of her life on North Brother Island. She was made comfortable, given a humble but pleasant cottage, and paid work to test medical samples at the hospital. A small fox terrier was permitted to her as a companion. As she settled into her quarantine life, the fight seemed to drain out of her. Eventually, her good behavior convinced hospital staff to allow Mallon to visit the city from time to time. She didn't attempt to flee.

Mary Mallon died on North Brother Island in November 1938, alone except for her pet dog, ten years before antibiotics were successfully used to treat the disease.

In the end, America's first known healthy carrier of *S. typhi* bacteria was responsible for at least 122 cases of typhoid fever, with five confirmed deaths . . . but the real number of people infected by Mallon was undoubtedly higher.

Perhaps much higher.

CONTAINMENT

Colored scanning electron micrograph of Penicillium notatum *spores*

1918 INFLUENZA
(aka Spanish Flu, Spanish Lady, French Flu, Purple Death)

Patient ZERO
Unknown

Cause: H1N1 influenza A (virus)

Symptoms: Fever, cough, malaise, and subsequent pneumonia causing purple face (heliotrope cyanosis), delirium, coma

Transmission: Airborne respiratory droplets

Impact: An estimated death toll of 50 million people worldwide

Where: Worldwide

When: 1918

Little-Known Fact: The flu almost certainly did not originate in Spain. Rather, the well-publicized flu case of King Alfonso XIII of Spain led to the unfounded belief that the virus came from there.

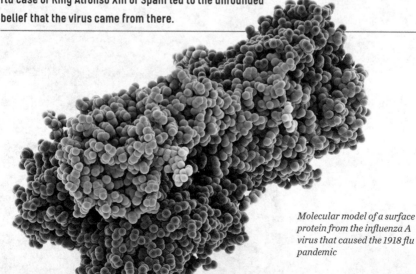

Molecular model of a surface protein from the influenza A virus that caused the 1918 flu pandemic

"These men start with what appears to be an ordinary attack of LaGrippe or Influenza, and when brought to the Hosp. they very rapidly develop the most vicious type of Pneumonia that has ever been seen. Two hours after admission they have the Mahogony spots over the cheek bones, and a few hours later you can begin to see the Cyanosis extending from their ears and spreading all over the face, until it is hard to distinguish the colored men from the white. It is only a matter of a few hours then until death comes, and it is simply a struggle for air until they suffocate. It is horrible. One can stand it to see one, two or twenty men die, but to see these poor devils dropping out like flies sort of gets on your nerves. We have been averaging about 100 deaths per day, and still keeping it up. There is no doubt in my mind that there is a new mixed infection here, but what I don't know . . ."

—September 29, 1918 Letter from Surgical Ward No 16 Camp Devens, Massachusetts

The wave of a new, deadly influenza virus that washed over the world in 1918 was the deadliest flu pandemic of all time. It killed an estimated 50 million people worldwide, while also decreasing the life expectancy of Americans by more than twelve years. Known as the "Spanish" influenza, la grippe espagnole, and the "Purple Death," the infection killed with a brutal swiftness, often within hours or days of exhibiting symptoms of the illness.

World War I, which ended officially in November 1918, played an outsized role in globalizing the pandemic, serving up soldiers and civilians in cramped, unsanitary quarters ideal for a flu virus to fester and spread. Another hallmark of this pandemic was how lethal it was to young adults age fifteen to thirty-four years—those who usually fare better during typical flu bouts. Even given the relatively advanced state of medical knowledge and care at the time (compared to just a century prior), the casualties of this strain were enormous. In fact, the particular strain of the 1918 pandemic influenza H1N1 virus has perplexed and frightened scientists ever since.

Stories from 1918 still haunt survivors and their families: Coffins piled up in cemeteries without enough men to bury them, daily funerals, a shortage of caskets, people fighting over mask mandates (yes, it happened back then, too). Oddly, schools were havens of cleanliness and safety for many children whose homes were often not as clean or well ventilated. They also served as sanctuaries of normalcy for many of those children who lost one or both parents to the flu.

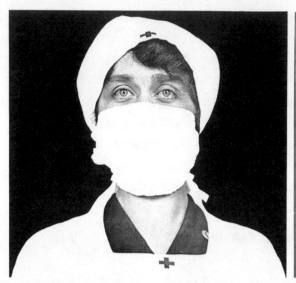

To Prevent

Influenza!

Do not take any person's breath.
Keep the mouth and teeth clean.
Avoid those that cough and sneeze.
Don't visit poorly ventilated places.
Keep warm, get fresh air and sunshine.
Don't use common drinking cups, towels, etc.
Cover your mouth when you cough and sneeze.
Avoid Worry, Fear and Fatigue.
Stay at home if you have a cold.
Walk to your work or office.
In sick rooms wear a gauze mask like in illustration.

A 1918 public health poster showing a Red Cross nurse wearing a gauze mask to prevent catching the flu

Reports of the sick included the classic flu symptoms of fever, cough, and malaise. For 95 percent of those who got it, their cases were unremarkable, their symptoms typical but not severe, and they recovered. For the other 5 percent, however, the symptoms were far worse. These patients would be stricken with a telltale purplish-blue hue to their faces called heliotrope cyanosis, a sign that they were suffocating to death. Some would vomit blood, which also poured from their nostrils, sometimes for more than twenty-four hours straight. Violent coughing tore abdominal muscles, and delirium and coma were commonplace.

A North Carolinian named Glenn Hollar recalled, "That was the roughest time ever. Like I say, people would come up and look in your window and holler and see if you was still alive, is about all. They wouldn't come in."

Philadelphia resident Anne Van Dyke described how the dead were handled: "They wouldn't bury 'em. They had so many died that they keep putting them in garages ... garages full of caskets."

As a testament to the terrifying viciousness of the 1918 flu virus, questions still remain about it. Why was it so lethal compared

INFLUENZA!

ALL PERSONS
Excepting Physicians and Nurses, are Forbidden, Under Penalty of Law, of **Entering or Leaving** This House, Without Written Permission from the BOARD of HEALTH.

to past flu viruses? Why did it kill so many young people? And the inevitable: Where did it come from?

We know that the virus itself is a brilliant speck of evolution. It's an RNA orthomyxovirus, sometimes filament-shaped and sometimes spherical. The main types are A, B, C, and D, as determined by their RNA material. Type C is relatively mild and generally don't cause epidemics. Type D only infects cattle and pigs and B only infects humans and seals. It's influenza A we worry most about.

Influenza A can infect humans as well as a disturbing variety of animals: pigs, horses, seals, whales, domestic fowl, and wild aquatic birds. Often it lives without a problem in wild birds or fowl. Also it doesn't jump willy-nilly between all animals, but rather dances a lethal two-step between specific couplets: birds to humans, for example, or in the case of the 1998 H3N2 virus, from humans to pigs. Pigs also sometimes work as an intermediary between us and other species.

As for those names (H1N1, H3N2), those refer to the surface proteins on the virus. H is for eighteen types of hemagglutinin and N for eleven types of neuraminidase. Hemagglutinin (HA) is responsible for influenza's ability to attach to a cell in your nose, throat, or lung, and then get internalized into that cell. It's called hemagglutinin because the protein causes blood cells (heme) to clump

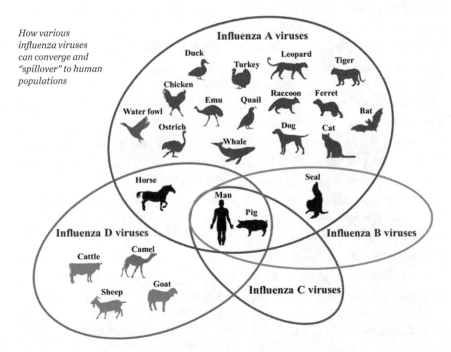

How various influenza viruses can converge and "spillover" to human populations

(agglutinate). Neuraminidase (NA) is the protein that allows newly created virus particles, called *virions,* to be released from the human cell that just made it. Both HA and NA are antigens, which means the human immune system recognizes them and can direct the body to launch an immune reaction against them.

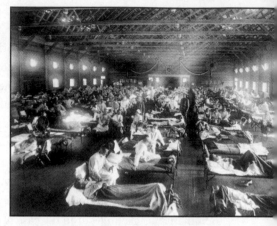

Influenza ward, Camp Funston, Kansas, 1918

The 1918 flu virus is so simple, and yet also achingly complicated. There are only a handful of proteins encoded in those RNA strands—eight to fourteen in general (in comparison, a human has an estimated 19,000 proteins encoded in our genes, though this number is still a strong subject of research). But those proteins, particularly the HA and NA, tend to change like clockwork.

In what we call antigenic drift, small changes occur as point mutations, which occur with a single nucleotide alteration in the encoding genes of the flu virus. These are minor changes, but if enough of them happen (and they do, within a year), then the resulting virus particles created by these altered genes can make them different enough that your body no longer recognizes it. All that immunity you received from catching the flu last year (or getting the flu shot) won't protect you against the new version. It's why you might get the flu twice in one season. And yet, one point mutation from you, jumping to your coworker, may not cause you to fall ill. Those two strains are probably close enough that you're still protected after your first infection. Every year, in a process called antigenic characterization, the CDC tests the antigens of two thousand influenza viruses circulating around to decide what next year's vaccine should include.

Now, antigenic *shift* is another issue altogether. Every ten to forty years or so, viruses make a huge mutational leap. When it happens, a whole new subtype is created, often after a recombinant exchange of genes from two or more different influenza A viruses. The "swine flu" in 2009 is a recent example, the result of viruses from humans, birds, and Eurasian and North American swine suddenly combining into a novel virus never before seen by humans. Without any immunity, it easily infects people, who spread it to others who are similarly unprotected. In short, antigenic shift produces a novel influenza virus.

It creates a pandemic.

Since the late nineteenth century, there have been antigenic shifts that created novel flu pandemics (beyond seasonal antigenic drift flu variations). These include the 1889 flu pandemic, the 1957–1958 (H2N2), the 1968–1969 (H3N3), the 2009 (H1N1, the so-called swine flu), and of course, the 1918 pandemic. A typical flu season usually results in 300,000 to 650,000 deaths worldwide annually. In a novel flu pandemic, doubling or tripling that number is usually the case. For perspective, the 1918 pandemic blew up that thinking, with a staggering 50 million deaths estimated.

◆ ◆ ◆

For decades after the 1918 pandemic, the question of "Where did the virus come from?" was answered with stories. Concepts arose from news articles, personal accounts, and records of mass human migrations that might have pushed a novel virus into immune-naïve populations numbering in the millions. Later, DNA evidence would come closer to the truth.

One common theory held that the origin of the virus was not in faraway China, but in Haskell, Kansas. American historian John M. Barry described it as a flat place where trees were scarce and horses were still used to deliver mail weekly. Haskell survived on agriculture and livestock, its farmers growing wheat and corn, its ranchers raising hogs, chickens, and cattle.

Loring Miner, the local physician, tended to patients within several hundred square miles. According to Barry's *The Great Influenza*, in late January and early February, Dr. Miner began noting some unexpectedly sick flu patients. The local *Santa Fe Monitor* reported a rash of flu cases as well: "Mrs. Eva Van Alstine is sick with pneumonia . . . Ralph Lindeman is still quite sick . . . Homer Moody has been reported quite sick." Dozens of young, healthy people in Haskell County were described as being rapidly struck down with the flu, and some were reportedly dying. By March, the local outbreak appeared to end.

Dr. Miner supposedly reported the epidemic to the health authorities, but there's no record of it, possibly because of wartime censorship. But it's also possible that the "Haskell epidemic" was elsewhere—at the Haskell Institute, a boarding school on the other side of Kansas, close to Camp Funston, a major army training center located near Manhattan, Kansas.

Nestled along the serpentine Kansas River, Camp Funston was a country unto itself. The largest of sixteen divisional training camps in the US, it housed up to 40,000 men during World War I, including the soon-to-be "doughboys" who would fight overseas in American General Pershing's forces. The sprawling 20,000-acre compound contained numerous bunkhouses holding 150 sleeping men at a time, plus infirmaries, libraries, a pool hall with seventy tables, a barber with forty chairs, schools, theaters, and even coffee-roasting houses to supply the beans for countless of gallons of coffee downed by the camp's soldiers.

American soldiers gargle saltwater to prevent the flu, Camp Dix, New Jersey, 1918

But Camp Funston also demanded a lot from its inhabitants. Summers baked the camp like an oven, and windstorms kicked up a choking haze of dried manure, ashes (from burned manure), and dust. The winters were bitterly cold, creating conditions in the cramped bunkhouses that were anything but healthy for the trainees. On March 4, the camp records note an uptick among a handful of men who'd fallen sick with fevers, body aches, and sore throats. But the numbers were low, so no one got alarmed.

By March 11, something was clearly afoot. Reports of men getting sick were streaming in all morning: Mess cook Private Albert Gitchell had a "bad cold" with a fever, sore throat, and muscle aches. Corporal Lee W. Drake walked in with the same complaints, his fever registering at 103 degrees. By noon that day, more than a hundred were struck with the same illness.

Notably, the aforementioned Private Gitchell often comes up in discussions of the 1918 flu origins. Gitchell, a cook, was a likely suspect. After all, who

Mask enforcement during pandemics is not a new thing.

better to serve up a helping of flu virus along with some army grub? Also, he probably had close contact with butchered animals at the camp, a theoretical location for a zoonotic spillover event. Nevertheless, it's very unlikely Gitchell seeded a pandemic, as cases were on the rise before his. Though his memorials refer to him as patient zero of the 1918 flu, he would survive not only this first wave of the flu, but World War I as well, ultimately dying at the respectable age of seventy-eight in South Dakota.

All told, 1,100 people would get infected at Camp Funston, though the spread wouldn't stop there. It also reached a boarding school in Lawrence, Kansas, less than 90 miles (144 km) away. The Haskell Institute, a boarding school for Native American children, was the largest of its kind. However, he school was sadly lacking in funds and good sanitation. There was also a decent amount of contact between the school population and those who lived and worked at Camp Funston. A week after the outbreak began at Camp Funston, the Haskell Institute was drowning in sick children, their hospital wards overflowing.

An alert was sent to *Public Health Reports*, later the CDC's Morbidity and *Mortality Weekly Report*. The Haskell Institute was most likely the "Haskell" mentioned in the *Public Health Reports* on April 5, referring to "the occurrence of 18 cases of influenza of severe type, from which 3 deaths resulted." Strangely, the 1918 death records from Haskell County reported only 11 deaths—none from influenza and only one from pneumonia (a complication that can be the cause of death in influenza).

If not from a small rural town in Kansas with more chickens and pigs than people, where did these soldiers catch the flu? Could it have truly originated at Camp Funston?

It seems to have a good setup for the creation of a novel virus. The trains that brought thousands of soldiers to the camp traveled through Kansas City, the location of the largest hog feedlot in the state. Raising poultry was a significant

American Red Cross workers transport an influenza victim, St. Louis, Missouri, 1918.

part of Kansas agriculture at the time. Because there was no refrigeration, live animals were likely kept at the camp in pens until slaughter—and close to soldiers. Further, the Kansas River, which runs along the southern edge of Camp Funston, is visited by migrating geese and nearly every species of American duck as well—all of which harbor avian versions of our influenza.

Camp Funston was also the perfect launching point for a pandemic. The camp hosted a buffet line of guests. On March 2, days before the first flu cases emerged, Camp Funston hosted several English and French officers, who conceivably could've unleashed a first wave of the flu from Europe.

But one key piece of information crumbles the Camp Funston theory. Epidemiological evidence from New York City in March 1918 shows a spike in deaths there above and beyond the norm for that time of year. The spikes were oddly large among fifteen- to forty-four-year-olds compared to other age groups, creating that "W" curve with a big spike in the middle age ranges that was typical of the 1918 pandemic. Because the New York City numbers were rising at the same time as the far less lethal outbreak at Camp Funston, it seems unlikely that the 1918 flu started in Kansas.

Other origin theories abound. A global search for bronchitis, pneumonia, or flu-like symptoms in an outbreak leading up to the fall 1918 pandemic reveals two interesting locations: Étaples, France, and Aldershot Garrison, England.

Antigenic Sin and the W-Death Curve

In searching for answers as to why the 1918 flu hit twenty- to forty-year-olds so hard, one possibility is something called "original antigenic sin." And while it might sound like the awkward title of a Hollywood B movie, it's an idea that informs our approach to influenza and viruses in general.

The theory holds that when young children are exposed to an antigen— in this case the flu— their immune system's B cells and T cells imprint the memory of the virus to confer lifelong protection to anything similar to it.

Evolutionary biologist Michael Worobey and his colleagues believe that older people in 1918 were already exposed to an H1 virus as children. Whereas those twenty- to forty-year-olds were exposed to an influenza strain that was an H3N8, not the right strain to imprint and provide some level of future protection. Also, because younger people tend to be healthier, their immune systems may have caused a "cytokine storm" during the initial infection, where a robust, healthy immune system goes into overdrive with high fevers, lung hemorrhages, and, in some cases, death. A study by David Morens, Jeffery Taubenberger, and Anthony Fauci reviewed data from over 8,000 autopsies and found most 1918 influenza deaths resulted from a secondary bacterial pneumonia, which may have also triggered a cytokine storm, making it harder to recover.

In a normal influenza season, a graph showing deaths on the y-axis and age on the x-axis results in a U-shaped curve, indicating more deaths in children and the elderly. This makes sense, given their weakened immune systems related to their ages. But in the lethal 1918 influenza pandemic, the normal U-shaped death curve became a W-shaped curve, with a big hump in the middle for younger adults. If there's a lesson begging to be learned from the 1918 pandemic, it's that new viruses don't necessarily conform to either our expectations or the behavior of past viruses.

In northern France, there was a British army base in Étaples, a former fishing village before the war located 15 miles (24 km) south of Boulogne-sur-Mer. Once a lovely spot that attracted artists and tourists, it was overtaken by a string of British Army bases, transforming it into "a dirty, loathsome, smelly little town," according to Lady Olave Baden-Powell, a volunteer at the time. In Étaples, soldiers received training in gas warfare, drilled endlessly with their bayonets, and marched on the dunes for weeks before deploying to the Western Front.

The barracks at Étaples were crowded, cramped, and cold. Its hospital received soldiers from the front, up to 23,000 at a time—two million in total—and experienced waves of dysentery, typhoid, typhus, and lung infections. Many came back from the front lines still recovering from the respiratory onslaught of gas attacks. The soldiers had done their best to ready themselves with inoculations against cholera, typhoid, and smallpox. But still, infection came.

During the winter in early 1917, a disease outbreak hit Étaples that was considered "almost an epidemic," thought to be brought on by "the influenza bacillus," according to the *Lancet*. (At the time, they still didn't know it was a virus.) Reported symptoms were respiratory and called a "purulent bronchitis" that was very different from a regular type of bronchitis. Victims of it suffered from cyanosis and pneumonias, and the mortality was a staggering 45 percent at its height across a seven-week period.

Étaples represented yet another perfect stew of factors that could bring about a novel influenza virus. There were masses of young people in crowded conditions, their immune systems depressed from stress and other infections, plus a population of live pigs regularly slaughtered to feed soldiers, who were also the beneficiaries of frequently purchased live fowl from the nearby villagers. The area was surrounded by sea marshes frequented by migrating birds, making it truly an ideal place to birth a novel virus.

Across the English Channel in southeast England, an army hospital in Aldershot was dealing with its version of purulent bronchitis. The hospital was experiencing an epidemic of a severe respiratory illness with identical symptoms to those suffering in Étaples, including the same "remarkable heliotrope cyanosis." The mortality rate was high, ranging from 6 to 25 percent.

After dealing with the second wave of the pandemic from August through December of 1918, a physician from the hospital at Aldershot looked back on their experiences in the cold months of 1917, remarking that, "in essentials the influenzo-pneumococcal 'purulent bronchitis' that we and others described in 1916 and 1917 is fundamentally the same condition as the 'influenzal pneumonia' of this present pandemic . . ." Which meant that, clinically at least, the doctors were seeing the same beast. And more than a year before the first spring wave of the 1918 pandemic.

◆ ◆ ◆

Clinical reports and documents from more than a century ago can chip away at the origins of the deadliest known flu pandemic in human history, but genetic evidence is the gold standard. The problem is in finding intact human tissue from that time with influenza RNA still viable enough to be sequenced.

RNA is single stranded, unlike double-stranded DNA, and the molecule is less stable and degrades easily. Natural enzymes that seek and destroy RNA, called RNases, are everywhere. They coat our desks, our fingertips, our laptops, our coffee cups. These RNase enzymes are a molecular biologist's worst nightmare because an ungloved touch or some non-autoclaved glassware could destroy a sample quickly.

So, finding a viable sample was the first challenge. Luckily (or rather, not), the pandemic hit everywhere around the globe, including isolated areas in places like Alaska. Brevig Mission, a village with a pre-pandemic population of eighty, was reduced to only eight people after the flu decimated the area. However, because the village was both remote and got overwhelmed fast, the dead were quickly buried and their bodies still frozen in permafrost, thus preserving their tissues—including the influenza viruses that killed them.

After receiving permission from the village elders in 1951, microbiologist Johan Hultin obtained lung samples from a little girl. In her place of burial, she was still well-preserved, in a blue dress and red hair ribbons. Unfortunately, techniques to preserve the samples were far from perfect in 1951, and Hultin failed to isolate the virus.

American microbiologist Dr. Terrence Tumpey working with specimens of the influenza strain from the 1918 flu pandemic

Influentia, "Spanish" Flu, and *La Grippe*

Spanish king Alfonso XIII got the 1918 flu, leading Spain to become its namesake.

The word "influenza" comes from the Italian word meaning "influence," derived from the Latin word *influentia*, which means "to flow into." The words have medieval astrological qualities to them, as illness was at times seen as influenced by the stars. But the term also described an outbreak, as in *influenza di febbre scarlattina*, which translates as "an outbreak of scarlet fever."

The French called it *grippe*, which translates literally as "seizure," first seen in 1776. It came into common parlance in the American lexicon to mean the flu and was referred to in the early twentieth century as the grippe.

But why did the "Spanish Flu" or the "Spanish Lady" become the moniker for the 1918 flu pandemic? Because of censorship, many countries didn't acknowledge the pandemic publicly. However, in World War I, Spain was neutral, so they spoke freely about the pandemic within their borders. (Interestingly, they called it the "French Flu," as they thought it was spread to Spain from France.)

In May 1918, when King Alfonso XIII came down with the flu, fears of "Spanish Flu" got even worse, though the origins of the pandemic likely had nothing to do with Spain at all.

In 1997, virologist Jeffery Taubenberger tried again, along with Hultin, then seventy-two years old. Taubenberger, biologist Ann Reid, and their team had already sequenced nine fragments of the 1918 virus genome from formalin-preserved blocks of lung tissue embedded in paraffin, belonging to a US service member from South Carolina who'd died within six days of being admitted to Fort Jackson's camp hospital. Once again, permission was sought and granted from Brevig Mission elders, and once again frozen lung samples were obtained, this time from an Inuit villager named "Lucy" who likely died in her mid-twenties and was found buried seven feet deep. Taubenberger and Hultin were able to isolate more segments of the viral genome, resulting in a full sequence of the original virus, with the help of other RNA bits from the South Carolina service member, as well as a flu victim from Camp Upton in New York.

Bit by bit, the entire genome of the 1918 virus was carefully sequenced until it was done in 2004. The entire genome of the infamous 1918 flu virus—all eight genes—was complete and ready to be studied.

Surely the genes would tell us why the virus was so very lethal? Only they didn't. There wasn't anything in the proteins that showed why the 1918 influenza strain was so virulent. And the sequences also didn't show any genetic changes to explain its virulence. It was still a puzzle.

The genes did reveal that an ancestor of the 1918 virus likely infected humans sometime between 1900 and 1915. The NA gene was avian in nature and seemed to jump to mammals just before 1918. Meanwhile, the HA gene was quite mammalian, closer to swine or human than bird flu HAs.

Now that the whole genomic sequence existed, the next step was seeing how the virus worked in real life. First, Peter Palese at Icahn School of Medicine at Mount Sinai had to perform a preliminary step, to reverse engineer the 1918 influenza genes from RNA into DNA using a technique he had helped pioneer to produce influenza viruses. The DNA was then inserted into plasmids (a circular DNA molecule that can replicate on its own inside cells, typically found in bacteria).

Finally, the moment to re-create the deadliest pandemic virus in recorded history arrived. Which, you can imagine, also came with significant concerns. Particularly the falling-into-the-wrong-hands kind of concerns capable of keeping scientists up at night. A careful decision was made by CDC Director Julie Gerberding to allow a single researcher to pursue the actual reconstruction. Terrence Tumpey, the branch chief of the Influenza Division's Immunology and Pathogenesis Branch at the CDC, was chosen for the task. Only Tumpey had access to the project at this point, complete with Biosafety Level 3 *Mission Impossible*–style iris scanners, biometric fingerprint locks, and, rounding out the safety protocols, a daily dose of antiviral flu medications (oseltamivir, or Tamiflu).

Tumpey placed plasmids with the pandemic viral DNA into human kidney cell cultures, where they directed the kidney cells to make the influenza virus. They do this by hijacking the cell's machinery, in a similar way to influenza. The kidney cells obeyed the plasmid DNA blueprints and started manufacturing flu virus.

The day that Dr. Tumpey knew he'd manufactured the virus, he sent an email to colleagues with a single line: "That's one small step for man, one giant leap for mankind."

Once he'd finished, Tumpey then worked alongside a group of researchers under stringent biosafety protocols to see how the virus would affect a population of mice. Because the genomes didn't show anything special about the lethality of the virus, perhaps it would reveal itself in action?

It did. The mice got sick, really sick. Within four days, the amount of virus in the mice lungs was 39,000 times higher than when they were infected with "ordinary" influenza viruses. The mice lost up to 13 percent of their body weight within two days and many died swiftly, within the same short amount of time. It also caused severe lung damage. In fact, the recombinant 1918 virus was 100 times more lethal compared to other flu viruses. When they swapped out the HA gene for a different one, the mice didn't die or get nearly as sick as they did with the 1918 virus. Researchers further learned that the polymerase—the enzyme that helps the virus replicate—was incredibly good at its job.

According to Tumpey and his colleagues, it wasn't just one special gene that made the 1918 influenza virus so deadly. It was "the constellation of all eight genes together [that] make an exceptionally virulent virus," they wrote. "This was evolution, nature, chance, and the globalization of the human species at work."

RNA viruses in particular have a telltale mutation rate, which evolutionary biologist Michael Worobey and his colleagues took advantage of using a molecular clock (see HIV, page 130) to do a phylogenetic analysis, or an evolutionary development study. It showed that, a few years before 1918, the human HA in the deadly flu virus acquired its partner in virulence, the avian NA and internal protein genes. This is well before the Étaples and the Fort Funston, Kansas, outbreaks. Furthermore, the phylogenetic analysis points to a possible North American origin of the NA component.

In time, we may have more precise molecular clock methods to estimate exactly when and how the 1918 pandemic virus truly came to be, or novel methods we haven't discovered as yet. In the meantime, we can at least benefit from a far better understanding of what's out there, waiting to emerge and take the world by storm, as we've painfully relearned with COVID-19 (see page 111). The latter will likely not just leave the world with a neat goodbye, similar to the 1918 flu, pieces of which are embedded within our seasonal flu, year after year.

VACCINES
From Variolation to Messenger RNA

Impact: Vaccines have prevented billions of illnesses and deaths from notorious diseases like smallpox, yellow fever, rabies, hepatitis A and B, and diphtheria

When: 1774 and 1796

Who: Benjamin Jesty and, more famously, Edward Jenner

What Happened Next: Technology is steadily creating novel vaccines, such as using viruses as a delivery mechanism, but incomplete immunization of populations means many diseases cannot be completely eradicated.

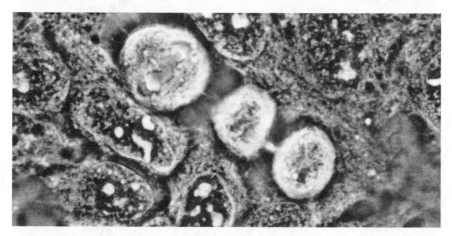

Cell culture using cells originally taken, without permission, from Henrietta Lacks

We used to take vaccines for granted, assuming they'd always be there when we needed them. Then came COVID-19, and their absence transformed the pandemic into a nightmare from which some countries couldn't seem to wake. Perhaps most people didn't imagine having zero immunity to a new, potentially lethal infection spreading globally. But for most of human history, vaccines weren't an option, and their existence was as inconceivable as is their absence now.

So when and from where did vaccines come?

An anti-vaccination cartoon warning against smallpox inoculation using cowpox virus, England, 1802

Vaccines didn't actually make their first appearance in the 1900s or even the late 1800s, but a hundred years before that. The history-making first vaccine arrived in 1798, for smallpox. There was, however, a predecessor.

Variolation.

Better known as smallpox inoculation, variolation is worth mentioning solely for its etymology. It comes from the word variola, which is the name for smallpox, but is derived from the Latin *varius*, meaning different, changing, or spotted. It doesn't end there: *varius* comes from the medieval Latin *variola*, for pustule or pock, which gets us to smallpox, an obvious combination of "small" and "pox," the latter originating from the old English word *pocc* for pustule, blister, or ulcer. By adding the "small," it distinguished smallpox from the Great Pox, or syphilis (see page 189).

Here's how variolation worked. The first step involved collecting liquid from a smallpox blister. If that wasn't available, dried-up scabs from someone infected would do. If you went with scabs, you might grind them up into a powder, then puff it up the nostrils of a smallpox-naïve (or never previously infected) person. You could also cut that person's arm and rub blister liquid into the wound. The least-abhorrent method (and least effective) involved wrapping a virus-infused blanket around someone. But regardless of how infection happened, the idea was to inoculate a person with a small, safer amount of virus, allowing them to get a less vicious case of smallpox, hopefully with a lower rate of scarring—and death.

Some of the first known instances of inoculation were seen in China in the 1500s, but undocumented practices were rumored to occur centuries earlier. However, the methods were clearer by the 1700s. When the Kangxi Emperor suffered from a bout of smallpox, he had his children variolated to protect them. In 1732, a medical text described two processes for variolation. (And be sure to note the nostril delineation, likely a reflection of the yin and yang concepts, the "right" representing the feminine or yin, and "left" being the masculine or yang nature.)

Method #1:
Use "fluid seedlings": Moisten the pox with clear water and break it into a wrapped cotton. Stuff it in the nostril. . . . Left nostril for a man and right for a woman. Then stuff up the nostril with clean dry cotton again.

Method #2:
If you use crust seedlings, wrap them up with a piece of paper, then grind up the crust into a powder with a smooth and hard gavel. . . . Wet the cotton slice and wrap six li (0.3 grams) of smallpox powder inside and make it into a ball. Don't be rough with it. Bind it with a silken thread and stuff it into the nostril, the left nostril for a man and right nostril for a woman. Don't do it carelessly.

The practice of variolation in the Ottoman Empire was thought to have originated in India, where inoculation was documented in 1768, but it had likely been occurring for centuries, if not thousands of years. Methods included using a sharp metal needle to pierce a pustule, then puncturing the skin of the upper arm in a circle (not unlike how smallpox vaccines were done in recent times). In central Sudan during the eighteenth and nineteenth centuries, the practice of "buying the smallpox," or *Tishteree el Jidderi*, was a method observed by the Sennar women. It involved tying a cloth around the limb of an infected child, after which parents of uninfected children would bargain over specific pustules on the sick child. After a deal was struck, the cloth was then brought home and tied on the uninfected child. Another approach, called *Dak el Jedri*, or "hitting the smallpox," was described earlier and included rubbing liquid from a pustule into a cut, often on the leg.

Lady Mary Montagu, née Pierrepont, an early advocate for smallpox variolation, ca. 1756

By the seventeenth century, variolation was a common practice in India and swaths of Africa and China, and eventually made headway in Europe and England later, thanks to Lady Mary Wortley Montagu.

Born May 16, 1689, Mary Pierrepont was an aristocrat praised for her beauty. Even at the tender age of seven, she was the "toast of the season" and had her name inscribed on a crystal goblet declaring her so. At the age of twenty-three, she married Lord Montagu and soon became a fixture at royal court in London. At age twenty-six, after losing a brother to smallpox, Lady Montagu contracted the disease as well. She survived, but the scars marred her famous beauty.

A year later she went to Turkey with her husband, then the British ambassador to the country. Terrified that her children would succumb to smallpox, she learned about variolation in Turkey, or "ingrafting" as it was called there.

Typically around the eighth day, variolated children would come down with fevers and a mild illness for two or three days, contracting fewer pustules on their faces compared to a full-blown infection. Lady Montagu didn't know of anyone who'd died this way, but later statistics would reveal that variolation had a mortality rate of up to 2 percent, versus up to 30 percent from natural exposure. Montagu's son was variolated this way and recovered well.

After returning to England, she allowed others to watch her daughter variolated during an outbreak in 1721. King George I's physician, Sir Hans Sloane, watched and later tested the inoculation on prisoners in exchange for their freedom, assuming they survived. They did. Variolation took off in popularity soon after.

In the United States, variolation had its beginnings in Boston, when an enslaved African man named Onesimus (his country of origin is not definitively known) explained that he'd been inoculated in his home country and was now immune. He told the man who enslaved him, a Puritan minister named Cotton Mather (he of the Salem witch trials infamy). Mather took down Onesimus's words (phonetically, and with odd capitalization) to describe the process: "People take Juice of Small-Pox; and Cutty-skin, and Putt in a Drop." Mather soon learned that the practice was widespread in Africa. Armed with knowledge to protect Bostonians from the 1721 epidemic, he raced to inoculate those he could. At the time, one in seven people died of smallpox; but only six of the 242 inoculated perished. The method spread. In 1775, the Continental Army was variolated by order of George Washington. By the 1800s, the technique was being used worldwide.

♦ ♦ ♦

But variolation was not vaccination. Variolation gave smallpox to people so they could experience a less-potent form of the disease, and thereby gain immunity.

By definition, vaccination induces immunity in someone by giving them a substance that is not the actual disease itself. Vaccination types (see sidebar, page 248) are varied, but none involve giving the actual pathogen, unchanged, to a new host. Variolation nevertheless paved the way for the first true vaccination.

Enter Edward Jenner.

Born in Gloucestershire, England, in 1749, just as variolation was gaining popularity, Jenner was then subjected to it as a child. Orphaned at age five, he was eventually sent to a boarding school that suffered a smallpox epidemic. Before he could contract the disease, Jenner was bled until his blood was "thin," and purged with medicines (commonly used methods at the time included mercury salts like calomel to induce diarrhea, or antimony compounds to cause vomiting) until he was horribly dehydrated. He was also given a low-calorie vegetarian diet. Thin as a skeleton, Jenner also stayed in an "inoculation stable" belonging to the variolator. He managed to survive the inoculation but was left with insomnia and anxiety—not a great trade-off, but better than death. His studies apparently suffered as a result, and a future

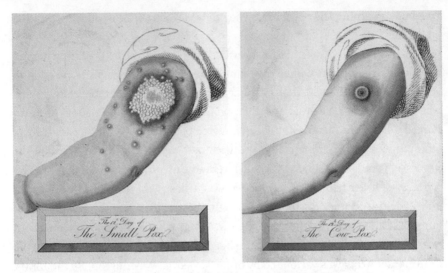

A comparison of smallpox (left) and cowpox (right) inoculation, twelve days later

in the law and the clergy were no longer an option. Instead he chose something more interesting and "easier"—a career in medicine.

After a move to London, Jenner was mentored by a famed surgeon, John Hunter. Jenner was eventually admitted to the Royal Society, a renowned academy of science, for (deep breath) research into the nesting habits of cuckoos.

But history recalls Jenner less for the cuckoos and more for this keen observation—that milkmaids who contracted *cow*pox didn't contract *small*pox, hence their famed "milkmaid complexion." Cowpox, a disease that affects the udders of cows and muzzles of nursing calves, occasionally jumped to the hands and arms of milkmaids.

But could it be protective? The pocks of both diseases sure looked alike.

An English farmer named Benjamin Jesty noted that two of his employees who had cowpox also cared for children with smallpox and yet never caught it. Jesty decided he would try to protect his family in a similar fashion. In 1774, he took a darning needle, pricked lesions from an infected cow, and scratched the arms of his wife, Elizabeth, and two sons, Robert and Benjamin, with it.

None of them contracted smallpox during later epidemics, and when variolated years later, they had no reaction.

Though Jenner is usually credited with producing the first vaccine, the honor should really be shared with Jesty. Jenner's famous first vaccine occurred later in

HeLa Cells

Henrietta Lacks (1920–1951), whose cells, taken without her permission, power cancer research to this day.

We can't talk about vaccines without giving thanks to one woman who never knew she'd be responsible for saving millions of lives.

Henrietta Lacks.

In 1951, Lacks, a Black mother of five from Baltimore, died of cervical cancer at the tragically young age of thirty-one. During her radiation treatment at Johns Hopkins Hospital, her cancerous cervical cells were taken and kept for research purposes without her knowledge or consent, a common if wildly unethical practice at the time.

In the early 1950s, efforts to grow human cells in a lab were not going well. Human cells were considered finicky at the time and died easily. But when researchers cultured Lacks's cancerous cells, they didn't just grow—they thrived, doubling in about a day. And they kept going, and going. Lacks's cells had three perfect qualities that scientists had wished for: They were durable, prolific, and immortal. The cells were named after the original patient from whom they were taken. Scientists used the first two letters from her first and last name: HeLa.

For the first twenty-five years HeLa cells were being used, they found their way into laboratories and research across the globe, but the Lacks family was unaware that HeLa cells even existed, as detailed in *The Immortal Life of Henrietta Lacks*, by Rebecca Skloot. Those cells are now responsible for more than 17,000 patents. They allowed American virologist Jonas Salk to develop the vaccine for polio. They have also been used in cancer research, cloning (HeLa cells were the first human cells to be cloned), zero-gravity research, and cosmetics testing, as well as hormone, infectious diseases, and toxicology research.

The question of who owns HeLa cells remains. Once a person has died, do the living cells they leave behind have their own legal rights? And will Lacks's family ever be able to have control over the cells, a portion of the millions of dollars of profit from research based on them, or even a proper accounting of their legacy? The answer regarding who truly owns Henrietta Lacks's cells may never truly be answered.

What we do know is that Henrietta Lacks didn't just unwittingly help humanity battle disease and teach us more about ourselves; she also opened up tangled and vital conversation about who we are, and who owns our very genes. We owe her and her family more than we can ever repay.

1796, when he took a fluid sample from a cowpox lesion on a milkmaid named Sarah Nelmes, who was infected by a cow named Blossom. Jenner then applied it to the scratched arm of an eight-year-old boy, James Phipps. After some mild illness and fever a week later, he recovered and did not come down with smallpox after two subsequent exposures. A paper detailing the breakthrough was submitted to the Royal Society but was rejected for being "incredible" (as in, not credible).

Jenner repeated the procedure in 1798 (he had to wait for cowpox outbreaks to do his work). Finally, he published a paper where he called his procedure *variolae vaccinae* (*vacca* is Latin for cow), meaning "smallpox of the cow." And from that was born the term *vaccine*, and of course *vaccination*. From cows!

As is common when the scientific status quo is shaken, Jesty and Jenner were ridiculed. Jesty was pelted by neighbors who feared humans would turn into cows from the procedure, while Jenner drew ire from angry physicians who feared something more realistic—losing the lucrative business of variolating patients.

A final twist to the story of the world's first vaccine involves a case of medical mistaken identity. It turns out that the vaccinia virus, which had been used for the ultimate vaccination-led eradication of smallpox, was in fact not cowpox but a related virus from the poxvirus family. Somewhere along the decades-long history of the vaccine's cultivation, vaccinia emerged. It may be the laboratory survivor of a virus that no longer exists in nature. Possibly, it was a genetic recombination of cowpox, smallpox, and a related horsepox created by hundreds of rounds of growth and generations via lab cultures, a process known as *passage*. Scientists aren't sure when the change happened exactly, but by 1939 it became obvious that the vaccinia used in smallpox vaccines and cowpox were two biologically distinct viruses. The vaccinia-based vaccine was able to confer short-term immunity to smallpox similar to Jenner's vaccine, with immunity lasting about five years, not lifelong, and was genetically closer to a horsepox virus than the original cowpox virus.

◆ ◆ ◆

Inspired by the breakthrough smallpox vaccine, famed French biologist Louis Pasteur made a landmark finding in the late 1800s. In the process of investigating chicken cholera (*Pasteurella multocida*), Pasteur was inoculating chickens with generation after generation of cholera grown in cultures made of chicken muscle decoctions, or "germ-sowings" as he called them, though today we call them the aforementioned passage. His assistant was supposed to inoculate a group of chickens with fresh bacteria grown from a culture but forgot, went on holiday, and returned a month later, when he proceeded to inoculate them with an old culture. Shockingly, the chickens didn't die. Instead they grew ill, recovered, and became immune to the cholera pathogen.

But how? Pasteur had unwittingly discovered attenuation, or the weakening of a bacteria or virus. In this case it was through germ-sowings, though Pasteur believed it was from oxygen exposure at the time. According to Pasteur, "By changing the mode of culture . . . we have a means of obtaining progressively decreasing virulences, and, finally, a true vaccine virus, which does not kill, but communicates a benign disease, preservative from its mortal form."

Pasteur would soon be hailed for creating a novel vaccine for anthrax (*Bacillus anthracis*), a disease humans can contract from infected animals.

Only this time it was a ruse.

One of Pasteur's contemporaries, Jean Joseph Henri Toussaint, was a veterinarian who had made strides at the time in creating an anthrax vaccine. When Robert Koch discovered anthrax bacteria in 1877 (see Anthrax & Biological Warfare, page 294), Toussaint went on to work on a vaccine. In 1880, Toussaint killed anthrax bacilli by cooking them at 55°C (131°F) for ten minutes, then vaccinated sheep and dogs with it, to excellent results—all the vaccinated animals survived, and all the unvaccinated ones perished. He then repeated the trial using bacilli that were weakened with carbolic acid, the same substance used by British surgeon Joseph Lister as an antiseptic in surgery.

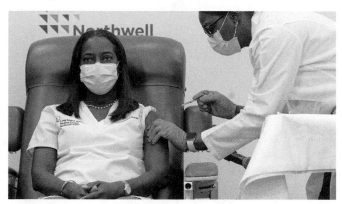

Sandra Lindsay, an intensive care nurse working in Queens, New York, received the first COVID-19 vaccine administered in the United States, December 14, 2020.

A Vaccine Buffet

Vaccines have come a long way from those pioneered by Benjamin Jesty and Edward Jenner. The following are examples of the varied types of vaccines now available around the world.

Attenuation: kind of alive

This type of vaccine uses a weakened version of the original pathogen (or sometimes a related one, as with the tuberculosis vaccine) that won't make a recipient sick. To attenuate the virus or bacterium, it is serially grown in culture after culture, until the strain changes and becomes less potent. The attenuated yellow fever vaccine was created by growing it via serial passage through chicken and mouse embryos, for example. The measles vaccine was obtained from a child in 1954 and passaged for ten years before a viable, weakened version was created. Other attenuated vaccines include the oral typhoid and polio vaccines, as well as the typhus vaccine.

Inactivation: not alive

Inactivated vaccines don't contain any live virus or bacteria. Instead they contain pathogens killed by heat or a chemical like formalin (formaldehyde). Rabies and hepatitis A vaccines are whole, inactivated viral vaccines.

Polysaccharide vaccines: the sugar-coat method

These vaccines consist of molecule chains made of sugar subunits found on the surface capsules of some bacteria. These are the antigens, or the part of the pathogen that the human body makes antibodies against, which protect a person from the whole pathogen itself. Examples include the pneumococcal (causing pneumonia in some), meningococcal (meningitis and septicemia), and *Salmonella typhi* (typhoid fever) vaccines.

Conjugate vaccines: a little help from a friend

Sometimes, vaccines alone don't provide a sufficient immune response. To overcome this, a protein called a conjugate is linked to a piece of the pathogen, which then elicits an immune response, allowing the body to gain a longer immune "memory" against a disease. *Adjuvants*, on the other hand, are substances added to a vaccine to elicit a stronger overall immune response. The latest shingles vaccine contains such an adjuvant made from the inner bark of the Chilean soapbark tree, *Quillaja saponaria*.

Recombinant vaccines: made just for you in a lab

Remember how viruses like to hijack our cells by using their DNA or RNA to tell our cells to make pieces of virus, thus making more of itself? Recombinant vaccines use a similar technique. Only this time, humans are the masters of the hijacking.

Our immune systems will respond to particular proteins on viruses and bacteria. So scientists place a piece of DNA that codes for a protein on a virus or bacteria inside, say, some yeast cells. Those yeast cells then follow those DNA orders to make that protein in large quantities, and those proteins are placed in vaccines to elicit an immune response. This is how the hepatitis B vaccine is made.

Toxoid vaccines: defanging toxins

Some pathogens, such as diphtheria or tetanus, create toxins or poisons that cause damage as part of the disease. Those toxins can be gathered, inactivated with heat or chemicals (at which point they're called toxoids and are no longer dangerous), then used as a vaccine. So even if you get diphtheria or tetanus, they can't harm you because the vaccine trained your body to disarm them of their most powerful weapon.

Protein-based vaccines: pieces of the whole

Some vaccines are created by growing a given virus (in eggs, for example), then breaking up the virus with detergents and isolating the component within it that allows the human body to develop immunity to the virus. The seasonal influenza vaccine is created this way. Protein-based vaccines, as well as some of the others listed above that contain pieces of pathogens, are also called subunit vaccines.

New ideas: nucleic acid vaccines and vector vaccines

Some new vaccines use only nucleic acids, like DNA or RNA, to tell the human body to make proteins that will elicit an immune response. Unlike previous vaccines that require months to create, these can be produced far faster.

In some cases, RNA vaccines release these pieces of code into your body to tell the cells to create proteins on the surface of viruses, like the surface spike protein on SARS-CoV-2. In other cases, the RNA might code for a human antibody specifically made to recognize a virus or bacteria and launch an attack— a completely different concept.

Vector vaccines use the "shell" of a virus, like a common cold adenovirus, but replace the genes inside with DNA or RNA that makes a surface protein. So the vaccine's created "virus" will enter your cells the same way a cold virus might, but instead of giving you a runny nose and sore throat, your body gets to work making a protein that elicits an immune response. A number of vector vaccines are underway for HIV, malaria, Ebola, and herpes, among others.

There are over a hundred SARS-CoV-2 vaccines in human trials as of the writing of this book, including novel RNA vaccines and vector vaccines, and no doubt these will pave the way for vaccines both in the fight against infectious diseases and other diseases—like cancer.

Illustration of RNA polymerase (orange) using DNA (purple) as a template to create a strand of RNA (red), which will itself be a template for specific protein

Pasteur then performed a successful trial of his anthrax vaccine for the Agricultural Society of Melun, claiming it was created similarly to his chicken cholera vaccines, via his "oxygenated" method. But in truth, Pasteur documented in his notebooks that his assistant, Charles Chamberland, creator of the Chamberland filter (see page 48), had used a chemical, potassium bichromate, to weaken the anthrax fashioned after Toussaint's methods.

Pasteur would go on to do more vaccine work, including a partnership with Pierre Paul Émile Roux, a French physician and immunologist. They dried the spinal cords of rabbits that had died of rabies, and Pasteur injected a small amount of liquid "bouillon" made from dried tissue into nine-year-old Joseph Meister, who'd sustained fourteen wounds from a rabid dog.

Pasteur suspected that the dried spinal cord contained two substances, "side by side, one which is living and capable of multiplying rapidly in the nervous system, and another, not alive, having the faculty . . . to inhibit the development of the first?" His theory was that this "inhibitor" was what prevented rabies from multiplying in the infected boy.

The theory wasn't correct, but luckily for the boy, the vaccine was a success (see page 315).

Pasteur demanded that his notebooks be kept private after his death. They stayed that way until they were donated to the French National Library by his last living relative in 1946, where they were restricted from viewing until 1971. Alas, they revealed that Toussaint deserved the credit, more than a hundred years later, for the first successful anthrax vaccine. Further, they showed his rabies vaccine, though effective, was partially grounded on an erroneous theory.

Since the time of Jenner, Jesty, Pasteur, and Toussaint, and all the attendant drama and controversy of their efforts, we've developed vaccines for a variety of diseases that once wreaked havoc on humans. These include yellow fever, rabies, hepatitis A and B, cholera, and diphtheria, to name a few. Most of us don't spend much time thinking about getting shots, aside from the seasonal flu, funny-sounding ones your doctor recommends (pneumo-what?), and that tetanus booster you get after stepping on a nail while making furniture for your new Etsy shop.

So when you get your next shot, wincing as the needle goes in, remember that vaccines prevent 2.5 million deaths a year, or close to the entire population of the city of Chicago. It might just make that little jab less painful.

POLIO
Patient ZERO
A very young Italian immigrant

Cause: *Poliovirus* (virus)

Symptoms: Neck pain/stiffness, headaches, fever, vomiting (related to meningitis), paralysis, asphyxiation

Where: Brooklyn, United States

When: June 1916

Transmission: Oral ingestion via unwashed hands, contaminated food or drink. Less commonly, ingested airborne droplets

Little-Known Fact: The threshold for herd immunity in polio is low compared to other infectious diseases; only 80 to 85 percent immunization in a population is required.

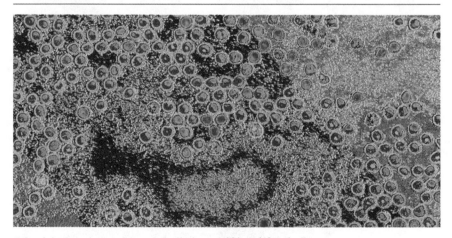

Colored electron micrograph image of polioviruses within an infected cell

S omething was wrong in Pigtown.

In June 1916, the children of recent immigrants started getting sick in the densely populated, largely Italian neighborhood in Brooklyn. Newspaper accounts of the time reported that Italian parents, scared of

Pigtown, Brooklyn, New York

the strange symptoms their children were exhibiting, approached doctors and priests "complaining that their child could not hold a bottle or that the leg seemed limp . . . and there had been a little loss of appetite and some restlessness." Feverish symptoms and muscle pains soon advanced to struggles with breathing and entire loss of feeling in the legs. For parents, it was absolutely terrifying, as their child could, within a couple of days, go from perfectly healthy to crawling across the floor, unable to move their legs.

Within a few days, the first deaths were reported and the New York City Health Department sent a team of investigators to Pigtown for an emergency house-to-house inspection. There they found the dreaded telltale signs of "infantile paralysis."

Or, as it's known today, poliomyelitis, typically shortened to polio.

After emerging in Pigtown that year, polio would go on to infect some 27,000 people across twenty-six states, resulting in about 7,000 deaths in America's first

"HELP ME WIN **MY** VICTORY!!~JOIN THE MARCH OF DIMES!"

A March of Dimes polio poster, ca. 1942

widespread epidemic of the disease. New York City had it the worst: Of its 9,000 cases, almost 2,500 of them died. Most were children. The disease, as its early name of infantile paralysis suggests, mostly impacted the young.

At the time, a diagnosis of infantile paralysis would lead to exactly that. There were no known treatments for the disease that could permanently paralyze a child—or even kill them. At the time, a wide variety of things were incorrectly blamed for the source of the disease: moldy flour (nope, but see Ergotism, page 2), gooseberries, poisonous caterpillars, summer fruits, soft drinks, ice cream, candy, maggots in the colon, garbage, dust, tarantulas crawling over bananas, too much rain, too much heat, bad air from sewage odors, and infected milk bottles. For a time, a lot of attention centered on houseflies as well.

Rather, polio is an infectious disease caused by the poliovirus, which can be ingested or inhaled before it settles in the tonsils, larynx, or ileum. From there, the virus can potentially move on to the gut, infecting feces and spreading from person to person via unwashed hands after defecating or contaminated water. The virus also can be spread through droplets from infected people after they sneeze or cough, though this is much less common. For 90 to 95 percent of cases, the virus never moves on from the lymph nodes, resulting in no symptoms at all. In less than 10 percent of infected people, they experience abortive polio about a week later, with symptoms similar to a cold, including headaches or feeling slightly feverish from hot weather. And that's where polio starts and ends for the great majority of people.

But for others, it gets worse. Much worse.

In a fraction of these cases, the polio advances and the virus attacks the central nervous system, which can lead to all manner of tragic results. After the neurons are infected, their corresponding cells die off. As the nerves supplying the muscles start to die as well, muscle degeneration is common. Then comes meningitis with symptoms of neck pain, headaches, fever, and vomiting. Paralysis, polio's trademark symptom, can arrive next. Polio victims at this stage frequently can't move their legs. Worse, in some cases the virus destroys nerve cells in the muscles that facilitate swallowing, talking, even breathing. Each breath then becomes a struggle, the victim feeling like they've forgotten how to breathe naturally. At this stage, polio can kill through asphyxiation.

But there was still hope for those who experienced paralysis. While it can become permanent, in some cases muscle function starts to return to normal after about six months, with a full recovery possible in approximately two years.

INFANTILE PARALYSIS

This Notice is Posted in Compliance with Law

"Every person who shall wilfully tear down, remove or deface any notice posted in compliance with law shall be fined not more than seven dollars."

General Statutes of Connecticut, Revision of 1912, Sec. 1173.

Town Health Officer.

One of the saving graces of polio is that if you catch it and recover, you'll be immune to the disease for the rest of your life.

Victims and their loved ones in the 1916 epidemic could do little more than watch, pray, and hope that the disease didn't advance to the paralytic stage. As mentioned, there were no viable treatments. Survivors of polio could potentially spend the rest of their days in "calipers" or braces to help them walk again.

Meanwhile, New York City health authorities stepped into the fray, implementing strict quarantine measures and a variety of ineffective restrictions in an attempt to stop the rapid spread of the disease. Roadblocks were imposed around certain neighborhoods. Vehicles with children sixteen years old or younger were not allowed to enter or leave New York City unless carrying a health certificate from a professional stating that there had been no polio in their homes. If a family had a member come down with polio, the "polio house" was quarantined, complete with warning signs to ward off visitors. Immigrant families were blamed for the spread of the disease and subsequently shunned.

Then there were the poor cats and dogs. A briefly held but popular theory maintained that dogs and cats were carriers. The result was a mass abandonment and extermination of New York City's pets. More than 75,000 cats and 8,000 dogs were killed in 1916 in response to the epidemic.

Wealthy New Yorkers who could afford to do so left the city, some for the entirety of the summer. The suburbs, however, had their own quarantine measures in place. Public beaches banned swimming by nonresidents. Vaudeville theaters could not entertain anyone under age sixteen. Sunday school classes were cancelled for the summer.

By the fall of 1916, the worst of the disease was over.

If there was one medical benefit to come from the 1916 polio epidemic, however, it was an incredible amount of data on the spread of the disease, amassed by a federal health team. It took the team two more years to complete their report, and, while damning, it was also enormously helpful for future fights against polio outbreaks. The report showed that New York City's quarantine measures had failed, that animals had nothing to do with the spread of the disease, and that polio was transmitted between people by as-yet-unknown methods. It further stated that most polio cases were not even recognizable by the classic symptoms (such as paralysis), while recommending the study of abortive cases to better understand how the disease spread.

The US was hit by another polio epidemic, again concentrated in New York City in 1931, resulting in 4,500 cases. The disease remained enormously scary to people at the time because there still was no effective treatment. It also primarily struck children, their plight becoming symbolic of paralysis. The New York City Health Department seemed to be tossing darts at a board in their efforts to contain the 1931 epidemic, recommending some common if ineffective advice, as well as a variety of head-scratching measures, such as eating salt, hanging camphor around your neck, and using nasal sprays.

The poliovirus, unfazed by anything the public health experts could throw at it, came back again even stronger in 1942. The last, worst, and longest polio outbreak stretched over eleven years, from 1942 until 1953. Indeed 1952 was the worst on record for the United States, with 58,000 people diagnosed with the disease.

♦ ♦ ♦

The curious thing about polio in the twentieth century, the thing that puzzled so many scientists investigating it, was: Why did it seem more virulent among largely affluent populations? And why did it continue to return? The United States, one of the world's wealthiest nations, and indeed one of the most sanitary, kept getting slammed with polio epidemics while impoverished nations, with significantly less resources for public sanitation, were largely spared.

The answer, ironically, was the public health reforms of the late nineteenth century. In the wake of epidemics of infectious diseases caused by contaminated water, such as cholera and yellow fever, America went on a cleaning binge. Public health reforms across the country significantly improved sanitation and kept water supplies much cleaner and healthier than they'd ever been. The problem, however, was that polio had been a mild endemic condition in concentrated human populations for centuries. Due to the lack of hygiene and contaminated water sources, infants were usually infected early in their first year of life, and maternal antibodies still circulating in their bodies helped keep infections mild as the infants acquired their own immunity. Furthermore, higher rates of child mortality were expected in previous centuries, and polio was just another in a long list of infectious diseases likely to kill you before you became a teenager. And of course most people didn't die of polio, contributing significantly to herd immunity, which occurs when a large portion of a given community becomes immune to a disease, significantly reducing the chances of the disease spreading from person to person. In this case, the result was that older children and adults gained protection, as the disease had limited opportunities to move between hosts.

With cleaner water and better sanitation, children weren't getting exposed to polio as babies. When they were finally exposed, it was often at an older age, when their maternal antibody protection had disappeared. More and more children were at risk as the years after sanitation reforms passed, as the critical threshold of immune individuals in the population dropped below the amount needed for herd immunity. It didn't take much for polio, in a comparatively sanitized and wealthy country, to become an epidemic. Which is why polio kept coming back, again and again.

The disease raged in the eleven-year stretch between 1942 and 1953, when the first half of the baby boomer generation was born. Since no treatments existed for the disease, the best victims could get at the time were devices to help them better cope with their symptoms. And those devices, in turn, became dreaded symbols of the disease.

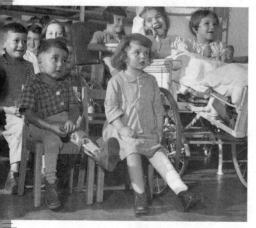

Children stricken with polio, ca. 1950

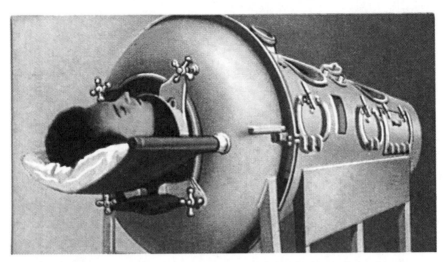

A person in an iron lung shown on a cigarette card, a collectible once sold with a pack of smokes

Those with paralyzed limbs were put into splints or braces to prevent muscle contractions, which, theoretically, allowed them to walk. Those who had problems breathing (as the virus attacked the nerves in the muscles controlling the lungs) were put into large, frightening metal respirators called iron lungs. Polio patients laid in them horizontally, while the devices, coffin-like tubes, did their breathing for them. The machine used negative air pressure in the chamber to expand the chest, which made patients suck in air; then it would put pressure back into the chamber, causing the lungs to push air back out again. Iron lungs did the work of breathing so a patient had time to heal and their own respiratory system could regain some of its vitality as the virus hopefully receded. For particularly difficult cases, however, iron lungs were the only way to keep a patient alive. Some were condemned to spend the remainder of their lives in them. Indeed, some victims of the disease still use them today.

◆ ◆ ◆

Photographs of children trying to walk in braces or breathe with the help of an iron lung created a great deal of public sympathy for polio victims—and a lot of pressure on scientists to develop a cure. Adding to the urgency was the towering and beloved presence of the nation's 32nd president, Franklin Delano Roosevelt, who was paralyzed from a polio infection he contracted at thirty-nine years old in 1921.

While it was not a secret that Roosevelt had polio, he tried to hide the impacts of his disease when he was in public. Through great force of will, strong leg braces, a walking cane, and some subtle support from those around him, he could force himself to stand and walk to some extent. Secret Service personnel would block him from photographers' view as he got in and out of cars with assistance.

From a very wealthy and influential East Coast family, Roosevelt caught polio as an adult bordering on middle age. His diagnosis defied much of the conventional thinking about polio at the time, demonstrating that it wasn't specific to children or to those in poverty.

Roosevelt first showed polio symptoms in 1921 while summering at a family property on Campobello Island in New Brunswick, Canada. It's not certain where or exactly how Roosevelt caught the disease, but the consensus is that he was likely exposed to the virus while visiting a Boy Scout camp in New York State just prior to traveling to Campobello.

After an alarming progression of symptoms over three days in August that year, Roosevelt suddenly could no longer bear his weight on his legs. On August 25, he was diagnosed with polio, a huge blow to both Franklin and his wife, Eleanor.

US President Franklin Delano Roosevelt, Warm Springs, Georgia, 1929

Roosevelt was a rising political star in New York state at the time, but he bowed out of politics to focus exclusively on rehabilitation for several years. After realizing that he could support himself in water, the very active Roosevelt took to regular swimming practice to build his muscles up again and stay healthy. By January 1922, he was fitted with braces for his legs. His car was also fitted with special hand controls so he could still drive it. Swimming and driving his own car were two things he could still enjoy.

Eleanor would later write in her autobiography, "Franklin's illness proved a blessing in disguise. He had to think out the fundamentals of living and learn the greatest of all lessons—infinite patience and never-ending persistence."

Although FDR remained partially paralyzed, he eventually made his way back into politics again, running successfully for governor of New York and president of the United States. He was the first—and so far only—president elected with a physical disability. The American public was very sympathetic to FDR's condition. What's more, Roosevelt was himself determined to always be publicly in "good cheer" about the disease and demonstrate to the American people a

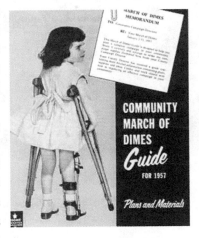

lasting commitment to his recovery. He also went out of his way to not appear physically limited by the disease, making a great effort to walk and stand at public speeches, with the use of leg braces and a cane. In the White House, Roosevelt used a special wheelchair he designed himself.

In 1938, Roosevelt made a massive impact on the race for a polio cure by launching the National Foundation for Infantile Paralysis, now known as the March of Dimes. The organization encouraged people around the country to send in dimes from their spare change in an effort to fight and find a cure for polio. (This is why Roosevelt is pictured on the dime to this day.) FDR's powerhouse foundation raised a significant amount of money and funneled it into promising research to develop a cure.

A team led by Jonas Salk finally did it in 1955. Salk, the director of the Virus Research Laboratory at the University of Pittsburgh, had been experimenting with a new approach to vaccines. At the time, the standard way to create a vaccine was to use a weakened version of the virus. This weakened, live virus would then be injected into a patient, where it would replicate and induce immunity to both the weakened version and the full-strength, wild-type virus, but without making the person sick.

Salk's approach differed in that he began experimenting with non-infectious, "killed" viruses to produce a polio vaccine. He had successfully created a killed vaccine for the flu in 1943 and thought the same principle might apply to polio. He began to publish his theory of a killed virus vaccine widely, attracting the attention of Roosevelt and the March of Dimes, which provided reliable financial support for Salk and his effort. After using formaldehyde to kill the poliovirus without destroying its antigenic properties, Salk then administered the resulting vaccine to a handful of willing volunteers, including himself and his wife. Encouraged by the success, Salk conducted a national study in 1954 with more than one million children. The results were incredibly successful. In 1955, Salk announced the vaccine was safe and ready for mass distribution. That year there were more than 20,000 cases of poliomyelitis in the United States. Two years later, the rate plummeted to 2,500.

Mobile distribution of the polio vaccine, California, 1960

Roosevelt passed away in 1945, ten years before the vaccine was finished. But since Salk's vaccine was a direct result of funding from the foundation Roosevelt established, we can add contributing to the eradication of polio to his incredible array of accomplishments.

Around the same time that Salk was developing the killed-virus vaccine, another polio vaccine was in development by Polish American medical researcher Albert Sabin, who worked at the University of Cincinnati. Sabin was developing an oral attenuated live-virus vaccine. He believed an oral vaccine would be superior to an injected vaccine as it would be easier to administer. After a great deal of testing, Sabin found three mutant strains of the poliovirus that stimulated antibody production in the human body, but without causing paralysis (or so it seemed). Sabin tested his vaccine on himself, his family, and prisoners from a facility in Chillicothe, Ohio, with great success. He couldn't get approval for a large-scale field test, as Salk's vaccine was already being used in the United States by that time, so Sabin turned to the Soviet Union, who conducted a successful field test in 1960. The US Public Health Service then approved the Sabin vaccine for manufacture and distribution in the United States, which eventually replaced the Salk vaccine because, as an oral vaccine, it was easier to administer.

There was a catch, though: The attenuated virus, after passing through the gut, can become infectious and pathogenic and, in a tiny fraction of recipients, actually cause polio. Which is pretty much the complete opposite of what you need from a vaccine. Sadly, it took a while to discover this possibility. It wasn't until 1999 that a US federal advisory panel recommended a return to the Salk vaccine. Ironically, it was too late to prevent the slow spread of some active strains of polio into the global community. While rare, on occasion the excreted vaccine-virus from the Sabin vaccine can continue to circulate in a heavily under-immunized population. After circulating and undergoing genetic mutations for 12 months or so, the virus can sometimes develop the ability to paralyze a victim again. This is known as "cVDPV" or circulating vaccine-derived poliovirus. It's one of the reasons polio has not yet been fully eradicated. Another problem is that globally, the injectable Salk vaccine was met with shortages and issues around teaching people how to give the injection. The Sabin vaccine, by contrast, is far simpler to administer, as it requires just two drops of oral liquid. For these reasons, the Sabin vaccine is still used in endemic areas of Afghanistan and Pakistan. But plans to phase in the injectable polio vaccine and phase out the oral vaccine are already under way.

Global mass vaccinations against polio were hugely successful between 1960 and 2016, by which time the disease was still circulating only in Afghanistan and Pakistan. In 2016, there were just forty-five cases of polio worldwide and the disease seemed on the verge of eradication. But it came back.

Vaccine drives were halted in Taliban-controlled parts of both countries, leaving thousands of children without protection. Rogue strains of polio then began showing up in sub-Saharan Africa. Eventually, a vaccine-derived poliovirus caused by remnants of the live virus used in the Sabin vaccine grew in strength, replicated, and became just as virulent as the original strains of the virus. Then COVID-19 hit, halting mass-immunization drives all over the world. In 2020, fifty million children in Afghanistan and Pakistan were unable to receive immunizations as a result. The WHO estimates that 200,000 cases will occur annually in ten years if we can't eradicate polio in these two strongholds.

Roosevelt Really Had Polio, Right?

Perhaps not. While the medical consensus has always been that FDR suffered from polio, recently medical historians such as Armond Goldman of the University of Texas in Galveston have put forth the idea that Roosevelt's symptoms more closely match those seen in Guillain-Barré syndrome

(GBS), an autoimmune disease linked to certain infections. Symptoms of Guillain-Barré syndrome are similar to those endured by Roosevelt: symmetric, ascending paralysis affecting the legs as well as the face; bowel and bladder dysfunction; and pain. Furthermore, GBS can strike at any age, whereas it was very unusual for someone to be stricken with polio at the age of thirty-nine, when Roosevelt fell ill.

However, polio is still a fair possibility given that Roosevelt also had fever (uncommon in GBS) and most people with GBS recover and do not need the use of a wheelchair, though it's possible in both diseases.

While an interesting historical curiosity, as a practical matter nothing would have changed in Roosevelt's treatment plan if he had been diagnosed with GBS, as effective treatment for that disease was still decades away. If anything, a misdiagnosis may have been a blessing in disguise as Roosevelt's polio diagnosis led to significant public attention and funding through the March of Dimes, which did a great deal to support the development of a polio vaccine.

A rare picture of FDR in a wheelchair, with Ruthie Bie, the granddaughter of his Hyde Park, New York, property's caretaker, and Fala, his dog, 1941

As Roosevelt said in 1932, "Can we . . . do less than our best in a crusade launched against the vast human loss inflicted by infantile paralysis? We all have glimpses, however vague of the whole picture; we understand at least a little of the human values involved in the stupendous problem presented by hundreds of thousands of polio victims. We have a gospel to preach. We need to make America 'polio conscious' to the end that the inexcusable case of positive neglect will be entirely eliminated."

HEPATITIS C
Patient ZERO
Unknown

Infectious agent: Hepatitis C virus

Location: Everywhere

Symptoms: Liver inflammation that can result in cancer and liver failure

Transmission: Blood exposure

Little-Known Fact: Called "the affectionate killer" because of its slow progression and lack of symptoms for years

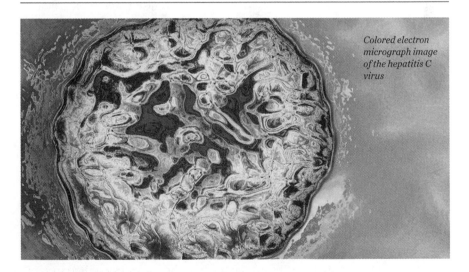

Colored electron micrograph image of the hepatitis C virus

n December 2017, a man in his sixties went to the emergency department (ED) at a hospital in Washington state with abdominal pain. To ease his discomfort, an injectable opioid was given to him by two nurses on duty. One month later, in January 2018, he was back in the ED with abdominal pain. Only this time, his laboratory tests showed something odd—his liver enzymes were abnormal, indicating inflammation in his liver. As part of the evaluation, a standard test for hepatitis viruses was ordered.

The man tested positive for hepatitis C virus (HCV), an infectious disease he'd never had before. He knew this for a fact because he had been tested for it in 2016, as was common practice by primary care providers treating people born between 1945 and 1965.

HCV is spread through blood exposure, so it's more common among those who use injectable drugs like heroin and also share needles. Healthcare workers are also at risk from accidental needlestick injuries while caring for HCV-positive patients. Serving time in prison is also a risk factor, as well as having received blood transfusions before 1982, when regular donor screening began for the virus.

The Washington state patient had none of these risks, so it was odd for him to have contracted HCV seemingly from nowhere. The Washington Department of Health was notified.

In March 2018, a second patient contracted HCV in Washington state, also with no risk factors. This patient was a woman in her fifties who sought help at the ED for neck pain in December. At that time, she too received injectable opioids for pain relief.

Investigators discovered that the HCV genotype in both infected patients was 96 percent the same, meaning they likely got HCV from the same source. Upon further questioning, the patients revealed that both received injectable pain medication in the same ED in December.

It didn't take long to track down the nurse who administered the pain medicine to both patients. Nowadays, doctors and nurses can't just grab medicine from a drawer and jot down on paper what they took and for which patient. In today's electronic medical records, everything can be easily tracked. It turned out that the nurse who administered the opioid-based medication to both patients had withdrawn it from the automated drug-dispensing system way more often than other nurses at the same hospital.

She quickly admitted to using the narcotics for her own drug addiction while also taking care of patients in the ED. How? Though she didn't explain exactly how she stole the meds, it's easy to imagine a few different ways. Injectable opioids could be drawn up into a clean syringe then partially injected into the person (the nurse, in this case) wishing to "divert" the drug elsewhere (hence the term "drug diversion"). Saline could also be pulled up into a syringe already containing medication to "top it off" and make the syringe look full. Or the medication could be injected directly into the patient, but not completely. The leftovers in

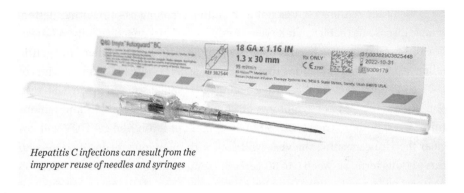

Hepatitis C infections can result from the improper reuse of needles and syringes

the used syringe could then be injected into the healthcare worker with the substance use disorder.

In the first scenario, the diverting healthcare worker would expose any infectious diseases they carried through their used needle into the patient. In the second scenario, a patient could expose the diverting healthcare worker with their own infectious diseases, like HIV, HCV, or hepatitis B.

After testing, they found that the nurse indeed had HCV, though she'd tested negative in 2013 while being screened for blood donation. Nevertheless, she had probably given it to those two patients. Investigators called the nearly three thousand patients who'd received injectable medicines while she'd been on duty. Several had HCV—and the ones treated by the nurse all had the identical HCV genotype.

Almost all had new diagnoses of HCV, but one had HCV all along—and that patient was treated twice by the nurse, in August and November 2017. The nurse had probably acquired HCV from this patient, then spread it to the other patients through her drug use.

Thanks to a slew of recent infectious diseases that have rocketed around the world, we are becoming accustomed to them making regular headlines, like:

Ebola Strikes Again.

Coronavirus Now Officially a Pandemic.

Flesh-Eating Bacteria Eats Away Limbs of Victim.

Brain-Eating Amoeba Kills Teenager After Swimming in Warm Lake.

HCV is not nearly so dramatic, but as a disease it can be devastating.

One of the reasons HCV is not as headline grabbing as some other better-known diseases is that it can take years or decades for its effects to fully manifest. For this reason, it's also been called, creepily, an "affectionate killer," or "gentle killer." Unlike other infections that can cause sickness quickly, many victims of HCV don't know they have it until a doctor tells them it showed up on a lab test. Sometimes, they'll have symptoms of jaundice, with yellowy skin and lemon-tinted eyes. Without treatment, 50 to 85 percent of those who get HCV will not clear the virus, meaning the virus will stay with them long term. Of those who carry it long term, between 5 to 30 percent (some studies cite a rate as high as 55 percent) will go on to suffer from liver disease, cirrhosis (late-stage scarring and poor liver function), and possibly liver cancer.

But how the disease operates inside the body can become secondary when a cluster of people with no risk factors suddenly come down with HCV. When it happens, that generally means an outbreak is afoot. And the most pressing question is how.

♦ ♦ ♦

HCV has probably been with us for a very long time, though we didn't actually name and identify the virus until 1989, practically yesterday in epidemiological terms. We're talking when the Berlin Wall came down, Madonna's "Like a Prayer" freaked out pearl-clutching moms everywhere, and *The Golden Girls* was a runaway TV hit. Before that, we discovered hepatitis C's cousin viruses, hepatitis A and B, in the 1970s. But it wasn't until 1989 when a "new" virus that had actually been found and known for a while (called NANB, or non-A, non-B) was officially called hepatitis C.

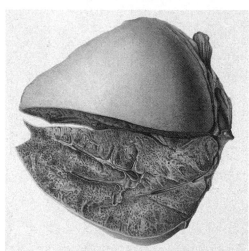

Lithograph of a diseased liver from someone who suffered from hepatitis, ca. 1897

It's a tiny virus, a little 60-nanometer ball of protein carrying within it a single strand of RNA. In new infections the virus attacks the liver, causing inflammation. About a third of those newly infected people show telltale symptoms that their liver is not happy. Sometimes their skin becomes an odd day-glow yellow and itches. The white sclera of their eyes may also turn yellow from jaundice. This happens when a pigment called bilirubin, a natural byproduct from the break-down of hemoglobin molecules inside red blood cells, accumulates at high levels in the body. With jaundice, the system of clearing bilirubin out by the liver isn't functioning properly. HCV sufferers tend to feel tired, they lose their appetite, their urine turns a shade resembling Coca-Cola, and their bowel movements are weirdly light-colored, like bread dough. But the majority of people won't have symptoms, or they'll be so mild they won't even notice they're harboring a virus.

Before we had names for them, all the hepatitis viruses were likely wreak-ing havoc throughout history. Around 400 BCE, Hippocrates described an illness called "epidemic" jaundice for the "fourth kind of jaundice," after the fourth humor (yellow bile). These "jaundice" epidemics sprouting at various points in history point to an infectious source, as opposed to other noninfectious causes of jaundice, such as liver failure from alcoholic cirrhosis, or pancreatic and liver cancer. When a likely viral hepatitis outbreak erupted in the eighth century, Pope Zacharias quarantined men in Rome. It may have affected Napoleon's army in Egypt at the turn of the nineteenth century, as well as during the American Civil War, when up to 70,000 cases were noted in the Union Army. Its preva-lence during war campaigns lent it other names, like "campaign jaundice" or "jaunisse de camps," as it was called during the Franco-Prussian war. Though we don't know definitively, these descriptions point to hepatitis A outbreaks, which are spread through the fecal–oral route, as in drinking water contami-nated by feces. Another major outbreak occurred in World War II (see sidebar, page 271).

Today, it would be strange to see HCV crop up in large numbers; we're more likely to see hepatitis A outbreaks due to contamination in produce, like the blackberry-associated outbreak in 2019, or associated with homelessness and IV drug use, which in 2017 resulted in 1,521 cases. When new clusters of HCV occur, healthcare professionals start searching for the reason. Sadly, drug diversion is sometimes the case.

If you are suffering from an opioid-use disorder or are trying to sell opi-oids on the black market, it's quite hard to procure prescription opioids without an actual prescription. Because of how powerfully they affect people, opioids are locked up everywhere—in pharmacies, emergency departments, and hos-pital wards. The rules about prescribing are stringent. But working within the healthcare system removes several barriers and brings you one step closer to the medications themselves. There have been a handful of HCV outbreaks related to drug diversion but they are still likely more that go unreported or unnoticed.

In 2013, a particularly horrifying HCV outbreak was tied to a dentist who put 7,000 people at risk and accidentally infected eighty-nine people in Tulsa, Oklahoma. It was the first HCV outbreak linked to a dental clinic in history. Some of the patients also contracted hepatitis B and HIV. The offending dentist allegedly had been reusing syringes and was found in possession of rusty dental tools.

Though the dental clinic appeared clean and well kept, the truth was far dirtier (on myriad levels). The dentist's drug cabinet contained a medication that expired ten years prior, and the cabinet itself wasn't locked. Morphine had also been used on patients, though the last morphine delivery to the office occurred four years prior. The dental assistants employed at the office performed intravenous sedation without licenses. At that time, the executive director of the Oklahoma Board of Dentistry saw the tools that came out of the autoclave, the high-temperature oven used to sterilize tools, and said, "I wouldn't let my nephews play with them out in the dirt. I mean, they were horrible. They had rust on them."

A separate incident in the small farming town of Fremont, Nebraska occurred in 2000 and 2001 at an oncology clinic, leading to the largest outbreak of HCV at the time. The citizens of Fremont initially were thrilled to get their own full-time oncologist. Before him, they only had part-time doctors in town and often had to trek an hour to Omaha for various problems and treatments. During the physician's short tenure in Fremont, a shocking 99 out of 857 cancer patients he treated turned up with a fresh HCV diagnosis.

Warning signs included improper waste management reports of blood vials found tossed into regular trash cans or randomly sitting on countertops overnight. Hypodermic needles and syringes were reportedly found in coffee cans instead of sharps containers. Witnesses described nurses with visible blood on their skin using bare hands to draw blood. Apparently, the doctor had further instructed his nurses to implement cost-saving measures that were blatantly dangerous, such as reusing syringes or employing a "community saline bag" for flushing chemotherapy ports. Anecdotal accounts detailed saline bags that contained pink-tinged liquid, instead of the expected clear saline. There was also an allegation that he used less chemotherapy than reported in order to save money by not opening second vials needed for treatment.

In a tell-all memoir, *A Never Event*, a victim named Evelyn McKnight teamed up with her attorney, Travis Bennington, to shed light on the scandal. There are mentions of the physician's dramatic suicide attempt, an affair, impersonating another doctor, as well as sexual misconduct and manipulation of a patient. The book reads like a lurid novel, a classic case of truth being far stranger than fiction. When the misconduct was exposed, the doctor and his head nurse lost their licenses, and he fled the country where he became a government health official in his home country. Multiple lawsuits were filed against him, and the trust he shattered in the medical profession in this small town will undoubtedly last for a very long time.

Another outbreak erupted in an endoscopy center in Nevada where syringes and vials were reused for patients undergoing procedures like colonoscopies. It was first noticed when the Southern Nevada Health District was alerted to two new cases of HCV in January 2008. Normally, that region sees two cases *per year*. But the medical practices at the endoscopy center had focused on patient volume and quick turnaround times. A nurse training at the center was told to record inaccurate times and sign off on paperwork done by other nurses. This allowed the center to double and triple bill insurance companies for anesthesia times. They would bill for things like a thirty-one-minute procedure, when in truth the colonoscopy only took seven minutes, and the upper endoscopy to the stomach only took a staggering two minutes. (In reality, an average colonoscopy takes fifteen to thirty minutes, and an upper endoscopy takes ten to fifteen minutes.) The nurse-in-training was also told to document that all patients were in good condition, even if they were not. "Every time it was documented exactly the same way, that they were in good health," she reported, before quitting three days later. Unlike the nurse-in-training, the vast majority of workers at the center didn't report anything wrong.

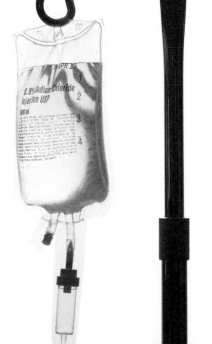

Improper use of IV fluids, including sharing of bags between patients, can lead to spread of hepatitis C infections.

The doctor in charge at the endoscopy center had allegedly urged nurse anesthetists to reuse single-use drug vials. A nurse would draw up an anesthetic agent called propofol, inject it into the IV port of a patient, and in doing so drew back some blood into the syringe. The needle would be replaced, but the same contaminated syringe would be used on another patient. Up to 63,000 people treated at the center were put at risk of exposure. At least nine of them contracted HCV and one died. And the aforementioned head of the endoscopy center was making over six million dollars in the year before the outbreak, much of it from egregious disregard for safety and outright fraud. He is currently in jail serving a life sentence after a jury found him guilty of twenty-seven criminal counts, including second-degree murder, insurance fraud, and negligence.

◆ ◆ ◆

According to the Centers for Disease Control (CDC), between 2006 and 2018 in the US, there were sixteen HCV outbreaks in outpatient and long-term care facilities, twenty-two outbreaks in hemodialysis settings, and four related to drug diversion by healthcare providers. In addition, there's a long list of single-case infections, too, and likely many more that went unreported.

Until 2015, there weren't great options for HCV treatment mostly because of the seriously unpleasant side effects from a regimen that included weekly injections for nearly a year and a cure rate of about half in genotype 1, which is seen most often in patients. Thankfully, there's now a new generation of direct-acting antiviral medications (DAAs) that are highly effective at inhibiting viral replication and infection, can be taken as pills, have high cure rates, and are far more tolerable. Unfortunately, they can cost upward to tens of thousands of dollars (some are $1,000 per pill).

The fact that a great treatment exists doesn't make any of the prior horrible lapses in care any better. Even with the seemingly endless lists of regulations for healthcare facilities to follow and a slew of organizations to enforce them, HCV outbreaks will undoubtedly still happen. But time and again, the likely reason for these outbreaks will be the same as they have been in the past: addiction, greed, laziness, cutting corners—all at the cost of human lives.

Hepatitis B Hiding in a Vaccine

Yellow fever was a significant concern for American troops in World War II, so much so that a vaccine was required for any personnel serving in tropical regions. However, in March 1942, multiple army troops came down with jaundice. Normally, when a group of people with such an obvious connection comes down with a symptom like jaundice, it indicates a disease outbreak in the group.

But army personnel *outside* of tropical areas were also coming down with jaundice, right here in the US. The Army began an intensive investigation into the cause and, after sending out a questionnaire, quickly found the common denominator: a yellow fever vaccine given to 425,000 army soldiers. Many vaccines, it turned out, contained human serum (blood plasma without a key clotting agent), donated by army soldiers. The virus for the vaccine was both grown and suspended in the serum, which likely contained the infective hepatitis agent. The serum was quickly removed from the vaccines, and no further cases were recorded.

Soldiers line up to get vaccinated before being sent overseas to fight in World War II, 1939

Not until 1987 were the soldiers able to find out what type of hepatitis they had (it was hepatitis B). Around 330,000 soldiers likely contracted it from the contaminated vaccine, and of those about 50,000 suffered from jaundice.

Hepatitis B, for which there is a vaccine but no cure, is usually spread through blood, by sexual intercourse, or "vertically" from mothers giving birth and infecting their newborns. Vertical transmission often results in children who have a chronic hepatitis B infection that can result in cirrhosis and liver cancer later in life. But horizontal transmission—through sex, blood, or in this case, vaccines later in life—usually results in a brief illness.

The three-part vaccine for hep B is very effective. Most healthcare professionals are vaccinated. And luckily, these vaccines no longer contain human serum, which, from a hygienic standpoint, is a very good thing.

THE INFECTION–DISEASE LINK

From Viruses, Bacteria, and Parasites to Cancer

Impact: One-fifth of all the world's human cancers are blamed on infectious causes

When: Since ancient times

Who: Can affect anyone, depending upon lifestyle factors

What Happened Next: Attempts to decrease global cancer burden with vaccines, early detection, and eradication of cancer-causing infections

Little-Known Fact: Chickens were the first animals that helped prove that viruses can cause cancer.

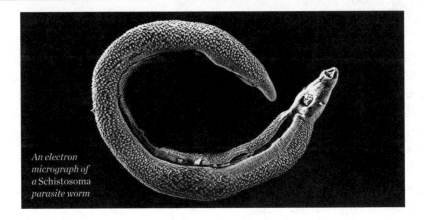

An electron micrograph of a Schistosoma parasite worm

W e tend to think of infections as foreign invaders that break through our body's defenses, temporarily wrecking us in clever, insidious ways, and then poof! Once our bodies finally mount an immune defense, they recede because we've destroyed them on our own or we've taken antibiotics to kill them off—if they haven't killed us first. But many infections don't always leave us with parting gifts like immunity or smallpox scars. Sometimes they never leave, squatting in our liver cells for the rest of our lives. And sometimes, they leave us with a far different type of legacy, a lethal one.

Cancer.

Most of us have a decent idea about why cancer happens. Carcinogens and risks abound, seemingly everywhere. Smoking can cause lung cancer while contributing to myriad other issues. Too much booze can cause liver cancer. Overindulgence in red meat is linked to colon cancer. A few surprising correlations exist—piping-hot drinks, being a french-fry cook, working night shifts, pepperoni—all linked to cancer. Then there's the stuff you're born with, such as inherited risks for breast, colon, and ovarian cancers. Sun worshippers who eschew protection and experience repeated, peeling sunburns are at greater risk for melanoma. Radiation exposure can cause thyroid cancer. The list goes on.

Food. Genetics. Environment. Increasingly, the warnings are everywhere. We scan packages of bacon to ascertain if they have nitrate preservatives, or read yet another article about the relative benefits or downsides to coffee, which, depending on the day, will save us from or kill us with cancer. We do DNA tests and map out family trees to determine if we're destined to inherit certain diseases.

But we tend not to consider other causes altogether, including a host of species that prey on us and can sometimes change our cells—and our lives—in irrevocable ways. We often talk about events that change our destiny, but disease can alter the very fabric of our selves, meaning our DNA. Carcinogens aren't just static, inanimate things, like cigarettes and radiation. They can be living things trying to survive, like you and me. Like bacteria, parasites, and viruses.

It took time to realize how and why these cancers occurred. In 1842, Italian surgeon Domenico Rigoni-Stern observed that nuns didn't seem to suffer from cervical cancer, unlike married women, the first hint that the sexually transmitted human papillomavirus (HPV) was both responsible for the malignancy and was transmissible. Also discovered in the nineteenth century was the transmissibility of a type of lung cancer in sheep (caused by the jaagsiekte virus, named after the Afrikaans words for "chase" (jaag) and "sickness" (siekte), describing animals who became breathless while being chased). In 1908, two scientists in Denmark, Vilhelm Ellerman and Oluf Bang, found that if you took chickens with leukemia, filtered out the cancerous blood cells and any bacteria, then inoculated the serum into healthy fowl, the cancer could be transferred. The filterable cancer-causing substance was later found to be avian leukosis virus. A few years later, a landmark finding established the basis of our understanding of the link between prior disease and the cancer landscape even further.

An American virologist named Peyton Rous, who would go on to win the Nobel Prize, conducted a groundbreaking though seemingly simple experiment that showed the relationship between infection and malignancy. In his lab Rous had chickens afflicted with a type of spindle cell cancer sarcoma. He took a piece of the chicken cancer and meticulously ground it up with sterile sand, spun it in a centrifuge, and passed it through a Berkefeld filter with pores small enough to remove individual cells. Those cells were then extracted and injected into the breasts of healthy chickens, who subsequently grew sarcomas at the injection sites.

Two more tumor-causing viruses were discovered in the 1930s. One was the Shope papillomavirus, found in the skin of cottontail rabbits, the other a mouse mammary tumor virus that was transferred from mother to baby via a cancerous agent, or "milk factor," from nursing the young.

More and more pathogenic discoveries would be made in the coming decades that would begin to expose the full role of infection in causing or otherwise setting the stage for malignancies. And all those discoveries, small and large, would come to fruition decades later in the creation of actual vaccines that prevent cancer.

◆ ◆ ◆

Perhaps the best-known cancer-preventing vaccine is for the human papilloma virus (HPV). These are a group of over 150 HPV virus subtypes that can infect people. In 2013–2014, about 42 percent of adults ages eighteen to fifty-nine tested positive for genital HPV. We tend to think of warts as a mildly embarrassing cosmetic irritation, but the kind you can't see are far scarier. You may never actually be able to view genital warts on yourself, but the virus can still be spread by skin-to-skin contact that even condoms can't always prevent (because HPV can infect areas not covered by a condom). And though it's only a select group of HPV subtypes that cause cancer, and most people with HPV won't get an HPV-related cancer in their lives, it's still a frightening proposition to know it's a possibility. In you. Waiting in place until the right set of factors converge, including time, luck, smoking, drinking, or other co-infections, like HIV, to transform your healthy cells into something far more sinister.

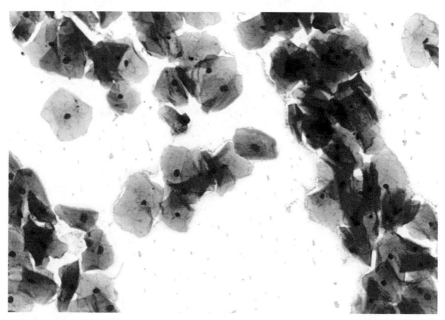

Normal cervical cells on a Pap test

HPV is often screened for during routine pelvic exams with Pap smears, because it's a known cause of cervical cancer. But strains of HPV can also cause throat, anal, and vulvar cancers. And yes, sexual transmission is how it happens— vaginal, anal, and oral. There is even a risk to gynecologists who do HPV-removing electrosurgical procedures that aerosolize the virus, and can result in them developing throat cancers. It's hard to imagine that this virus, which looks like a golf ball under an electron microscope and only encodes eight to ten proteins, can cause such damage. But HPV is incredibly clever. Two of its genes encode proteins E6 and E7, which do a subversive job of inactivating the human proteins responsible for keeping cells nice and orderly.

In healthy skin and mucous membranes, epithelial cells are programmed to do very specific jobs, like make keratin or lipids. Cells also have tumor-suppressor proteins (like p53 and retinoblastoma gene product pRB) to prevent them from becoming tumors. And HPV's proteins screw up how p53 and pRB do their jobs by interfering with DNA repair, stopping apoptosis (programmed cell death), and messing up normal cell growth, all necessary for the proper function-ing of our bodies. When the HPV virus gets involved, the outcome is that DNA replication goes unchecked and cells proliferate at abnormally high rates. And when those cells proliferate, too many DNA mistakes occur, they don't get fixed, and they can start growing out of control. Which is pretty much the hallmark of a cancer cell that stopped listening to its internal regulation mechanisms and adopted a god complex.

And then they wait. From the time of infection to cancer, it can be ten to twenty years, though in some cases it can be as short as two.

The first HPV vaccine was approved in 2006 and recommended for adolescents typically eleven to twelve years old and younger adults, but available to adult patients as well. Now there are three vaccines, and they are some of the few known cancer-preventing vaccines we have today. It can prevent 90 percent of HPV-related cancers in people. This bears repeating: We have a cancer prevention vaccine.

And not just one. Other known cancer-causing viruses include hepatitis B (HBV), and hepatitis C (HCV). After chronic, long-standing infections of the liver, people with these infections can succumb to liver and other cancers, and in the case of HCV, it may also be linked to non-Hodgkin's lymphoma. There is now a vaccine against HBV—the first anti-cancer vaccine to arrive before the HPV vaccine—but not yet one for HCV. Someday, perhaps, there will be even more cancer-preventing vaccines, but so far these are all we have.

Another virus that can cause malignancy is Epstein-Barr virus, or EBV, which most people know about if they've had mononucleosis, or mono, also informally referred to as the kissing disease. EBV is a type of herpes virus and one of those, once caught, that we carry forever, like chickenpox (which erupts later in life as shingles). Most of the time EBV is not a problem, but it can increase your risk of getting nasopharyngeal (head and neck) cancer, stomach cancer, and lymphomas including Burkitt lymphoma (though these tend to be more common in places like Southeast Asia and Africa).

We know that people living with HIV are more likely to get Hodgkin's lymphoma, lung cancer, oral cancers, Kaposi's sarcoma, liver cancer, and anal cancer, though HIV is not directly the cause of the malignancy, but rather a means to weaken the immune system so that cancers can occur more easily. Human herpesvirus 8 (HHV-8) causes Kaposi's sarcoma, a cancer that tends to appear in older men of Mediterranean descent and in immunocompromised patients with HIV and AIDS.

There's also a Merkel cell polyomavirus that causes Merkel cell carcinoma, a rare type of skin cancer, to which 63 to 75 percent of the population has been exposed. Human T-lymphotrophic virus type 1, a retrovirus like HIV, can cause types of leukemia and non-Hodgkin's lymphoma called adult T-cell leukemia/lymphoma.

Now, before you panic and call your doctor to demand a slew of tests, remember that despite those huge numbers in the case of Merkel cell polyomavirus exposure, less than 0.35 percent out of 100,000 people will get cancer related to it. With HTLV, there is often a long latency period between infection and cancer, on the order of decades, and even so only a small fraction of people will go on to develop a malignancy.

Even with HPV, there are often cofactors and other risks that play into the development of cancer, such as use of hormonal contraceptives, chlamydia (see below), herpesviruses, smoking, and nutrition, so it certainly isn't destined to happen. As with HIV (see page 276), immune suppression plays a large role in the development of Kaposi's sarcoma. For cancers to develop, there is a highly complex process affecting cells every step along the way in the transformation from healthy to cancerous, including the body's inability to clear the pathogen, inflammation, increased mutation rates in normal cells, genetic instability, and alteration of signaling pathways in how cells grow, proliferate, and die (or don't die). A mortal cell doesn't become a malignant, immortal one by a simple infection alone.

◆ ◆ ◆

Viruses aren't the only pathogens linked to cancer—bacteria and parasites have their place as cancer-causers, or abettors, as well. One of the best known is *Helicobacter pylori*, bacteria that appear to thrive in the severely acidic environment of the human stomach. Just the discovery of *H. pylori*, the helical-shaped (hence the name) bacterium, was groundbreaking. First observed in the late-nineteenth century in the stomachs of dogs and people, it wasn't possible to culture it at first, given its highly acidic needs. Not until the 1980s was it definitively found in the stomachs of patients suffering from peptic ulcer disease and also proven to be a carcinogen that caused stomach cancer.

We tend to think that ulcers are annoying side effects of stress or too many dashes of Tabasco in our food. When they're really bad, they bore an

Molecular model of the protein coat of an HPV particle

excruciating hole in your stomach and, if left untreated, can cause you to bleed to death. In the past, some physicians scoffed at the idea of ulcer-causing bacteria. Their advice: People just needed to lighten up, pop more Pepto-Bismol, and that pesky stomach ulcer will go away.

It wasn't that simple. A physician and researcher in Australia, Dr. Barry Marshall, was convinced of this as well, and he suspected that *H. pylori* was the true culprit. So he drank a concoction of broth and *H. pylori* that he'd gathered from a patient and waited to see what would happen. (It was a level of dedication to one's work that bordered on the fanatical, but also deserved some kind of award.) What followed was a lot of stomach pain, nausea, and vomiting. In short, he'd successfully sickened himself with gastritis—inflammation of the stomach—the precursor to stomach ulcers. He then tested his stomach acid and found the *H. pylori* thriving there. Along with his collaborator Robin Warren, they would go on to get a rather big award in the end—the 2005 Nobel Prize in Physiology or Medicine.

H. pylori may cause stomach cancer by a variety of mechanisms. The complicated inflammatory cascade it encourages in the stomach cells may set the stage for cancer, along with gene mutations that prime stomach cells to transform into cancerous ones. When an *H. pylori* infection is eradicated in a patient with a cocktail of antibiotics, their risk for stomach cancer and *H. pylori*–related lymphoma decreases.

Other bacteria have been suspected of causing cancer as well, such as chlamydia, the sexually transmitted infection that appears to just keep on giving. While it doesn't cause cancer directly, women who have active or past chlamydia infections are more likely to get cervical cancer than those without. (So far, no physicians have volunteered to give themselves chlamydia to test this one out.)

Still other bacteria have been considered carcinogenic, though it's not definitive, including *Salmonella typhi* (the cause of typhoid fever) for gallbladder cancer, and *Chlamydia pneumoniae* (not related to the sexually transmitted kind) for lung cancer. It's only a matter of time before more research yields discoveries around how these bacteria contribute far more than what came of the original infection.

◆ ◆ ◆

Last and certainly least, for how much they make our skin crawl, are parasites. And yet getting one is rather easy: Imagine eating some undercooked shrimp or carp in a less-developed region of East Asia. In the fish flesh there might be some metacercariae—the cyst-encased form of *Clonorchis sinensis*, the Chinese liver fluke. A type of flatworm, the fluke cyst casing dissolves in the acid of your stomach, and the metacercariae go on to penetrate your intestines before making the one-to-two-day journey to your bile ducts within the liver. There, they make themselves comfy, gobble up their favorite food, bile, and grow into adults in about a month. Then they start laying eggs that are excreted into your bile ducts and finally out of your body through your feces. Unfortunately, your bile ducts can become cancerous related to those wormy squatters, who can live up to thirty years. (*Opisthorchis viverrini* is another liver fluke in Southeast Asia that can cause the same result.)

Next up is *Schistosoma haematobium*, which has the dubious distinction of being the first parasite known as a Group 1 carcinogen by the International Agency for Research on Cancer and the US National Toxicology Program. Apparently, the earliest recorded case of schistosomiasis was recorded by a German physician named Theodor Maximillian Bilharz, who found one of these little flukes inside a dead soldier in Cairo, Egypt, in 1851. (Napoleon's soldiers

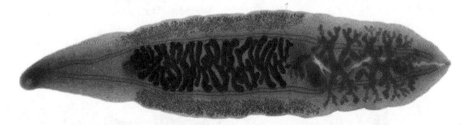

A Chinese liver fluke (C. sinensis*) parasite*

Fecal Transplants

Yup, poop transplants are a thing.

In the medical world, fecal microbiota transplants are becoming old news, but not many people have heard of them. After a chapter full of doom and gloom from all the ways that pathogens are trying to off us, it's perhaps ironic that a beacon of hope comes in the form of poop.

This is how fecal transplants work: Feces from healthy donors with no communicable diseases are processed and given to a recipient who is suffering from *Clostridioides difficile* (formerly *Clostridium difficile*), or *C. diff*, colitis. An inflamed colon is nothing to laugh about; just ask anyone who's ever had *C. diff*. The disease occurs when natural bacteria of the gut are eliminated after a course of antibiotics, and *C. diff* steps in to fill the void.

C. diff creates spores that are very hard to kill and can even live for long periods on surfaces. It's ubiquitous in the environment and can sneakily find its way into sick patients. Spread via the fecal--oral route, *C. diff* can, in severe cases, cause the colon to dilate to massive sizes, something called "toxic megacolon," which scares most medical students the first time they hear it. And it should, as the condition has a high mortality rate and can pop a life-threatening hole in your intestine.

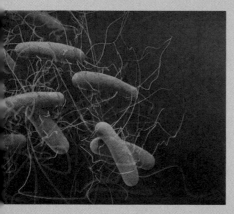

An illustration of Clostridioides difficile *bacteria*

Despite the existence of antibiotics that cure *C. diff*, it has a nasty habit of returning. And, unsurprisingly, the people who happen to get *C. diff* are usually dealing with a lot of other medical issues already.

Enter (no pun intended) the fecal transplant.

It's a brilliant idea, actually. We tend not to think of the bacterial contents of our intestines as an organ, but in many ways, it is—a complex network of trillions of microbes working in concert with the rest of our body to aid our immune system, help with digestion, and provide us vitamins (such as vitamin K). Many of us try to support this microbiome by eating probiotics: kimchi, sauerkraut, pickles, and such. With fecal transplants, it's taken a step further. The donor feces are processed and delivered via capsules that are swallowed. Sometimes it's given via a tube running from the nose down the esophagus and past the stomach. Retention enemas are also employed, as is colonoscopy delivery. Cure rates can be as high as 90 percent and appear to be long lasting. Fecal microbiota transplants are also being investigated for treatment of non-infectious diseases, such as insulin resistance, a precursor to type 2 diabetes.

Just one of those slightly icky things happening in the world of bacteria that, for a change, is really awesome.

reputedly suffered from schistosomiasis during the Egyptian campaign in 1799–1801; their main symptom was urinating blood.) Schistosomiasis is also called bilharzia in the German physician's honor.

Schistosomiasis is only second to malaria in terms of global burden of disease, death, and disability due to parasitic infections. In several parts of the world where fresh water is found, the parasite breeds in snails and then leaves to swim in the water, until it can find unsuspecting human victims wading while bathing or washing laundry. Then, it penetrates the skin, maturing within the human and eventually laying eggs that cause inflammation in the liver, intestines, blood, lungs, and bladder, which can become calcified. The inflammation from the eggs results in bloody urine and eventually bladder cancer.

Worldwide, about one-fifth of all human cancers are blamed on infectious causes, which is staggering. Perhaps the good thing about knowing that so many pathogens can cause such horrible disease is just that: We know. Which means we can develop more treatments, detect cancers earlier, even when they are precancerous (such as via Pap smears or anal versions of the same, which may be helpful for detecting early anal cancers), and develop more vaccines to prevent the entire process from the start.

In the meantime, there are things we can do to prevent what may feel like the inevitable. A healthy lifestyle actually does have an effect on our fates, particularly if we avoid tobacco products, alcohol, and unhealthy eating habits. Safe sex is a good thing, and getting cleared of infections when you have them makes a difference, like in *H. pylori* ulcers and HCV.

Take care of your body. After all, scientists have been doing crazy things like drinking *H. pylori* shakes in order to save lives. Being better to ourselves is really the least we can do.

CHOLERA
Patient ZERO
The Lewis Baby

Cause: *Vibrio cholerae* (bacteria)

Transmission: Ingestion of infected fecal matter, generally through water and/or food

Symptoms: Vomiting, uncontrollable diarrhea, severe dehydration, blue skin

Impact: Cholera can spread and kill with devastating speed

Where: Broad Street neighborhood, London, England

When: 1854

Little-Known Fact: The discovery of how cholera spread in London's Broad Street outbreak spawned the field of epidemiology.

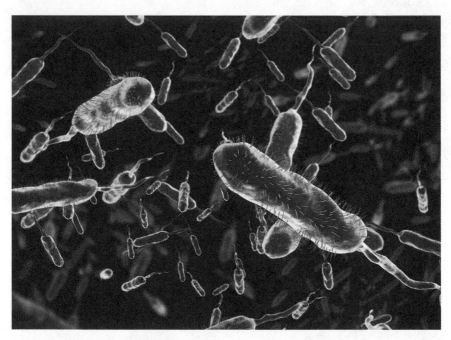

Illustration of Vibrio cholerae *bacteria*

n March 1849, James Polk, age fifty-three, America's youngest president in its short history, stepped down from office, fulfilling his promise to serve only one term. A Democrat from Tennessee and protégé of Andrew Jackson, Polk was very popular at the time, widely hailed as a successful president. During his four-year tenure, America had defeated Mexico in the Mexican-American War and expanded its territory by a million square miles, adding much of present-day California, the American Southwest, Washington, and Texas to its borders. The expansions came at substantial cost, however, as the debate over whether to admit new states into the Union as free or slave states directly laid the groundwork for the coming Civil War. Polk, a plantation owner who profited from the work of enslaved people, relied on income from those people to fund his

First Lady Sarah Polk and US President James Polk, ca. 1848

early retirement from political office. Despite his political success, Polk kept to his campaign promise of only serving one term and stepped aside for Zachary Taylor to become the 12th president of the United States.

Soon after leaving office, Polk and his wife set out on a tour of the southern states, heading down the Atlantic seaboard to Alabama, Mississippi, and Louisiana before traveling back up the Mississippi River to their home state of Tennessee. The trip went splendidly all the way until New Orleans, with cheering crowds greeting Polk at every stop.

But in New Orleans, Polk caught an illness that sickened him on the journey north up the Mississippi River. What began as abdominal distress advanced to diarrhea, vomiting, and dehydration. Then those symptoms got worse, with extreme fatigue setting in as severe, relentless diarrhea took hold.

Polk's trip was cut short and he retreated home to Nashville to try to recover from his illness. But the symptoms continued, and his health further deteriorated. On June 15, 1849, a scant three months after he left office, Polk died from the illness. Owing to Nashville city laws designed to prevent the spread of disease, the 11th president of the United States was buried for a year in a mass grave in a Nashville cemetery.

So what was the gastrointestinal illness that claimed the life of a US president? Cholera.

The word alone was enough to strike fear into the heart of anyone living in the nineteenth century. Cholera is a hideous disease caused by infection from the bacterium *Vibrio cholerae*. If we ingest the bacteria, it creates a special toxin that tricks our intestinal cells into spewing out water virtually nonstop. What starts with convulsions of pain quickly advances to violent vomiting and uncontrollable diarrhea.

The effects go well beyond the typical "stomach flu" symptoms most people have experienced. Those stricken with cholera pass quart upon quart of a milky white liquid termed "rice-water stools" by doctors. This liquid has a characteristic fishy odor and is full of electrolytes the body squeezes out of its intestinal cells. It also contains legions of cholera bacteria.

The resulting dehydration from the onslaught of diarrhea is often so intense that, if not treated, it will make a patient's skin look blue (a trademark of cholera), send them into shock, and then death. Quickly. Needless to say, it's not a pleasant way to die. The only small mercy of cholera is that it tends to kill fast.

◆ ◆ ◆

Cholera bacteria live in fecal matter and spread through contaminated food or water. If the bacteria in the milky white liquid expelled by the body get into food or, worse, a public water supply, an epidemic can spread with shocking

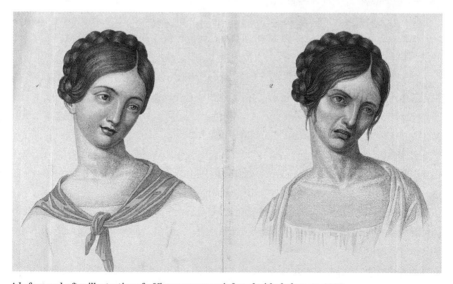

A before-and-after illustration of a Viennese woman infected with cholera, ca. 1830

speed. But the need to protect water supplies from cholera exposure wasn't realized until nearly the end of the nineteenth century, after several devastating global waves of cholera pandemics.

And while it has probably been with us for many centuries, it's a disease deeply associated with the nineteenth century, when five successive cholera pandemics happened in one hundred years. The 1800s could rightly be labeled "The Century of Cholera."

People the world over were intensely afraid of cholera outbreaks, as virtually every country on the planet was hit hard by one at some point in the nineteenth century. Cholera was particularly brutal in urban slums, where people lived tightly packed, in deep poverty, and where mortality rates were already sky-high. (For example, in Bethnal Green, London, the average age of death in the mid-nineteenth century was . . . sixteen years old.)

Before cholera, people who lived in the industrial slums of the world were already ravaged by multiple infectious diseases prone to spread mercilessly in crowded, dirty environments. In fact, a major reason why life expectancy in such areas was so brutally low was because of rampant typhus, typhoid, smallpox, tuberculosis, measles, dysentery, diphtheria, scarlet fever, rickets, whooping cough, bronchitis, and pneumonia. A shockingly prolific list.

Since the rise of capitalist economies and the division of people into broad socioeconomic classes with disproportionate access to health care, sanitization, and clean running water, disease epidemics have always hit harder among the poor and disenfranchised. By the time of the cholera epidemics of the nineteenth century, however, the industrial revolution had made the division between the rich and poor substantially wider. (Those socioeconomic divisions, now global in scale, are often why diseases considered eradicated in wealthy Western countries are still ravaging populations in other parts of the world.)

So when cholera arrived in nineteenth-century England, it found an impoverished working-class population rife with a variety of other diseases and living in squalid, close quarters. It was the perfect setting to seed a pandemic.

Between 1817 and 1823, the first cholera pandemic of the century is believed to have sprung from pilgrimage sites in Bengal, India, before finding its way to the Indian subcontinent, and from there into the Middle East and Asia before subsiding. Medical historians largely agree that the disease's origins lie in the brackish, warm water of the Indian Ganges-Brahmaputra Delta, where cholera has probably been endemic for centuries. Pilgrims traveling to sacred Hindu

Illustration of a grim reaper bringing death by cholera, 1912

sites in the region appear to have enabled its subsequent spread along trade routes into the Middle East and Asia.

The second cholera pandemic of the century, which raged from 1826 to 1837, tore through Asia, North Africa, Europe, and the Americas. With a devastating mortality rate of 50 percent, it spread across the world via trade routes, catching rides on merchant ships, as well as in the digestive tracts of soldiers, religious pilgrims, itinerant travelers, migrants, and refugees.

Cholera felt particularly at home in slums, most notably those in England. The deeply unsanitary living conditions in England's industrial cities were infamous at this time. Writing about Manchester in 1845, the German political theorist Friedrich Engels described the people in poor areas of the city as "pale, lank, narrow-chested, hollow-eyed ghosts" who lived in homes that were "kennels to sleep and die in." Delving deeper, he wrote, "Passing along a rough bank, among stakes and washing lines, one penetrates into the chaos of small one-storied, one-roomed huts, in most of which there is no artificial floor; kitchen, living, and sleeping-room all in one . . ."

Nor were there near enough toilets.

Sometimes several hundred people shared one or two hastily constructed outdoor toilets. At other times, they just used a privy installed over a hole in the ground in the basement of the slum house. A civil engineer recorded in a survey of two houses in the 1840s, "I found whole areas of the cellars of both houses were full of nightsoil to the depth of three feet, which had been permitted for years to accumulate from the overflow of the cesspools." If

Crowded row houses with back-to-back terraces in Victorian London, 1872, about when indoor plumbing began to be adopted widely

anyone living in such conditions at the time came down with cholera, there was virtually no way to stop the wildfire-like spread of the disease.

The third cholera pandemic of the century raged across the world from 1846 until 1863. In England alone, it killed approximately 50,000 people between 1848 and 1850. (This was the pandemic wave that took US president James Polk.) In retrospect, it was easy to see why. Reporting on the spread of cholera in 1848 London, journalist and reform advocate Henry Mayhew commented on the atrocious conditions: "As we passed along the reeking banks of the sewer, the sun shone upon a narrow slip of the water. In the bright light it appeared the colour of strong green tea, and positively looked as solid as black marble in the shadow . . . and yet we were assured this was the only water the wretched inhabitants had to drink. As we gazed in horror at it, we saw drains and sewers emptying their filthy contents into it; we saw a whole tier of doorless privies in the open road, common to men and women, built over it; we heard bucket after bucket of filth splash into it; and the limbs of the vagrant boys bathing in it seemed by pure force of contrast, white as Parisian marble." Considered the worst year for cholera in history, 1854 was also the first year that the tide began to turn in the fight against the disease.

♦ ♦ ♦

For years, cholera's mode of transmission was a genuine medical mystery. The disease seemed to rise up from nowhere, kill a swath of people quickly, then recede again almost overnight, only to emerge somewhere else a few days later. No other known disease at the time was quite so sporadic and quixotic. Its patterns confounded physicians to what, in retrospect, should have been obvious: It was contagious, spreading from person to person, showing up anew in places where people arrived from a previous outbreak. To be fair, most people at the time were unaware of what "contagious" even meant.

Enter a bona fide medical hero, John Snow, who proved to the rest of the world how cholera spread, pioneering the use of disease mapping and founding the science of epidemiology.

In the mid-nineteenth century, medical opinion of how cholera spread coalesced primarily around bad air, or "miasma" (see Germ Theory, page 38). All you have to do is imagine the stink

English physician John Snow, who discovered that cholera spread through contaminated food or water, not bad air or "miasma"

Portrait einer Cholera-Präfervativ-Frau
von M. G. Saphir.

A humorous illustration of a woman over-prepared with dubious protections against cholera, ca. 1832

of English slums, the people crammed together in tiny rooms, defecating in the basements of their tenements, to understand where this idea came from. Breathing in that stink would certainly make anyone unfamiliar with it feel sicker, but it wouldn't give them cholera.

To catch cholera, you need to ingest contaminated food or water. Snow suspected this in 1854, when London was in the throes of its worst cholera outbreak ever. He proposed that water might be the culprit of the disease, pointing to the fact that cholera outbreaks tended to cluster among populations who share the same water supply. To prove his hypothesis, Snow analyzed an outbreak in

Dr. John Snow's "ghost map" of London's Broad Street cholera outbreak in 1854

the Soho district of London. On the night of August 31, 1854, two hundred people came down with cholera in this West End neighborhood. Within ten days, the death toll was five hundred, a full 10 percent of the area's population.

Snow mapped the addresses of those who died from the disease, noticing quickly that many deaths were near a specific water source: the pump on Broad Street. Snow's now-famous "ghost map" showed the concentration of cholera deaths in that specific area. And though the map indicated a problem localized there, Snow still needed to sort out some troubling details in his data.

Among them were that some deaths in the Soho outbreak happened far away from the pump. And there were a lot of people in the neighborhood who weren't getting sick at all. So Snow began a thorough, door-to-door examination of the area, with one principal question: Where did you get your water?

It turned out that those who died in parts of the neighborhood far from the Broad Street pump were nonetheless getting their water from it. One mother sent her sons to the pump specifically because she preferred the taste of its water. Some children in other parts of Soho who were sick with cholera had drunk from the Broad Street pump on their way to school. Which solved one of Snow's two mysteries.

He then investigated why some residents in the district weren't getting sick at all. He learned that all seventy of the workers in the local Broad Street brewery were still healthy. Why? Because the brewery had its own water supply. And, to be more accurate, the brewery offered a generous perk to its workers: free beer. Many of the brewery workers said they never drank any water, just beer. All the time. Snow also discovered that, in a large workhouse (where those unable to support themselves were offered food and shelter in exchange for work), only a few of the 535 residents had come down with cholera, as the workhouse also had its own supply of water.

The evidence was overwhelming. Snow presented his case to local health authorities and offered a simple but brilliant solution: temporarily remove the handle on the Broad Street pump. Afterward, the cholera outbreak slowed to a halt.

Shockingly, the General Board of Health, a central administrative authority established a few years earlier specifically to combat cholera outbreaks, was still unconvinced that water was the true culprit. They conducted an investigation as well and concluded that, once again, "bad air" was the cause of the disease. Such was the relative strength of the miasma paradigm (and the relative strength of the stink that pervaded the neighborhood). So, when the General Board of Health pointed out that all the houses where people died were clearly unsanitary, they weren't technically wrong.

The story might have ended there. But Snow showed his map to the assistant curate of St. Luke's Church on nearby Berwick Street, Reverend Henry Whitehead. Whitehead was a popular figure in the neighborhood and was deeply concerned about the cholera outbreak. At first, Whitehead didn't believe Snow either; in fact, he was so opposed to Snow's assertions about the Broad Street pump that he decided to launch his own investigation, drawing on his extensive neighborhood connections to interview families impacted by the disease and disprove Snow's theory. He questioned many residents of the Broad Street area, including those who had fled the neighborhood at the first sign of a cholera outbreak, asking if they had consumed Broad Street pump water.

His investigations ultimately led him to the patient zero of the Broad Street outbreak. The first person to die of cholera in the 1854 London outbreak was the baby daughter of Thomas and Sarah Lewis, who lived in the parlor room of

The Great Chicago Cholera Outbreak of 1885

One of the worst cholera outbreaks in US history occurred in 1885, when a heavy storm carried sewage from the Chicago River far enough into Lake Michigan that it reached the intake system for the city's drinking water and contaminated the water supply. A full one-eighth of the city's population (90,000 people) died in the resulting outbreak of cholera, which ultimately led to a new large-scale sewage project that would reroute the city's sewage away from the Lake Michigan intake plant.

It was a terrible outbreak—or it might have been—if it had ever happened. This oft-reported story was an urban myth concocted in the mid-twentieth century to sell Chicagoans on a flood control scheme. While a large storm, which sent polluted water into Lake Michigan, did indeed occur in 1885, the prevailing winds actually pushed the contaminated water away from the city's intake system. No cholera outbreak occurred.

The seeds of the legend were planted in a 1956 publication from the Metropolitan Water Reclamation District. By 1976, the story included the deaths of "90,000 Chicagoans," and was often recorded in books, newspapers, and magazines. Libby Hill, a professor at Northeastern Illinois University, finally asked the question no one thought to ask: Did it actually happen?

Turns out, the answer was a resounding no.

40 Broad Street, quite near to the infamous water pump. Around 6:00 a.m. on the already hot morning of August 28, Sarah Lewis had discovered her daughter was ill, vomiting and expelling a green, watery diarrhea with a "pungent smell." Sarah washed her baby's diapers, then threw the dirty water into the cesspool drain located a scant three feet away from the Broad Street water pump. Whitehead learned that the very next day the upstairs neighbors of the Lewis family came down with cholera. The reverend urged the General Board of Health to excavate the cesspit and, sure enough, the pit had an obvious leak into the water supply.

The source of the outbreak was confirmed at last. And Snow's theory about contaminated water was validated, a major milestone in our attempts to understand and fight a disease that spread with a deadly quickness via a resource essential to keeping us alive.

As for the Lewis baby, patient zero of the 1854 outbreak—she died tragically four days after contracting cholera, her name lost to history. Her father died two weeks later. During that time, Sarah Lewis continued to dump water contaminated with infected stool into the cesspool, unwittingly contributing to the spread of the epidemic.

A terrible irony of the cholera outbreaks of the nineteenth century is that the disease is actually simple to treat. In fact, of all the diseases that have led to historical epidemics, cholera stands out as the easiest to defeat. Simple home remedies that cost mere pennies can help someone recover from cholera and reduce mortality rate to less than 1 percent. The answer lies in oral hydration therapy. In one liter of clean water, you dissolve half a teaspoon of salt and five or six teaspoons of sugar (modern rehydration solutions usually also have potassium salts), then encourage the patient to continually drink this mixture. Basically, cholera is treated by staying very hydrated until the disease runs its course. Had this simple fact been known to residents of Broad Street and everywhere else around the world hit hard by cholera epidemics in the nineteenth century, millions of lives could have been saved.

Reverand Henry Whitehead, a local curate for the Broad Street neighborhood, who came to believe John Snow's theory about how cholera spread.

Snow and Whitehead became friends in the wake of the 1854 outbreak. Their efforts laid the groundwork for the burgeoning science of epidemiology, combining scientific inquiry with demographic investigation, fundamentally changing the way we study the patterns of disease. Whitehead outlived Snow by forty years, but he always kept a portrait of Snow in his study to remind him "that in any profession the highest order of work is achieved, not by fussy empirical demands for 'something to be done,' but by the patient study of eternal laws."

ANTHRAX & BIOLOGICAL WARFARE

The Weaponization of Disease

Cause: *Bacillus anthracis* (bacteria)

Impact: Deadlier than heaving plague-infested bodies at enemies, but part of a long history of disease weaponization.

When: First used (ineffectively) by the Germans in WWI. Heavy favorite of the Japanese in the 1930s. Illegally investigated and stockpiled by the Soviet Union during the Cold War. Last (known) stockpile of weaponized anthrax found in Iraq in 2003.

Who: Imperial German Army, Kwantung Army, Red Army, Iraqi Ground Forces

What Happened Next: The Geneva Protocol prohibited biological warfare in 1925, followed by the Biological Weapons Convention, prohibiting development, production, and stockpiling of biological weapons in 1972. Both have been violated.

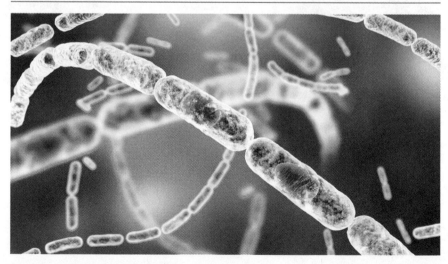

Illustration of Bacillus anthracis *bacteria*

n 1979, a woman named Anna Komina worked in a ceramics factory in the Soviet city Sverdlovsk (now Yekaterinburg). She came home from work one day complaining that it was difficult to breathe. She also felt faint and dizzy. After a few days of bed rest, her symptoms gradually improved. However, shortly after she got up and started moving around again, she suddenly collapsed.

Komina's son called her doctor, who in turn called an ambulance. It took five hours for the emergency responders to stabilize her blood pressure enough just to transport her to the hospital. But their efforts were in vain. Komina died the next day.

The Soviet cloak of secrecy quickly fell over her mysterious illness and abrupt death. Komina's family was forbidden from taking her body back from the hospital to be prepared for a proper funeral. Instead, police were assigned to guard her coffin as it was transported to the cemetery. Her death certificate listed her cause of death as "bacterial pneumonia."

Yekaterinburg (formerly Sverdlovsk), Russia, 2004

Komina wasn't the only person to suddenly take ill and die in Sverdlovsk. Nearly every employee at the ceramic factory where Komina worked also got sick, and with similar symptoms. Within a week, nearly all of them were dead. At least sixty-six workers from the ceramic factory died amid an outbreak of mysterious illness, though the actual death count remains unknown. All hospital records related to the Sverdlovsk ceramic factory illnesses and deaths were subsequently destroyed by the KGB.

The actual cause of all those deaths: anthrax. A disease caused by infection from the *Bacillus anthracis* bacteria, anthrax is typically associated with livestock animals and is rare in humans. Most people who catch anthrax do so from working with infected animals, such as shearing the wool of infected sheep. What makes anthrax scary, however, is its ability to spread via a tiny amount of airborne spores that linger for a really long time in the environment where they are introduced. That's also why it has such potential as a biological weapon.

Anthrax infects people (or animals) when its bacterial spores enter their body, typically through breathing them in, but also from ingesting contaminated food and water, or the introduction of spores to the bloodstream through an exposed cut or wound. Once inside a body, the bacteria produce exotoxins—toxins secreted by the bacteria into the body. The exotoxins are absolutely brilliant at suppressing the immune system. They prevent infection-fighting cells from recruiting more help, stop macrophages from consuming and destroying bacteria, and cause some cells to simply die. While the bacteria multiply in the lymphatic system, the toxins can reach levels within human cells that result in a kind of point of no return. Even if you kill off the bacteria with antibiotics, it's sometimes too late. If the remaining toxin can't be reached, the victim can die of shock. Anthrax anti-toxin and rapid medical care can make a big difference. Thankfully, anthrax is not contagious between humans, a small saving grace of the disease.

Anthrax symptoms can manifest anywhere from one day to two months after initial exposure. The mode of transmission (via a skin wound or through inhaling spores, for example), determines which symptoms display. The skin form includes a small blister followed by local swelling and eventually a painless but unsightly black ulcer on the skin. When untreated, skin-based anthrax results in death about 20 percent of the time.

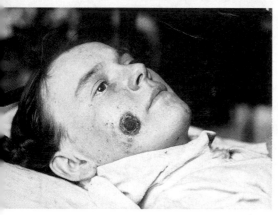

A sore from an anthrax infection

It's the respiratory or inhalation form that's particularly deadly. Fever, chills, headaches, difficulty breathing, and drenching sweats are all symptoms of a lung-based anthrax infection. If enough spores are inhaled, the bacteria can overwhelm the body and kill its host. Even with treatment, inhalation anthrax survival is rare—around 80 to 90 percent of people die once infected with it.

The disturbingly high mortality rate of the disease is another reason it has such allure as a biological weapon. Indeed, according to the Centers for Disease Control (CDC), anthrax is among the most likely agents to be used in a biological attack. Why? Because anthrax spores can be found naturally in soil and are also easy to produce in a lab. They also last for a long time once released into the environment, increasing the odds of someone eventually breathing them in. Finally, and insidiously, they can be unleashed on the sly, with the spores easily disguised in powders, sprays, food, or water.

Which brings us back to Sverdlovsk in 1979. After Komina and most of her coworkers died from anthrax, the Soviet government was compelled to provide an official explanation about what happened. They decided to put the blame on the black-market trade in livestock meat, claiming there was an outbreak of anthrax among local sheep and cattle. Those who bought their meat from black marketeers (rather than government-approved vendors) had caught anthrax and died.

But news of the deaths managed to escape the closed city and make its way to international media outlets. And the story exploded. Was the USSR secretly experimenting with anthrax? How? On whom? If any of this was true, it would be a violation of the Biological Weapons Convention treaty the country signed in 1972.

The Soviet government suddenly found itself in the uncomfortable position of having to document an anthrax outbreak to world governments paying close attention to the situation. Soviet scientists were then dispersed widely to prove that the anthrax infection had indeed come from meat contamination. They pointed to local livestock deaths that had occurred just days before the outbreak in Sverdlovsk, providing gory slideshows to demonstrate that the infections had taken hold in the intestines of their human victims. (To prove, they hoped, that it came from contaminated food.) Suspiciously missing from the slides, however, were any images of the patients' lungs.

Global suspicions remained and international pressure on the Soviet Union intensified. In 1986, Harvard biologist Matthew Meselson and his anthropologist wife Jeanne Guillemin were allowed into the Soviet Union for a limited investigation of the now seven-year-old outbreak. In the later view of the CIA, the Soviet-sponsored invitation was issued to "establish a propaganda position in the public domain and a version plausible to persons without access to intelligence information." Meselson and Guillemin interviewed a collection of people connected to the outbreak. Despite strong suspicions from the CIA that the anthrax release was connected to a biological weapons program in the USSR, Meselson was forced to conclude, from the limited evidence he reviewed and limited sources he spoke with, that the official Soviet contaminated meat explanation was "plausible and consistent with what is known from medical literature and recorded human experiences with anthrax."

But the American government remained suspicious of the incident. After the collapse of the Soviet Union in 1991, Boris Yeltsin became the first president of Russia. Anxious to improve American–Russian relations, Yeltsin surprised everyone by declaring in 1992 that he agreed with American suspicions about the Sverdlovsk incident and that the anthrax release was indeed the result of a secret USSR biological weapons program. At the time of the incident, Yeltsin had been the head of the Communist Party in the Sverdlovsk region and felt he had been personally lied to by the KGB about what really happened. He pledged to set the record straight.

Yeltsin promised pensions to the Sverdlovsk families who had lost members to the outbreak, issued a decree that Russia would adhere to the 1972 Biological Weapons Convention, and then invited Meselson to return to Russia for an unrestricted investigation. This time, Meselson was allowed to see autopsy slides of the victims' lungs, which confirmed what had been suspected all along: The Sverdlovsk anthrax outbreak was airborne. Meselson's final report laid the blame for the anthrax outbreak not on contaminated meat, but on the inhalation of anthrax spores. The location of the ceramic factory, where so many workers got sick, was also noted. It was directly downwind of a Soviet military building ominously named Compound 19. Meselson concluded that the source of the outbreak was the military installation, which was engaged in biological weapons research and had inadvertently released a small amount of anthrax into the air. How much? Meselson determined that the amount of anthrax released could have been as little as a gram, perhaps even less. If a scant gram of airborne anthrax can infect hundreds of people, imagine what a pound of it dropped in a major city could do.

It later turned out that Meselson still didn't learn the whole story. Anthrax was indeed being studied by the USSR for weaponization in Compound 19 in violation of international agreements. Unrevealed until years later was that the entire outbreak, and the deaths that resulted, were all due to an avoidable work accident.

The real story is that a worker in Compound 19 removed a filter from the compound's exhaust system. Before doing so, he turned off the exhaust system so the filter could be cleaned and reinstalled, a job he did not finish, meaning the filter was not in place when the worker's shift ended. He did leave a note explaining this for the incoming supervisor—who never saw it. When the new supervisor noticed the exhaust system off inside Compound 19, he simply turned it back on. The mistake was later discovered, and the filter was quickly reinstalled.

But it was too late for the hundreds of people downwind of Compound 19 who breathed in the deadly spores.

◆ ◆ ◆

The Soviets were by no means the first, or only, global power to consider the potential of anthrax for biological weapons. The Germans were actually the first to use anthrax in a military conflict. In World War I, German spies poisoned Allied livestock and animal feed with anthrax. The Germans attempted to spread it in Argentina, Finland, France, and the United States.

The covert anthrax mission in America came during the early years of the war, when America was still officially a neutral country but was busy supplying Allied efforts with livestock and munitions. German secret agents established a small laboratory in the basement of a house in Silver Spring, Maryland, where they grew anthrax cultures and distributed the bacteria along a covert network of operatives in several port cities. And that's where the historical record gets murky. It remains unclear how effective these attacks were, though it appears they only had a minor impact.

After the chemical warfare horrors of World War I, where tear gas, hydrogen cyanide, chlorine, phosgene, and mustard gas were all used, most countries signed on to the Geneva Protocol in 1925, which banned the use of chemical and biological weapons in war. While thirty-eight countries ultimately joined the treaty (and many more later ratified it), the scope of the

Japanese naval troops in gas masks, Shanghai, China, 1937

protocol was limited. Its rules were only binding in conflicts between countries who'd participated in the treaty. So, there were no restrictions on using chemical or biological weapons for internal conflicts, such as civil wars, nor were there any prohibitions on using such weapons against countries who had not signed the treaty. Furthermore, there was no restriction on possessing biological weapons. Many countries used the various loopholes in the Geneva Protocol to assert a right for retaliation in kind should they ever be attacked with chemical or biological weapons. As a result, the Geneva Protocol was effectively watered down to a "no first use" agreement. (An intriguing historical aside is that, while many world powers signed the treaty, notably absent were China, Japan, and the United States, who finally ratified the treaty in 1952, 1970, and 1971, respectively. The United States was the last major world power to agree to the Geneva Protocol, fifty-three years after the end of World War I.)

Japan began experimenting with weaponizing anthrax during their occupation of Manchuria, China, in 1932 as part of a large, state-sponsored initiative to investigate a variety of biological weapon possibilities. To experiment with dosage levels, Chinese prisoners were exposed to anthrax. The Japanese then attacked at least eleven Chinese cities with anthrax by dropping spores from airplanes.

The United States and Great Britain turned their attention to anthrax in the early years of World War II. Both countries experimented with anthrax as a retaliatory weapon (exploiting the Geneva Protocol exception) against attacks by Germany or Japan. The US conducted tests at sites in Mississippi and Utah, eventually stockpiling 5,000 anthrax-laced bombs filled for possible deployment after an attack by Germany. They were never used.

Great Britain, meanwhile, tested anthrax weapons by stocking a small Scottish island with eighty sheep and then detonating a couple anthrax bombs. Unsurprisingly, all the sheep died. What's more, the island remained uninhabitable for forty years, until 1986, when Britain decontaminated it by soaking the island in a mixture of formaldehyde and seawater. Britain's experiment proved to them (and the world) that anthrax lasts a very long time in the environment, its longevity and hardiness possibly scarier than its lethality.

In 1972, the Biological Weapons Convention, a new international treaty, was signed by more than one hundred countries, prohibiting the development, possession, and stockpiling of biological weapons. Signatories included Great Britain, the United States, Iraq, and the Soviet Union. All the signing countries were required to destroy their current stockpiles (which is what happened to those 5,000 anthrax bombs developed by the US). The Soviet Union almost immediately violated the treaty with their anthrax research at Sverdlovsk. This is also why the 1979 outbreak drew such international attention.

Anthrax didn't surface in the news again until 2001, when the United States endured multiple anthrax attacks on the heels of September 11. Letters filled with white powder containing anthrax were mailed to the offices of Senator Tom Daschle (Democrat-SD) and Senator Patrick Leahy (Democrat-VT), as well as several news and media outlets on the East Coast. After the attack was brought to light, forty-three people tested positive for exposure to anthrax, twenty-two got sick with the disease, and four died. More than 10,000 people were deemed at risk for exposure while the letters wound their way through the US mail system.

A biohazard team responds to an anthrax-tainted letter sent to Senator Tom Daschle, Washington, DC, November 7, 2001.

The ensuing FBI investigation was one of the largest and most complex in the history of law enforcement. Ultimately it focused on Bruce Edwards Ivins, a senior biodefense researcher at Fort Detrick in Maryland. After learning criminal charges were about to be filed against him, Ivins killed himself with an overdose of acetaminophen. But no direct evidence of his involvement was ever definitively established, and the 2001 anthrax attacks remain officially unsolved.

◆ ◆ ◆

Weaponized anthrax was hardly the first biological weapon used in war. Other forms of biological weapons have been used by, and against, humans for centuries, possibly even stretching back into prehistory. Thankfully, a lack of understanding of how disease truly spread prevented large-scale biological warfare for most of human history.

Nevertheless, poisoned arrows have long been a favorite tool of war, used by cultures around the world. Melanesians in present-day Vanuatu contaminated their arrows with tetanus by covering them in the contents of crab burrows. (Crab burrows can contain *Clostridium tetani*, the organism that causes tetanus.) Scythians in present-day Ukraine had an elaborate recipe to create an arrow poison that involved mixing the blood of decomposed vipers with dried-out human blood. Poisoning wells was also a favored tactic of invading armies. The Greeks poisoned the water supply of Kirra with hellebore, a toxic herbaceous plant in 590 BCE, seeding a mass outbreak of diarrhea and allowing the Grecian army to conquer the city. Animal cadavers were often dumped into wells, and human corpses as well, though the first documented use of corpses to foment disease didn't occur until Frederick Barbarossa, the Holy Roman emperor, attempted to conquer Italy in 1155. After successfully besieging Tortona, his army poisoned its wells with human remains and razed the city to the ground.

Botanical illustration of black hellebore, a poisonous plant believed to have been used in ancient warfare.

In 1346, when the Tartars set siege to the Crimean city of Caffa, they catapulted victims of the bubonic plague into the city. And it may have even seemed like it worked: An outbreak of the plague took down the defenders of Caffa shortly thereafter, and the Tartars conquered the city. (But the fleas from the pervasive rats were a more likely culprit.) The tactic was a bit shortsighted, though. Upon moving into a plague-infected city, the Tartars soon came down with the disease themselves. The plague-inflicting technique was used several times throughout history. The most recent example took place in 1710 during the Great Northern War, when the Russians besieging Reval (present-day Tallinn, Estonia) used the same tactic.

War makers eventually got more sophisticated than poisoning wells or catapulting corpses. In 1495, the Spanish contaminated barrels of wine with blood from leprosy patients, then snuck the wine into the supply lines of the French army they were fighting in southern Italy. But as leprosy can have an incubation period of anywhere from one to twenty years, it wasn't the most effective strategy. In 1650, the Polish army, under the advisement of artillery expert Kazimierz Siemienowicz, experimented with firing hollow shells containing the saliva of rabid dogs at their enemies.

Napoleon took biological warfare a step further, demonstrating an impressive understanding of the environmental conditions necessary for the spread of disease. While besieging Mantua in 1797, he arranged for the nearby plains to be flooded to encourage the spread of "swamp fever" (i.e., malaria) among the Austrians occupying the city.

Distributing disease-infected clothing is another time-tested way to engage in biological warfare. The most notorious example comes from North American history, when British soldiers gave smallpox-infected blankets (see sidebar, page 149) to two emissaries from the Indigenous Delaware tribe who were besieging Fort Pitt at the time (the only documented incident of such a tactic). Tunisians also tried to disperse plague-infected clothing among their enemies in 1785 while attacking La Calle (present-day El Kala), Algeria. During the American Civil War, Confederate doctor Luke Blackburn traveled to Bermuda in 1864 to help contain outbreaks of yellow fever and smallpox. While there, he found his Hippocratic oath challenging to maintain in the face of his Confederate sympathies and came up with an elaborate plan to spread disease among the Northerners. He acquired the clothes of victims who had died of smallpox or yellow fever, and then shipped trunks of the clothing back to the northern United States to be sold in consignment stores. Thankfully, the tactic didn't work. Yellow fever requires mosquitoes for transmission, and smallpox virus particles would have been too old to be contagious, after being shipped all the way from Bermuda. (Far from being ostracized for his attempts at biological warfare, Blackburn went on to become the governor of Kentucky.)

◆ ◆ ◆

As microbiology became an established scientific discipline in the late nineteenth century, international concern grew over its potential for large-scale biological warfare. Two international declarations on warfare sought to ban the use of poisoned weapons in 1874 in Brussels and 1899 in The Hague. A byproduct of international treaties regulating biological warfare, however, is paranoia. Most of the countries that participated in the accords feared other countries were developing a biological warfare program in secret, so they sometimes did exactly that themselves. Therefore it wasn't a huge surprise when the Germans used anthrax in covert operations against Allied troops in World War I.

But no country embraced the potential for biological warfare with as much enthusiasm as Japan between World Wars I and II. Radical nationalist Shirō Ishii, who would go on to become Japan's surgeon general, undertook government-sponsored research into biological weapons in 1930, later becoming head of Japan's bioweapons program during World War II. At its peak, the program employed more than 5,000 people. The Japanese freely experimented

Shirō Ishii, head of Japan's bioweapons program during World War II, was responsible for thousands of deaths.

with some twenty-five types of biological agents on prisoners, killing an average of six hundred people per year. They also unleashed a variety of biological agents during the Second Sino-Japanese War (1937–45), poisoning Chinese wells with typhus and cholera, and dropping plague-infested fleas over Chinese cities. These techniques worked, causing outbreaks of disease in China that outlasted the war itself, sometimes by many years. By some estimates, tens of thousands of Chinese died as a direct result of Japan's biological warfare program.

During the Cold War, Americans and Soviets conducted extensive biological weapons research. Both countries conducted tests on animals, volunteers, and even unsuspecting civilian populations. The most notorious American example was the intentional contamination of the New York City subway system by US Army scientists in 1966 with *Bacillus globigii*, a bacterial strain that has similarities to anthrax. While the bacteria are not known to be harmful to people, the test was intended to study how an anthrax attack on the US might spread among the population. The United States finally abandoned its biological warfare research efforts by executive orders of President Richard Nixon in 1969 and 1970, ratifying the Geneva Protocol and the Biological Weapons Convention in 1975, and destroying its stockpile of biologi-

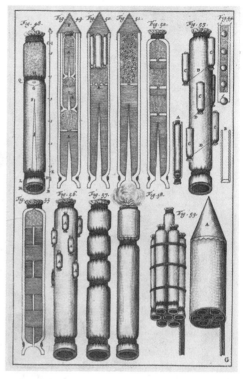

Illustrations of early rockets from Artis Magnae Artilleria, *a book by Kazimierz Siemienowicz, an early pioneer of rocketry, who supposedly fired artillery containing the saliva of rabid dogs in an 1650 battle*

cal weapons. The Soviet Union, while also officially signing off on the Biological Weapons Convention, continued its research until its collapse in 1991.

Today, no country is openly pursuing biological weapons; however as recently as 2003, in Saddam Hussein–controlled Iraq, remnants of an active biological weapons program were revealed. Iraq had investigated and stockpiled weaponized anthrax, botulinum, and aflatoxin, which were discovered by UN inspectors after the Gulf War in 1991. Although Hussein never deployed biowarfare, his regime did not destroy the weapons in the wake of the war and kept laboratories dedicated to biological weapons going until 2003. It appears now that all the countries of the world are at least nominally following the rules of international treaties against the development and use of large-scale biological weapons. But are there still secret labs researching biological agents and how to weaponize them?

Almost certainly.

Cult Attack

Rajneeshees line up to greet Rajneesh driving by in one of his many Rolls Royces in 1982.

The largest bioterrorism attack in US history didn't occur as part of international military conflict, but in an attempt by a cult to win a county election. In the early 1980s, the Rajneeshee cult, led by its charismatic leader, Bhagwan Shree Rajneesh (later known as "Osho"), bought a 64,000-acre ranch in eastern Oregon, then established Rajneeshpuram, a commune of 7,000 adherents that was incorporated into a city. The Rajneeshees wore red, engaged in communal work and open relationships, and used guards armed with Uzis and a jeep with a mounted machine gun to patrol their compound. Needless to say, their relationship with the locals was rife with conflict. When various building permits for Rajneeshpuram were denied by Wasco County officials, the cult tried new tactics: political intimidation and biological warfare.

The Rajneeshees fielded their own slate of candidates for local governmental in the 1984 elections, then decided that poisoning a group of other Wasco County residents would suppress (read: scare) local voters enough for their candidates to prevail. Cult members armed with homegrown salmonella visited a variety of local salad bars and dumped the contaminant into salad dressings. More than 750 people got sick with salmonella poisoning, forty-five seriously enough to require hospitalization. Fortunately, no one died. The cult had almost settled on trying to spread typhoid fever instead of salmonella, which would have certainly resulted in fatalities.

In the end, their strategy failed miserably. Local residents turned out en masse to vote against the Rajneeshee candidates, who didn't win a single seat. The salmonella outbreak itself was initially attributed to improper restaurant hygiene, but after a cult member fled and revealed details of the attack to the authorities, the US government launched an investigation. They found a fully operational bioterrorism lab on the cult's property, complete with intact salmonella cultures. Arrests followed and the Rajneesh movement collapsed, much to the relief of Wasco County residents. And salad bar enthusiasts everywhere.

RABIES
Patient ZERO
Unknown

Cause: *Rabies lyssavirus*

Symptoms: Headaches, fever, fear of water, brain swelling, convulsions, seizures, hallucinations, respiratory paralysis

Where: Worldwide

When: Dating back at least until the nineteenth century BCE

Transmission: Most often by bite from a rabies-infected animal

Little-Known Fact: Absent treatment, rabies is the most lethal virus on the planet, killing nearly 100 percent of its victims.

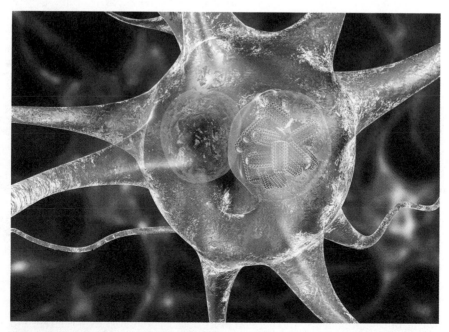

Illustration of rabies virus particles within a neuron

O n a balmy fall day in eastern France in 1885, a shepherd boy named Jean-Baptiste Jupille sat on a grassy hillside gazing out at the pastoral scene before him. Jupille and a handful of other young shepherds were enjoying a brief respite, chatting among themselves while casually watching their flocks as the sheep grazed calmly on the surrounding fields.

Without warning, a stray dog crested the nearby hill and came charging down at the shepherds, a thick, white foam clinging to its mouth. The foam was a telltale sign of rabies, a vicious disease with no known cure at the time, which meant a bite from that dog was paramount to a death sentence.

Jupille called out to the other boys to run—and run they did—as he reached for his whip to confront the mad dog. Jupille got in a whip strike before the charging dog was on him, biting down hard on his arm. In the ensuing brutal struggle, Jupille was able to pry the dog's jaws open, freeing his arm. He then used his whip to tie the dog's muzzle shut, after which the dog continued to snarl, rave, and claw. Jupille knocked the dog senseless by repeatedly banging its head with his wooden shoe. He then dragged the dog to a nearby brook and drowned it.

Jupille's brave actions undoubtedly saved his friends, but at great cost. With multiple deep bite wounds on his arm, Jupille would likely contract rabies. If you were bitten by a rabid dog, it wasn't guaranteed that you'd contract the disease—you'd have to wait for a few tense weeks, maybe even a month to be sure, all the while wondering if you'd live or die.

Maybe the bite wasn't deep enough? Maybe the saliva wasn't infectious enough?

However, if you began to feel a tingling in the wound followed by headaches and fever, well, it was a death sentence almost without exception. Next would come an inexplicable fear of water, followed by the swelling of your brain, and finally seizures and convulsions. Then, ultimately, coma and death.

Luckily for Jupille, the summer of 1885 brought the first glimmer of hope in several millennia of struggle against rabies. Because Louis Pasteur, a French scientist and a renowned pioneer in the development of modern bacteriology, had just discovered a rabies vaccine.

♦ ♦ ♦

Rabies is an ancient disease. It's also the most lethal known virus in the world, once killing nearly 100 percent of its hosts. The first known reference to rabies dates to the nineteenth century BCE, when it is mentioned in the Laws of

A woodcut of a rabid dog scene from the Middle Ages

Eshnunna, two cuneiform tablets dictating legal code in ancient Mesopotamia. The code dictates that an owner of a "mad dog" had to pay a fine if the dog caused the death of a person. (Note: This rabies death fine was on a sliding scale according to class position. For most people, you'd need to shell out 40 shekels of silver, but if your dog killed an enslaved person, you'd only need to hand over 15 shekels.)

Rabies is an acute viral infection of the central nervous system. The only way for a human being to catch it is through the saliva of an infected animal. Because of our ancient associations with domesticated canines, most of the time, this means a dog bite. But any warm-blooded mammal can carry the disease. Raccoons, bats, skunks, and foxes are the most common wild reservoirs.

Rabies is caused by *Rabies lyssavirus*, present in the saliva of infected animals. Once the virus enters the human body it avoids the bloodstream, bypassing much of our body's built-in immune system. Instead, rabies travels within the axoplasm, or cell material within nerve fibers, following a pathway to the brain. An incubation period as short as nine days or as long as several years follows, during which victims don't know whether or not they've contracted the disease. If it takes hold in the brain, the horrific symptoms of rabies begin to appear.

Caricature depicting the chaos caused by a rabid dog on a London street, 1826

And those symptoms in humans are almost identical to those in rabid dogs, one of the reasons the disease has so unnerved people for centuries. Like their canine counterparts, human victims of rabies will foam at the mouth, become unpredictably vicious and angry, and suffer from seizures, convulsions, and hallucinations. It's little wonder that folktales of werewolves and vampires haunting the European countryside may have originated with rabies.

Perhaps the most bizarre symptom of a rabies infection is a crippling fear of water. Cruelly, victims suffer from extreme thirst, which they aren't able to quench because severe muscle spasms in the throat prevent them from drinking. The spasms are so severe and painful that the mere sight of water leads to horrible hydrophobia. And herein lies the ruthless yet brilliant design of the rabies pathogen: As the virus accumulates in the saliva, all that foaming at the mouth and the inability to swallow equates to a huge population of rabies virus in the mouth, just waiting to spread via a bite. Finally, the victim will fall into a coma followed shortly by death, usually from respiratory failure or uncontrolled seizures.

There have been only a handful of times in recorded history where a person survived rabies. But for just about everyone else who contracted the virus before we had a vaccine, it resulted in a certain and grisly death. With its telltale symptoms and near–100 percent mortality rate, rabies was one of the most easily recognizable and feared diseases of the ancient and medieval world.

◆ ◆ ◆

According to legend, St. Hubert (656–728 CE) cured a man bitten by a rabid dog, which led to him becoming known as a protector against rabies. After his death, people made pilgrimages to an abbey that housed his remains in the Ardennes Forest in Belgium hoping to acquire protection from the disease, a demonstration of just how scared people were of getting rabies.

A St. Hubert's Key, used to cauterize bites from rabid animals until the 1880s.

Like most fears, the worst wouldn't come to pass for most. However, the terrible symptoms of the disease and the lack of any kind of treatment frightened huge numbers of people in medieval times. It wasn't unusual for those who'd been bitten by a rabid dog to kill themselves, or to have a friend do it for them.

The monks at the Abbey of Saint-Hubert began issuing pieces of iron to pilgrims, which they dubbed St. Hubert's Keys. Shaped like a nail, cross, or cone, these keys were intended to be used for cauterization, but were sometimes hung on the walls of houses to ward off the disease (which didn't work quite as well).

For many centuries, cauterization was the only possibly effective treatment following a rabies bite. By burning the wound, you'd potentially disinfect it, giving yourself a chance at least to destroy the rabies virus (along with your own flesh). Indeed the practice may have saved the life of a young Emily Brontë, allowing her to eventually write the classic *Wuthering Heights* (heavy itself with rabies symbolism). As a child, Emily was bitten by a rabid dog on the street while trying to give it water. Unfazed, the brave girl calmly walked home, stuck an iron in the fire, and burned her own wound, saying nothing of the incident to anyone else in her family until months later when danger from infection had passed.

All manner of laws were put in place to contain rabies in the nineteenth century. With the rise of the middle class, dogs suddenly found themselves in a new position. The pampered household dog, formerly exclusive to the domain of the upper classes, became increasingly common. By the 1840s in Paris, there are believed to have been about 100,000 pet dogs. Concurrent with the rise in popularity of the pet dog, however, was an out-of-proportion fear of rabies outbreaks. Newspapers regularly carried florid stories of "mad dogs" at loose in the city.

In Victorian England, it was decreed that all dogs had to be muzzled in public. The penalty for not doing so was that your dog was shot. Other countries passed restrictions on dog teeth, mandating that they be filed down to prevent them from inflicting deep bites.

And while dogs were rightly dreaded as the most common carrier of rabies throughout history, perhaps the most feared attack came from the significantly less common but substantially more horrifying rabid wolf. Wolves acquire rabies less frequently than their domesticated counterparts, but more quickly enter the "furious" stage of the disease. When it happens, they tended to launch themselves on mindless rampages of random attacks, displaying none of their usual cautiousness and careful calculation in their quest for prey. When wolves were more common in Europe and America, and people lived in closer proximity to them, these attacks were greatly—and understandably—feared.

A first-person account from Captain Albert Barnitz of the 7th Cavalry, stationed at Fort Larned, Kansas, in 1868 reveals the horror of a typical attack: "Quite a serious affair occurred at the Post on the night of the 5th. Colonel [Wynkoop], the Indian Agent, was sitting on his porch, with his wife and children, and Mrs. Nolan and Tappan, I believe, and Lt. Thompson of the 3rd Infantry— and others, when a mad wolf—a very large grey wolf—entered the Post and bit one of the sentinels—ran into the hospital and bit a man lying in bed—passed another tent, and pulled another man out of bed, biting him severely—bit one man's finger nearly off—bit at one woman, I believe and some other persons in bed, but did not bite through the bed clothes—passed through the hall of Captain Nolan's house, and pounced on a large dog which he found there, and whipped him badly in half a minute—and then passed the porch of Col. [Wynkoop]—and springing in upon Lt. Thompson bit him quite severely in several places—he then passed on to where there was a sentinel guarding the haystacks and tried to bite the sentinel, but did not succeed—the sentinel firing and shooting him there on the spot!"

Perhaps the most eminent person in history to die from rabies was Charles Lennox, the 4th Duke of Richmond, and the Governor-General of colonial Canada in the early nineteenth century. Lennox was a great lover of dogs. When his pet dog, Blucher, got into a fight with a fox while he was inspecting a Québec garrison in 1819, Lennox tried to separate the quarreling beasts. In the course of the struggle, Lennox was bitten by the fox. The injury healed and Lennox resumed his tour of Canada. However, in August that year the telltale sign of hydrophobia

Lithographs of Louis Pasteur, who created the first rabies vaccine by using an attenuated form of the virus to achieve immunity

set in—the fox had been rabid. In prophetic words, Lennox remarked to one of his officers at dinner, "I don't know what it is, but I cannot relish my wine tonight as usual. I feel that if I were a dog, I should be a shot for a mad one!" Two days later, he was dead.

◆ ◆ ◆

In 1831, when Louis Pasteur was nine years old and growing up in the Jura Mountains of eastern France, he witnessed a rabid wolf rampage through his home village. The wolf mutilated and killed several people—and infected several more with rabies via bite wounds—before it was finally put down. For the rest of his life, Pasteur was deeply terrified of rabies, carrying with him always the vivid memories of that childhood incident.

By 1885, Pasteur was already a pioneer and a force in the medical science world. A founder of modern bacteriology, Pasteur championed the "germ theory" of disease (see page 38), replacing the theory of spontaneous generation that previously held sway. He'd also developed a process of removing microbes from milk, leading to the eponymously named practice of pasteurization. If that wasn't enough, Pasteur also developed vaccines for fowl cholera and anthrax. His canon of work, however, still didn't include a lifelong goal: to develop a treatment for rabies.

Meanwhile, the disease was on the rise in France. In 1878, there were more than 500 cases of rabid dogs reported to the French authorities, leading to a mass slaughtering of stray dogs. By some estimates approximately 4,000 were killed that year alone.

Working with his team in Paris, Pasteur theorized that if a person was exposed to a weakened version of a disease, it would stimulate the body to produce defense mechanisms specific to that infection (known as the attenuated vaccine theory). Experimenting with this theory on rabbits and dogs, Pasteur's breakthrough came when he harvested spinal cords from rabid rabbits, dried the tissue out, then injected a mixture of boullion and dried spinal cord into dogs. By passing through rabbits and drying out first, the virus had become weakened (attenuated), and by the time it was injected into dogs it was weak enough that their bodies could fight off the disease. What's more, they became completely immune if exposed again to it. A rabies vaccine had been developed for the first time ever and a major

A sculpture of Jean-Baptiste Jupille, the first person to be inoculated with a rabies vaccine, wrestling with a rabid dog milestone had been reached, despite the fact that Pasteur's detailed theory of why the vaccine worked wasn't technically correct (see Vaccines, page 239). But Pasteur had yet to undertake a human trial of the vaccine when, in July 1885, an opportunity presented itself.

That month, Joseph Meister, a nine-year-old boy in Alsace, France, was bitten fourteen times by a rabid dog, which meant he would most certainly die. The boy's doctor cauterized his terrible wounds with carbolic acid, but there wasn't anything else he could do. In desperation, the boy's mother brought him to Paris to find Pasteur, though she didn't know his name and had only heard there was a microbiologist in the city who was working on a cure for rabies. She nevertheless found Pasteur, who needed to act fast.

After consulting with two leading Parisian physicians, Pasteur took a chance and administered the vaccine to the boy, the first human subject to receive it. Over the course of the next three weeks, Meister received thirteen vaccinations, each from increasingly "fresher" dried spinal cords, with more virulent virus. After each progressively more virulent injection, Pasteur and the Meister family tensely awaited the results. The last vaccination came from the most virulent sample, and still nothing happened. Amazingly, Meister didn't catch rabies. Enormously relieved and satisfied with the results, Pasteur announced that Meister had developed an immunity thanks to his vaccine, making him the first person in recorded history to become immune to rabies.

In October that same year, the courageous shepherd boy Jupille was also brought to Pasteur for treatment. Again the vaccine worked, saving the young boy's life. Impressed with the shepherd's courage, Pasteur went on to arrange for the Institut de France to give Jupille a 1,000 franc prize for heroism, allowing him to return home much better off both physically and financially. Jupille went on to become something of a minor celebrity in France as a brave survivor of rabies. A bronze statue of the boy vividly fighting off a dog still stands outside the Institut Pasteur in Paris.

Pasteur's vaccine was mass-produced and distributed as quickly as possible. By the end of 1886, two thousand people in Europe infected with rabies were treated with it. Today, rabies is extremely uncommon in developed countries and is on the path toward complete eradication. If given within the first ten days of exposure to rabies, the vaccine is 100 percent effective—an impressive turnaround for the world's most fatal virus, which, until very recently in human history, was also 100 percent effective at killing us.

The Milwaukee Protocol

A British public health advertisement against importing unlicensed pets, 1994

If left untreated, a rabies infection is always fatal. Or is it? A recent medical mystery has scientists scratching their heads.

In 2004, Jeanna Giese, then fifteen years old, picked up a bat outside her church in Fond du Lac, Wisconsin. The bat promptly bit her before flying off. Giese and her family didn't think much of it. The bite wound was very small and there wasn't much blood. They thought if the bat was healthy enough to fly, surely it didn't have rabies. As a result, they didn't take Jeanna to see a doctor.

The next month, however, Jeanna was admitted to the hospital, suffering from fluctuating consciousness, slurred speech, and other symptoms that indicated a full-blown rabies infection. If Jeanna had seen a doctor after the bite, she would have been prescribed a rabies vaccination as a matter of course. But now it was too late. The vaccine isn't effective once advanced symptoms develop. The verdict was as clear as it was tragic: Jeanna was going to die from rabies.

The doctors at the Children's Hospital of Wisconsin in Milwaukee, however, weren't ready to give up quite that easily. Led by Dr. Rodney Willoughby, an infectious disease specialist, the team developed what has since become known as the "Milwaukee Protocol" to treat Jeanna. As Willoughby put it, they decided to "shut the brain down and wait for the cavalry to come," by inducing Jeanna into a coma and giving her own immune system some much-needed time to build up antibodies against the virus. The doctors also injected her with antivirals. They had no idea if it would work, or, if it did work, whether or not Jeanna would wake up with permanent brain damage.

After a week in a coma, tests came back indicating that Jeanna's body was indeed fighting off the infection. Remarkably, she survived, recovered most of her cognitive functions within a few months, and went on to study biology at Marian University in Fond du Lac, Wisconsin.

But is the Milwaukee Protocol replicable? Since it was first used in 2004, the technique has been successfully applied twice more, once in 2008 and once again in 2011. However, it has also been attempted unsuccessfully multiple times as well. Post-exposure vaccination, rabies-immune globulin, antivirals, and supportive therapy, rather than inducing coma, are the prevailing treatments now. But, for the first time in thousands of years of recorded history, rabies is not a death sentence 100 percent of the time.

TUBERCULOSIS
The "All-Consuming" Disease

Cause: *Mycobacterium tuberculosis* (bacterium)

Symptoms: Fevers, shortness of breath, fatigue, night sweats, cough, weight loss, lung cavitations (holes)

Impact: Tuberculosis, still considered a pandemic, is possibly the most prolific pathogenic killer in human history.

When: As long as humans have walked the earth

Who: The disease is airborne, so it can infect anyone in (unprotected) contact with a victim. TB infections can also lay dormant for years.

What Happened Next: There is a vaccine that provides some protection to children, but it is not widely used in many countries, including the US.

Little-Known Fact: Women are half as likely to become infected with TB as men, a possible survival adaptation of the disease.

I should like to die of a consumption . . . Because the ladies would all say, "Look at that poor Byron, how interesting he looks in dying."

—*Lord Byron*

I n the rural town of Marion, Alabama, things are usually quiet. The center of town is a tiny grid of three by five streets, beyond which homes and scattered buildings are tucked into masses of trees and lawns. The population is less than 3,600 people, small by just about any standard. People tend to know each other's business, like it or not.

So in 2014, when cases of tuberculosis (TB) spiked in Marion, word got around fast. The case numbers—showing up at a rate 100 times higher than anywhere else in the US—shocked health officials. These were rates higher than those seen in India, Haiti, and Kenya, where tuberculosis is considered poorly

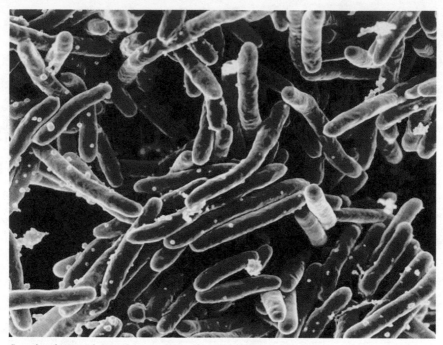

Scanning electron micrograph image of Mycobacterium tuberculosis

controlled. And while twenty people diagnosed with tuberculosis might not sound like a big number, in Marion it indicated a serious outbreak.

Tuberculosis is spread through respiratory droplets from coughs. Those most at risk tend to be in lower-income groups and without much access to healthcare, which describes many who live in Marion. But it was also hard to get a handle on how far TB had spread in the community, as not everyone has symptoms when they first catch it, and folks in Marion weren't eager to get tested or reveal their contacts for tracing purposes.

It didn't help that Marion is but a morning's drive from the Tuskegee Institute, the site of one of the worst ethical transgressions in medical history (see Tuskegee sidebar, page 200). So when medical authorities tried to coax contacts and other information from people in Marion, along with offers of testing and treatment, they were met with a wall of distrust. But without help, TB was poised to march right through that small Alabama town, spreading along a fault line of fear and distrust, infecting and killing at will.

Tuberculosis is an ancient bacterium, one that's survived about as long as we have. The disease might sound so familiar because it's deeply woven itself into our history and culture. As such, you're probably familiar with some of the symptoms, if only through what you've read in history books or seen in movies.

People with active TB infection can suffer from recurrent fevers, drenching night sweats, shortness of breath, and unflagging fatigue. Sometimes, victims have a productive cough with phlegm, and some cough up flecks of blood. Patients with TB will also lose weight no matter how hard they try to keep it on. They tend to get weak and their skin color changes—initially flushed from fever and later pale from anemia and chronic illness. In darker-skinned individuals with anemia, mucous membranes can turn pale, instead of a healthy pink. In 90 percent of people who become infected, TB can become latent, going inactive in the body for years only to reemerge later with symptoms that can kill.

Without treatment, active TB disease can wreak havoc on a body, causing cavitations, or holes, inside the lungs. The holes happen because knots of infected lung tissue die in their centers, described as "caseous necrosis" because of the cheesy substance of dead cells found in them. TB can further invade the spine and cause Pott's disease, a classic finding of untreated TB with arthritis and spinal malformations. Scrofula is another complication, where TB bacteria invade lymph nodes in your neck, causing swelling to the point of lesions eroding and draining through the skin.

Once these active infections take hold, three in twenty can die around the world, according to 2019 statistics by the WHO. Sometimes it's due to respiratory failure and sometimes from coughing up massive amounts of blood, typically after the TB lesions in the lungs erode into a major blood vessel. Other diseases, like the opportunistic infections associated with HIV and heart disease, do the final deed. When TB causes a victim to waste away, it's as if the disease consumes its victims into terminal debility, allowing other diseases to swoop in like vultures.

Hence, TB's other name: consumption.

◆ ◆ ◆

Given its age, the disease has achieved a certain cultural status. As long as we've been able to write, tuberculosis has had roles in literature and stories of old. In the movies, often a beautiful, tragic figure coughs up blood, speckling a handkerchief with a telltale crimson spray at a pivotal moment in the story. The afflicted waste away slowly until they are able to say their last goodbyes before dying in their sleep. The portrayals often seem nearly romantic—as romantic as death can get, anyway—but they're also fairly accurate.

A composite photograph of a young woman dying of tuberculosis, 1858

Through the years, TB has been known by a number of names, including the "romantic disease" and the "White Plague," named after the pale countenance of the afflicted or possibly the white lesions on the lungs seen in post-mortems. It was also called "phthisis," of Greek origin, meaning to decay or waste away, and the "king's evil," from a time when the literal "king's touch" was believed to heal the disease. Another nickname was "robber of youth," with its symptom "the graveyard cough." In classical Chinese, it was called *lào*, for the disease, and also described as *haifu*, translated as "destroyed palace." In the Song Dynasty, it was chillingly called *shīzhài*, or "disease which changes a living being into a corpse."

Around 1860, the term tuberculosis emerged from the Latin word *tuber*, as in root vegetables. *Tubercle* would describe a diminutive version of that, in reference to the small masses found on the lungs in TB sufferers. In the nineteenth century, it was also referred to by a rather dramatic and yet quite accurate moniker: "Captain of these Men of Death."

One might have gout, bowel problems, and dropsy (leg swelling and shortness of breath, often due to heart failure or severe liver disease), but it was TB that could wrest control of your ship and make it sink once and for all.

Because tuberculosis is an old disease, you might think modern technology, vaccines, and medicines (if they were universally accessible), would have done away with it by now. But that's far from the case. Tuberculosis is a simmering pandemic not close to ending—and it might, in fact, be the biggest pandemic we've ever known.

To wit: Tuberculosis has killed more than a billion people in the last two hundred years. Between the seventeenth and nineteenth centuries, it killed 20 percent of the human population. It currently infects about 2 billion people worldwide—fully a third of the world's population. It newly infects about 10.4 million people per year and is continuing to kill 1.5 million people annually.

Tuberculosis is seen throughout the ancient world in images, literature, DNA remnants, and human remains. Indian and Chinese texts describe the disease as early as 3,300 and 2,300 years ago, respectively. The deformed spines indicative of Pott's disease can be seen on terracotta figurines in the Temple of the Seven Dolls in Maya ruins. Mummified remains in northern Chile from the first millennium BCE had samples taken from within typical cavities of the lungs, left behind after TB disease. The Chilean samples showed bacilli dyed bright red under modern staining techniques (termed "acid-fast bacilli," or AFB to professionals, it means the bacilli retain their red color after rinsing with an acid solvent, and it's a classic hallmark for TB and certain bacteria). Egyptian mummies and artwork also show evidence of Pott's disease, while TB DNA has been found in mummy tissue. The term *schachepheth*, meaning wasting disease in Hebrew, is further mentioned in the Bible in the books of Deuteronomy and Leviticus. Finally, in Classical Greece, Hippocrates noted that "phthisis makes its attacks chiefly between the age of eighteen and thirty-five."

In literature, tuberculosis is everywhere. Think of Ruby Gillis, the stunning, beautiful friend in *Anne of Green Gables*, dying of "galloping consumption" as she grew thinner and paler. Emily Brontë and her family were stricken with tuberculosis, while her book *Wuthering Heights* is peppered with victims and casualties of the disease. There's Doc Holliday, gunfighter and friend of Wyatt Earp, a skeletal figure in constant battle with it.

Emily Brontë was stricken with what was likely tuberculosis in 1848, the year of her death.

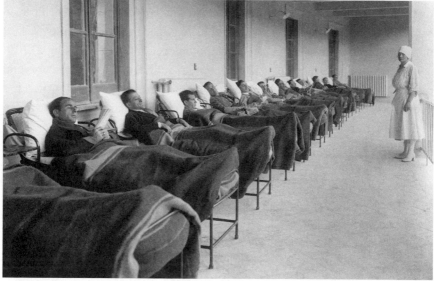

(Top) New England Sanitarium, Stoneham, Massachusetts, ca. 1930; (Bottom) An Italian sanitarium for TB patients, 1920

So many, fictional and real, have suffered or died from TB. Mimi in *La Bohème*. Fantine in *Les Misérables*. Frédéric Chopin. Vivien Leigh. Eleanor Roosevelt. George Orwell. President Andrew Jackson. Tina Turner. Nelson Mandela. Cat Stevens. Ho Chih Minh. Ringo Starr. Desmond Tutu.

♦ ♦ ♦

In addition to being perhaps the most prolific killer of all known pandemic diseases, TB may also be the oldest. Though sister diseases to tuberculosis, such as *M. bovis,* are found in cows and can spread to humans, it's more likely that TB originated with humans themselves, as opposed to arriving from a zoonotic spillover event (see page 23) years and years ago. Meaning TB lived within our species as we co-evolved with other primates, or it colonized us very early on in our evolution. We're talking about origins around three million years ago. Prior to that, certain *Mycobacterium* species may have existed during the breakup of the supercontinent Gondwanaland (more or less: Antarctica, Africa, Australia, and South America) in the early Jurassic period around 150 million years ago. So the TB family is *old.*

We know this because tuberculosis mutates somewhat slowly, which allows researchers to track how it has changed over time. The most recent common ancestor of the bacterium is thought to have originated around 70,000 BCE, which is also right when modern humans (*H. sapiens*) began significantly expanding and migrating out of Africa. This meant a larger global population, including clusters of humans congregating around fires for food, warmth, and protection—the perfect configuration for people to spread disease to one another while also inhaling soot, weakening both respiratory and immune functions.

Mycobacterium tuberculosis is one of those special bacterium-fueled infections called an obligate human pathogen—meaning, it can't live without us. There is no known animal reservoir where TB can hide for a time before jumping back to kill us. Fascinatingly, that translates as TB being forced to evolve with us. Models show that TB's lineage split around 70,000 years ago (when that large migration out of Africa took place), and then split again around 46,000 years ago (when humans continued spreading throughout Eurasia). The disease probably became more dangerous around that time, but luckily for us our population grew sufficiently large enough to prevent it from killing us all off.

The ironic balance about infectious diseases is that if a pathogen is too effective, it kills its host, and thus loses its ability to infect others and survive. If that were the case, TB wouldn't still be around today. But somewhere along the way, perhaps 20,000 to 30,000 years ago, TB evolved, gaining the ability to become latent, or dormant. This way, it could later emerge to successfully attack its victims, which were then older, weaker, sicker hosts. Some evidence further shows that women are less susceptible to TB, another insurance policy against killing off a whole species (and thus themselves).

Vintage TB public health posters

◆ ◆ ◆

It wasn't until the 1800s that scientists and doctors began to understand what tuberculosis was and why it made people sick. Descriptions in historical postmortems were helpful to understanding how the disease manifested. In 1843, when German physician Philipp Friedrich Hermann Klencke was able to infect rabbits with human and bovine TB, the disease was proved to be infectious, as opposed to something people were born with, as was believed by some at the time. By 1867, German-Swiss scientist Edwin Klebs was able to see the mobile bacilli in egg white cultures, and then used them to infect lab animals. Around the same time, German physician and microbiologist Robert Koch was able to isolate the bacillus and stained it with the help of his colleague Paul Ehrlich (see page 349).

Koch also developed tuberculin, a glycerol extract containing proteins from the disease's bacilli. He hoped it would help cure the disease—it didn't—but tuberculin ended up being extremely important in diagnosing it. Charles Mantoux, a French physician, developed the Mantoux test, in which tiny amounts of tuberculin were injected into the skin. If the injection site grew red and swollen,

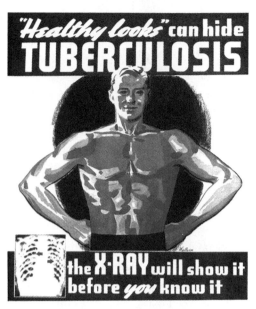

Christmas Seals Fight Tuberculosis

it was proof that you were infected with TB. The test, now known as the PPD (for purified protein derivative) skin test, is still used widely to identify TB infections, among other options like blood tests.

Later, Albert Calmette, a French physician, and Camille Guérin, a French veterinarian and bacteriologist, helped to culture a weakened version of TB—and that strain was used as a vaccine, called the BCG vaccine (after Bacillus Calmette-Guérin). When given to infants, the BCG vaccine can help protect them from getting pulmonary and other types of TB, though it's far less effective in older children and adults. So it wasn't the magic bullet against the disease. (As a fascinating aside, BCG is now also used as a treatment for bladder cancer.)

Because we still lack a universally effective vaccine, tuberculosis continues to quietly rage in countries throughout the world. Making matters worse, TB has mutated into multidrug-resistant forms that can cause devastating outbreaks that are hard to get under control. As a reportable disease in the US, TB outbreaks are aggressively managed by public health officials and workers. This means contact tracing, testing, and giving meds to people with infections. It also means isolating cases at home. And sometimes, these public health efforts involve a deeper invasion of privacy and control. DOT, or directly observed therapy, requires patients to come in daily to healthcare facilities to ensure they are actually swallowing their antibiotics.

In rare instances, if a patient is a threat to public health because of a high chance of spreading infection while also refusing to comply with medication requirements and isolation, states are legally allowed to issue emergency Department of Health "holds" on people. This means they are locked up in healthcare facilities to take their medications until they are no longer an infection risk. Because folks like this are not able for whatever reason to respect or consider the good of the greater public health, it's considered a necessary step.

TB remains a problem in countries around the world for a host of reasons. In the US, particular members of the population are more at risk: poor and low-income people, including those with inadequate access to medical care, as well as those with weakened immune systems related to factors like living with HIV, cancer, being a young children, or from taking immune-suppressing medications. Yet TB also finds ways into unexpected places (and people). In 2008, for example, several members of a Uruguayan basketball team contracted TB, with one of them dying from the disease. In a country with low rates of TB combined with it finding a way into wealthy, healthy athletes, it was a shock.

The same can be said when news of outbreaks in the US—like the one in Marion, Alabama—reaches the general public. It came as a surprise that a small, quiet town in rural Alabama was where TB gained a foothold and methodically spread. Although in the case of Marion it actually made sense: Healthcare was far from accessible to its residents; there were only two ambulances to service the entire town, and the nearest hospital, located some twenty minutes away, was overburdened. The fact that health authorities felt it necessary to offer financial incentives ($20 for TB testing, $20 for TB appointment follow-ups, $20 to get a chest X-ray, and $100 for completing treatment) underscored the profound economic disparities of the town.

In a world where people like to keep their lives private and avoid infections (who wants to go to a clinic where everyone in the waiting room might have TB, except maybe you?), the problems in Marion make more sense. After a lot of effort by healthcare workers and residents to ease tensions and frustration, the outbreak was finally brought under control. As of 2019, there were no new cases of TB reported in Marion.

If only we could say the same for the rest of the world.

Vampires or TB?

The term *consumption* once had much more dread attached to it than the clinical name, tuberculosis. After all, the term has a dark, macabre side to it: to be consumed, to be . . . someone else's meal. In the past, the deterioration of TB victims as they grew thinner and ashen brought on fears that something even more insidious was at work: vampires.

TB victims and so-called vampire victims, who were allegedly being fed upon by the undead on a regular basis, had a lot in common. They were wasting away without an obvious cause (as TB was not formally discovered until the mid-nineteenth century). Often, the afflicted had a close relative

Originally titled Love and Pain, *Edvard Munch's* Vampire, *was considered scandalous when it was first unveiled in 1894.*

die of the same thing, someone who could conceivably awaken from their graves to feed on their kin at night.

Occasionally, those dead were dug up and found to be curiously well-preserved. In Exeter, Rhode Island, at the turn of the nineteenth century, a man named Stukeley Tillinghast had fourteen children, six of whom died of consumption. When a seventh became ill, they exhumed the bodies of the children and found the first to die, Sarah, in "remarkable condition." Her heart was then cut out and burned to cinders. The seventh child died anyway.

About a century later, the belief was still prevalent when a nineteen-year-old woman named Mercy Lena Brown died of consumption, after which her younger brother fell ill. Surely, the family thought, he was being methodically fed upon by his deceased and vampiric sister. Mercy was exhumed and, though buried only two months, appeared to have liquid blood in her heart. The brother was given the ashes of her heart to consume as a talisman against further illness. He also died.

The vampire myth persisted for several reasons. First, victims buried in colder weather decompose slowly, giving the appearance of being "fresh" if exhumed before warm weather arrives. Normal decomposition can also result in the seeming appearance of nails and hair to grow longer, as the skin around them shrinks. Further, some TB victims do recover from acute illness and go on to be healthy for years afterward, making it seem as if they've had miraculous recoveries. Sadly, people would have to wait centuries longer before antibiotics could actually cure TB, as opposed to the misguided beheading and burning of corpses.

The Open-Air Cure

Imagine having tuberculosis: You're coughing, you're thin and weak, and your disease is eating away at you, pound by pound. You're told the only hope is the "open-air cure." Fresh air, rest, and good food is the best you can do—if you can afford it.

Since TB seemed less prevalent in mountainous areas, some European doctors in the 1800s thought that TB patients kept away from crowded villages and in open-air retreats called sanitariums, or sanitoriums, would cure them.

A 1940s diary by a sanitorium patient (written before the anti-tubercular medicine streptomycin was regularly used) provides a sense of what life was like in such a facility:

The rules are:

1. Absolute and utter rest of mind and body—no bath, no movement except to toilet once a day, no sitting up except propped by pillows and semi-reclining, no deep breath. Lead the life of a log, in fact. Don't try, therefore, to sew, knit, or write, except as occasional relief from reading and sleeping.

2. Eat nourishing food and have plenty of fresh air.

The Ringling Brothers and Barnum & Bailey Circus perform outside of Bellevue Hospital in New York, ca 1941.

It wasn't a technical cure for tuberculosis, but what it did was provide a means of better general care for the ill. In some cases, people with consumption pursued a similar approach, called "the prairie cure," as seen in Laura Ingalls Wilder's lightly fictionalized pioneer books. In other cases, such as in New York City, TB patients lined the balconies of New York City's Bellevue Hospital, breathing in fresh air—even if it meant lying in cots outside during the winter months. (Lucky for them, they were entertained by the Ringling Bros. and Barnum & Bailey Circus in the courtyard.)

None of these methods actually cured people, but they did manage to isolate sick individuals with a highly infectious disease away from family members and crowded, confined, indoor settings where the infection could be spread more easily. If they were lucky, they survived long enough for the TB to become latent, so they could resume a normal life again.

Meanwhile, the view was nice.

THE LAST SMALLPOX CASE
Patient Anti-ZERO
Janet Parker

Cause: *Variola major*/ *Variola minor* (virus)

Symptoms: Fever, vomiting, sores in the mouth, skin rash, blindness

Where: Birmingham, England

When: 1978

Transmission: Airborne

Little-Known Fact: Smallpox is one of only two eradicated diseases in the world; the other is rinderpest, a similar disease to smallpox that affected cattle.

A colored transmission electron micrograph of a smallpox (Variola major*) virus particle*

On Friday, August 11, 1978, Janet Parker, a forty-year-old medical photographer in Birmingham, England, woke up feeling sick. Her fatigue and fever worsened over the next couple of days until she abruptly developed a large number of red spots all over her body. Parker lived with her husband, Joseph, and two dogs in a modest home in Kings Norton, a suburb of Birmingham. She was an only child and remained close with her parents who lived nearby. After working as a crime scene photographer, Parker had recently taken a new position as a staff photographer at the Birmingham Medical School, where she primarily photographed tissue sections on slides.

British medical photographer Janet Parker, the last person to die from smallpox, Birmingham, England, 1978

On Monday, with no sign of improvement in her symptoms, a doctor was called for a home visit. The doctor diagnosed her symptoms as chickenpox, but Parker's mother, Hilda Witcomb, had doubts. Witcomb had nursed Parker through that disease as a child and, in her view, the marks on Parker's skin were different. They were larger for one thing, and they were blistering, unlike chickenpox.

The disease didn't improve on its own, so Parker was admitted to a Birmingham hospital nine days after she began showing symptoms. The first member of the hospital staff to examine her later recalled that Parker was doing "very poorly" and "had a very dramatic rash." But no one had yet considered that dreaded diagnosis: smallpox. Because smallpox was widely believed to be eradicated at the time.

Smallpox is a poxvirus, along with myxomatosis and cowpox, thought to have zoonotically transferred to humans from animals sometime around or before the third century BCE. Caused by the variola virus, the disease produces a high fever, body aches, vomiting, and its trademark outbreak of a progressive skin rash, where flat, red spots spread over the whole body, eventually transforming into large pustules. The disease is highly contagious, spreading through shed virus from the respiratory tract via sneezes and coughs, as well as lingering on surfaces. While most people recover from smallpox, it kills about 30 percent of its victims. Those that do survive often are left with permanent scarring.

The last naturally occurring case was recorded in 1977, the previous year, in Somalia. That single case led health authorities around the world to breathe a collective sigh of relief, believing that, after more than 3,000 years battling one of the world's worst infectious diseases, it was finally eradicated. In the nineteenth century alone, smallpox killed an estimated 300 million people, a shockingly high number unmatched by any other infectious disease. However, thanks to an extraordinary worldwide vaccination campaign spearheaded by the World Health Organization (WHO) beginning in 1959, no one in the entire world had been diagnosed with smallpox for an entire year.

But it wasn't long before doctors arrived at the seemingly impossible diagnosis, largely because there was no other possible explanation: Parker had smallpox. The disease was back. And it was particularly shocking that it appeared in Britain, a developed country that hadn't seen a case in more than five years.

The public health response was swift. Within a few hours, Parker was admitted to the Catherine-de-Barnes Isolation Hospital in nearby Solihull. By late that evening all of her close contacts, including her parents, were placed in quarantine. The next day, the diagnosis was confirmed by Dr. Henry Bedson, the head of the smallpox laboratory at the Birmingham Medical School.

And then all hell broke loose. The WHO and media organizations were alerted—both descended quickly and in numbers on the city of Birmingham, and with the same question: How could smallpox suddenly reappear? It was *the* great fear shared by medical practitioners around the world.

The entrance to the infectious disease ward at the Birmingham Hospital where Janet Parker was treated, 1978

But also: Parker's case simply didn't make medical sense. Smallpox was eradicated—and if that wasn't actually true, why did it strike again in Birmingham, England, of all places?

It wasn't long before the source of the infection—if not the method of transmission—was identified. The Birmingham Medical School, where Parker worked as a medical photographer, was one of less than twenty in the world approved by the WHO to have their own smallpox laboratory to study the highly dangerous and infectious disease. And Parker had recently been assigned to take photographs in the smallpox lab.

The next puzzle to sort out, however, was how she caught the disease there. What safety precautions had failed? And who else might've been exposed? The microbiology department was housed on the floor beneath Janet's office. Inside the microbiology department was the poxvirus laboratory, a small, 8-foot-square room with a sealed window and a freezer stocked with smallpox vials. Smallpox is typically transmitted via direct contact or shed virus in coughs and sneezes, but there have been rare instances of spread within a building via an airborne route. Could there have been a leak?

In the meantime, officials acted quickly to trace and contain the outbreak. Under the leadership of Dr. Surinder Bakhshi, medical officer for environmental health in Birmingham, an ambitious contact-tracing and quarantine effort was launched. Bakhshi and his team sought to trace and quarantine all of Parker's contacts, provide vaccine jabs to anyone who hadn't been recently vaccinated, give antibiotic injections to people in vulnerable health, and distribute an antiviral drug to everyone that had been exposed to the disease. More than 260 people who'd been in contact with Parker, including medical staff at the first hospital where she was admitted, were all quarantined. A further 500 people who may have been in contact with Parker were given vaccinations against the disease. Parker's home and car were even fumigated for good measure.

Parker's mother, Hilda Witcomb, came down with a "very minor" case of smallpox after nursing Parker through the early days of her illness. Thanks to the swift response of medical officials, Witcomb was the only other case in the Birmingham outbreak. She recovered fully after a week in the hospital.

Parker was not as lucky. As the days passed in isolation at the hospital in Solihull, the disease took hold in her. Before long, she was rendered nearly blind from the sores around her eyes and then went into kidney failure as well. She soon developed pneumonia and stopped responding verbally. By then, it was only a matter of time. When smallpox kills, it can do so by making the victim more vulnerable to other diseases and infections—such as pneumonia in Parker's case—as well as cause severe inflammation of the lungs and kidneys.

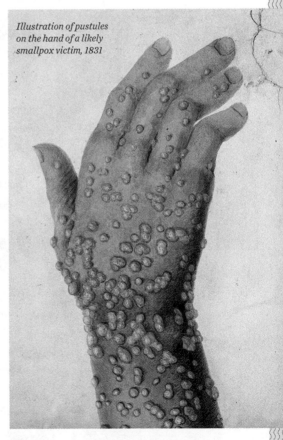

Illustration of pustules on the hand of a likely smallpox victim, 1831

On September 11, 1978, one month after she first showed symptoms, Parker died, becoming the world's last smallpox victim. Though she wouldn't be the last person to die as a result of the outbreak.

The WHO officially declared Birmingham free of the disease on October 16 that year, though other tragedies were in the offing. Parker's father passed away while in quarantine, a week before she did. His official cause of death was listed as cardiac arrest, the underlying cause believed to be stress over his daughter's plight.

The Future of Smallpox

After the tragic events of 1978, when it became clear that Birmingham had narrowly escaped a smallpox outbreak, the global scientific community began a debate that continues to this day: Since the disease has been eradicated, should laboratories retain samples of the virus? Before Janet Parker's death, just shy of twenty laboratories around the world were authorized to retain and study the smallpox virus. After Parker died, all known stocks of smallpox virus were destroyed or transferred to one of two WHO-approved facilities: the CDC in the United States and the State Research Center of Virology and Biotechnology VECTOR in Russia. In 1986, however, WHO recommended that all smallpox samples be destroyed, advocating that there was no longer any scientific reason to retain smallpox samples that overrode the extreme safety concerns inherent in stockpiling a dangerous virus. The destruction date was initially set for December 30, 1993; however, due to resistance from Russia and the United States, the date was pushed back to 1999, and then again to 2002. In the face of continued resistance, the World Health Assembly eventually agreed to permit both countries to temporarily retain smallpox vials for research purposes.

Kathleen Sebelius, former US Secretary of Health and Human Services

The argument in favor of keeping smallpox is that the virus may be useful down the road in the development of new vaccines or antiviral medicine. Also, the US Army was quick to point out that almost certainly there were unknown repositories of smallpox from countries that have never disclosed their holdings to the World Health Organization.

Kathleen Sebelius, US Secretary of Health and Human Services during the Obama administration, wrote an op-ed for the *New York Times* arguing that if a new outbreak of smallpox were to occur, having ready access to the virus would significantly speed up the ability to produce "new antiviral drugs" and "highly effective vaccines." She argued that destroying the last vials would be merely "a symbolic act." The debate continues to this day.

More terrifying, perhaps, is the possibility that smallpox might emerge again because of climate change. As the earth heats up, deep permafrost layers will gradually melt, potentially releasing ancient frozen viruses. Dark, cold, and oxygen-less, permafrost can act as a powerful preserver for microbes, bacteria, and viruses.

In 2016, an anthrax-infected reindeer corpse emerged from melting permafrost in Siberia and infected at least twenty people, killing one.

In 2016, a remote part of Siberia suffered an anthrax outbreak that killed one boy and hospitalized at least twenty people. The prevailing theory is that seventy-five years earlier, a reindeer died of anthrax and its body became trapped under a layer of permafrost. A summer heatwave hit in 2016, melting the permafrost and exposing the carcass, which then released anthrax spores into the surrounding environment. This, in turn, infected a nearby herd of reindeer grazing in the area, and then spread from there to humans.

Scientists are keeping a close eye on some known burial sites for smallpox victims in Siberia, where bodies were interred under the upper layers of permafrost. As those corpses have gradually been exposed again, scientists have tested tissue samples for evidence of surviving smallpox virus. So far, they have not found any but, worryingly, they have found fragments of smallpox DNA. So it is a real possibility that smallpox might return again.

Then there was Dr. Henry Bedson, the head of the smallpox laboratory at the Birmingham Medical School, where Parker caught the disease. Bedson was present the night Parker was first diagnosed with smallpox. Another doctor, Dr. Alasdair Geddes, who was with Bedson as they examined test samples from Parker, recalled the moment in a subsequent interview with the BBC.

"I said, 'Can you see anything Henry?', and he never answered," said Dr. Geddes. "So I gently moved his head aside so I could look down the microscope and there were brick-shaped particles that are characteristics of the smallpox virus. (Poxviruses such as smallpox have the unique advantage of being visible under a regular light microscope, rather than needing a scanning electron microscope to be seen.) He was horrified, because there was little doubt that in some way the smallpox virus had escaped from his laboratory and had infected Mrs. Parker."

The media quickly latched on to the narrative that unsafe conditions in Bedson's smallpox laboratory had allowed Parker to be infected with smallpox. Journalists camped outside his house, where he was quarantined inside with his wife and three children.

Bedson was increasingly distraught by the global scrutiny of his lab and felt personally responsible for Parker's death. He was an acknowledged global expert on smallpox, and yet the first case of the disease in five years in England came from his own lab.

The pressure and shame eventually overwhelmed him. On September 6, 1978, Dr. Bedson walked into the garden shed behind his home and slit his own throat. He died from the wound a few days later in a Birmingham hospital. His suicide note read, "I am sorry to have misplaced the trust which so many of my friends and colleagues have placed in me and my work."

Bedson's official biography in the Royal College of Physicians states, "Harassed as the chosen 'villain' of the tragedy, Henry Bedson's normally stable personality broke down and he took his own life. It could be said that he was a victim of his own dedicated conscientiousness, and of his extreme sense of responsibility."

Professor Henry Bedson, who headed the smallpox lab where Janet Parker was exposed, eventually took his own life over the ordeal.

After Parker died on September 11, no one else passed away from either smallpox or the events surrounding the lab breach in Birmingham. Extreme safety precautions had to be observed, however, in handling Parker's remains. The ward at the hospital where she died was literally sealed off for five years, with all the furniture and equipment left inside gathering dust.

Meanwhile, Parker's body was not allowed to be stored in a fridge for fear the virus could multiply. Ron Fleet, who was employed by the funeral home retained to handle her remains, later spoke of the gruesome results in a newspaper interview: "I was expecting to retrieve the body from a fridge in the mortuary, but . . . it was stored in a body bag that was kept on the floor of a garage away from the main hospital building. She was in a transparent body bag packed with wood shavings and sawdust. There was also some kind of liquid and I remember that I was frightened that the bag would split open. The body was covered in sores and scars—it was quite horrific. I was on my own and I needed help to lift the body . . . People from the hospital were very wary of helping me . . . When the day of the funeral arrived, the cars were given an escort by unmarked police vehicles just in case there was an accident . . . The body had to be cremated because there was a chance the virus could have thrived in the ground if Mrs. Parker had been buried. All other funerals were cancelled that day and the Robin Hood Crematorium was thoroughly cleaned afterwards."

◆ ◆ ◆

The question of how, exactly, Parker caught the disease remained a mystery. In December 1978, the Health and Safety Executive prosecuted the University of Birmingham for breach of safety legislation. At the time, blame centered on faulty ductwork in the poxvirus laboratory room. The leading theory of transmission was that Parker was infected while working in a telephone room serviced by ducts connected to the laboratory. But experts condemned this theory as highly unlikely. They demonstrated that it would take 20,000 years for a particle of smallpox to travel from the laboratory to the telephone room at the rate at which the smallpox-infected fluid was aspirated in the lab. A year later, three judges at Birmingham's Magistrates' Court dismissed the charges against the University of Birmingham.

The ruling should have exonerated Dr. Bedson.

But in 1980, the British government commissioned an official inquiry into Parker's death. Led by microbiologist R. A. Shooter, the eponymously named

WORLD HEALTH
THE MAGAZINE OF THE WORLD HEALTH ORGANIZATION   MAY 1980

smallpox is dead!

Shooter Report found there was "no doubt" that Parker had been infected at the smallpox laboratory. The inquiry found that the lab breached numerous protocols. Bedson had failed to inform British authorities about changes to his research that impacted lab safety. He had further misled the WHO about the volume of work he was conducting at the lab. He'd reported that his smallpox research was declining, when in fact the opposite was true—it had risen substantially as a result of Bedson wanting to finish his work in case his lab was closed. The WHO had also informed Bedson that the physical facilities of the lab did not meet WHO standards, though work continued there anyway. Worse still, several staff members who worked at the smallpox lab were found to be improperly trained to handle the pathogens.

Finally, the investigation also found out that while Parker had been vaccinated against smallpox, it had not been done recently enough to allow her to work in close proximity to the lab. The report concluded that the method of transfer must have occurred by one of three routes: on an air current, by personal contact, or by contact with contaminated equipment. Which of these was the real culprit, however, could not be answered by the investigation and remains unsolved to this day.

Thankfully, Parker was indeed the last person to die from smallpox. In 1980, two years after her death, smallpox was officially declared eradicated. It's difficult to overstate the importance of that moment. For more than 3,000 years, smallpox was one of the world's most feared infectious diseases, responsible for millions of deaths. It had almost single-handedly depopulated the entirety of the North American continent of its Indigenous population (see Indigenous Peoples & the Columbian Exchange, page 147). But inoculations against the disease that began in the eighteenth century, followed by a massive worldwide vaccination campaign orchestrated by the WHO in the twentieth century achieved the nearly impossible: complete eradication. To this day, smallpox remains the only major infectious disease that humans have successfully eradicated.

The Other Last Patient

While Parker was the last person to die from smallpox, she caught the disease in an artificial setting, related to her work as a medical photographer. Now that smallpox has officially been eradicated, such highly regulated laboratories that study the disease are the only potential sources of infection for anyone. Before its eradication, however, smallpox spread from person to person in an unbroken stretch heading back at least 3,000 years. So who was the last person to naturally catch smallpox? That "honor" goes to Ali Maow Maalin, a cook from Somalia, who caught the disease on October 22, 1977.

Maalin was exposed to the virus ten days earlier when he was driving a vehicle with two smallpox-infected children in it. In addition to working as a cook in the port town of Merca in south Somalia, Maalin was also sometimes employed by WHO as a vaccinator for, ironically, smallpox. Maalin, however, had managed to avoid being vaccinated himself, even though it was mandatory for all WHO workers. In a later interview he confessed, "I was scared of being vaccinated then. It looked like the shot hurt."

Ali Maow Maalin survived the world's last recorded case of naturally occurring smallpox.

When he fell ill with a fever, the first sign of smallpox, a little over a week after transporting the children, Maalin did not inform local officials because he did not want to be sent to an isolation camp. However, one of Maalin's friends reported his symptoms to the authorities in order to collect a reward offered by Somalian health officials for tips about smallpox cases. The country took smallpox eradication seriously, especially by 1977 when the goal to eradicate the disease was tantalizingly within reach.

After the WHO went to great lengths to track down any possible risk of spread by Maalin, it was determined that no one else caught the disease. Maalin entered the history books as the last person to naturally contract smallpox before it was officially declared eradicated on December 9, 1979. Thankfully, he recovered from smallpox and went on to become a passionate advocate for WHO vaccination campaigns for polio in Somalia. In a later interview, Maalin said, "Somalia was the last country with smallpox. I wanted to help ensure that we would not be the last place with polio too."

When Maalin died of malaria in 2013, he was hailed by WHO officials as having been one of the "true heroes" of their vaccination campaigns.

MEDICAL ADVANCES
From Pandemics to Progress

Impact: Each disease outbreak makes us more prepared for the next and often leads to groundbreaking medical discoveries.

When: Throughout known history

Who: Wu Lien-teh (layered masks), Emil von Behring (diphtheria antitoxin), Gerhard Domagk (Prontosil), Alexander Fleming (penicillin), Albert Schatz and Selman Waksman (streptomycin), James Lind (clinical trials)

What Happened Next: Less death and human suffering

Little-Known Fact: Bubble masks, like N95s, were the idea of a former women's magazine editor who was inspired by the shape of bra cups.

I n many ways, the history of humanity has been written in the ingenuity we've summoned to survive all kinds of horrors. The war between humans and infectious diseases is certainly among the oldest of these, one that has resulted in countless defeats and heartache. But it has also spawned incredible innovations and scientific discoveries that have saved hundreds of millions of lives. Proving once again that what separates us from other animals is our ability to adapt and overcome in the face of unimaginable adversity.

Some of the greatest medical advances we've made have grown directly from our desire to not repeat our experience of past pandemics. The most obvious examples of this are big discoveries like vaccines (see page 239) and germ theory (see page 38). We

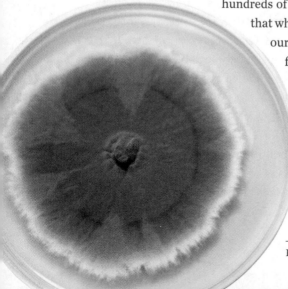

Penicillium *fungus growing in a petri dish*

Mask-wearing workers collect corpses in a painting of a plague outbreak in Marseille, France, 1720.

take for granted other advances that save innumerable lives, like face masks and hand sanitizer. That is, until we realized how precious they are when suddenly none are to be found at the start of a pandemic (see COVID-19, page 344).

Before we launch into what feels more like science fiction than real science, such as genomic sequencing, resurrecting historical pandemic viruses, and monoclonal antibody therapy, we should pay homage to the simple but crucial things that were born as a result of past pandemics. Like masks. Perhaps the best defense against airborne illnesses, facial coverings have a curious beginning—they were originally used not to block infection, but odors. In particular, the scent of rotting garbage or the stink from stagnant water and sewage thought to be the source of miasmas, or odiferous clouds of noxious gas once believed to cause disease, a concept that dates back to the fifth century BCE. During early epidemics, people didn't know that certain diseases like the plague, for example, were spread most often via flea-infested rats and not invisible disease-infested "clouds." (That said, a quite dangerous and rare form of pneumonic plague with a high death rate was, in fact, spread through the air.)

The early use of masks can be seen in an oil painting by Michel Serre from 1720 that depicts the removal of plague corpses in Marseille, France, in which workers with cloth masks carry away the dead to prevent themselves from catching the disease. Plague doctors once wore those surreal and creepy long-beak masks, the insides of which were actually stuffed with fine-smelling herbs (the wearer's nostrils were plugged with incense for good measure). Not only did the plague masks supposedly prevent miasmas from being inhaled, but the long beaks prevented doctors from getting too close to victims. A long stick or wand was employed to aid in examining patients to feel a pulse, lift garments, or even push the sick away if they strayed too close.

By the late 1800s, as germ theory made its way into medical consciousness, doctors began wearing masks during surgery to prevent cough or sneeze droplets from ending up in their patients. The masks in these cases were one-way protection, more for the patient than the wearer. Rather than blast the surgical field with antiseptic chemicals, more effort at the time was put toward keeping the area aseptic, or absent of bacteria. In 1897, Jan Mikulicz-Radecki, a Polish surgeon and inventor of surgical techniques and tools, created a mask that was "a piece of gauze tied by two strings to the cap, and swooping across the face so as to cover the nose and mouth and beard." By 1923, two-thirds of surgeons in the US and Europe used masks; by the 1940s, the practice was universal. In the 1960s, single-use surgical masks were found all over the world.

Tarbagan marmot

In 1910, a new kind of mask—technically a respirator—arrived, with the idea of protecting the wearer as well. In Manchuria, now northern China, a Tarbagan marmot (a sort of fancy groundhog) spread pneumonic plague to a hunter. Unlike typical plague transmitted by flea bites, pneumonic plague spread by respiratory droplets—meaning it spread quickly, and with a startling lethality of 100 percent. A doctor named Wu Lien-teh improved on the then-flimsy surgical masks, creating one with several layers of gauze and cloth. Wu's masks were easy to construct, not terribly expensive, and used by medical providers, soldiers, and citizens. They worked. Though this Manchurian pneumonic plague started by the marmot did kill 60,000 people, it would have run even more rampant without the protection of these new masks.

There were doubters, of course. One French doctor insulted Wu, belittling his masks with a racist comment of "What can we expect from a Chinaman?" The

French doctor then went maskless into a plague hospital to take care of the sick, promptly caught the plague, and died within two days. Death, in his case, was the ultimate rejoinder.

When the 1918 influenza pandemic arrived, public health officials had taken note of the relative success in Manchuria. Cloth or gauze masks were mandatory for some police forces and healthcare professionals. Even people in the general public were using them. In the open-air Camp Brooks Hospital in Brookline, Massachusetts, frequent mask changes, gowns and head coverings, and disinfectant hand washes were incredibly effective. Of the camp's 154-person medical staff, only eight contracted influenza, and five of those caught it while off duty outside the hospital.

In the 1950s, a woman named Sara Little Turnbull made a discovery that would change respirator masks forever. A former decor editor at *House Beautiful*, she was consulting with 3M on ways to produce gift-wrap ribbons out of melted

Seattle police officers wearing masks during the 1918 influenza pandemic

polymers. Turnbull thought the fibrous material made for great padding beneath jackets and blouses, as well as cups for molded bras.

By the late 1950s, Turnbull, who was mourning the death of family members after many hospital visits, struck on another idea. What if you stuck that molded polymer-fiber bra cup on your face? (At the height of the COVID-19 pandemic's first wave, people were photographed wearing rigged masks out of bra cups, likely unaware of the ingenious history behind the practice.)

The "bubble" surgical mask was created by the early 1960s and found to miraculously prevent the passage of pathogens through its material. Turnbull also helped with the later creation of the disposable N95 mask we know today, introduced by 3M in 1972.

Disposable surgical and respirator masks have been the norm for decades now. But the shortages during the COVID-19 pandemic and the desperate measures to reuse and sterilize these masks showed us that the throw-away culture in healthcare is problematic. Hospitals and other institutions frantically created UV sterilizers to reuse N95 masks during shortages, and the efficacy of washable cloth masks was hotly debated. Which fabrics? How many layers? Do we really need to wear them? The lessons learned could mean a future of non-disposable, sterilizable masks. It's not just a matter of smart environmental stewardship, but could mean the difference between life and death on the front lines of epidemics and pandemics.

◆ ◆ ◆

A respiratory barrier between the sick and the healthy wasn't the only simple, sensible way to prevent illness. Long before complex scientific breakthroughs came ideas even simpler than masks. For millennia, the separation of the sick from the healthy was the primary approach to preventing disease spread. Early examples of the practice can be found in the Bible and all those rules in Leviticus, including: "If the person who discharges the body fluid spits on someone who is clean, that person must wash his clothes and bathe in water; he will be unclean until evening." In the fourteenth century, Syrian physician Ibn Qayyim al-Jawziyyah recommended steering clear of the sick: "Flee from the one with leprosy as you flee from a lion."

In practice, the separation of the sick from the healthy happened in all kinds of ways depending on the era and the disease. People who had Hansen's disease were sometimes made to wear bells to warn others of their approach or used wooden clappers, the latter which also accompanied pleas for food or money. The healthy got in on the act, too. As we learned during the COVID-19 pandemic,

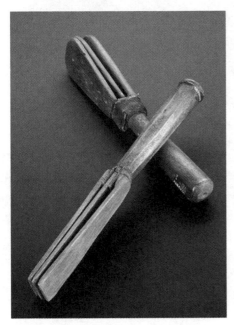

social isolation for safety's sake only makes you crave fresh air and freedom the longer you're stuck indoors. After being shut in for so long, adventurers during the Renaissance plague epidemic would leave their homes carrying poles eight to ten feet long called *batons de Roch* (St. Roch being the patron saint of dogs, bachelors, the falsely accused, and the plague). The poles were a handy way to keep people at a comfortable, plague-free distance. In the Victorian era, large bell-shaped skirts puffed out with hoops and crinolines likely helped with social distancing, though that wasn't the intended purpose. During the COVID-19 pandemic, X's marked where to stand safely in grocery store

A wooden leper clapper, seventeenth century

lines, high school graduations moved from indoor gyms to outdoor stadiums, Zoom meetings were ubiquitous, and drive-by birthday parties became the norm.

Quarantine and isolation practices (see Plague, page 53) were developed as a result of plague epidemics. The term "quarantine" originated in the fourteenth century, when the Venetian-controlled Mediterranean town of Ragusa (now Dubrovnik, Croatia) passed a radical law establishing a *trentino*, from the Italian word *trenta* for thirty, in which ships and people from plague-infested lands would be isolated for a month to prove they would not spread illness to Ragusa. Those thirty days would soon change to a *quarantino*, or forty days' separation.

In the fifteenth century, the Venetians set up isolation wards called "lazarettos" or "Nazarethums" (from *lazar*, Latin for a diseased person or leper, after the Biblical reference to Lazarus). Often the sick were separated from populated areas by natural barriers like rivers or the ocean or by setting up a remote colony, as was the case on the Hawaiian island of Moloka'i (see Hansen's disease, page 180). While crude, these were effective means of stopping outbreaks from

THE QUARANTINE QUESTION.

An illustration from an 1858 issue of Harper's Weekly, *the caption for which decries the arrival of disease via ports and those who brought them ashore from boats*

becoming epidemics. However, it wasn't until the 1918 influenza pandemic that science and public health practices often converged to make giant strides in slowing mass infection, resulting in several practices and treatments still used today.

One simple but remarkably effective approach for airborne illnesses is the understanding that better ventilation prevents rampant infectious spread (the "prairie cure" or "open-air" treatment for tuberculosis (see page 328) wasn't necessarily a cure, for example, but was certainly a step toward better hygiene and prevention of transmission). Another useful change after the 1918 influenza pandemic included hospital beds made of metal instead of wood so they could be more easily sanitized. Respiratory hygiene in the form of "no spitting" signs helped, but many of these actions were put in place too late and often with much resistance (in 1918, people apparently fought mask mandates, too; dissenters in Italy at the time called face masks "muzzles"). Anti-spitting campaigns and personal spittoons (they looked like little glass flasks) popped up internationally as well, to reduce the spread of tuberculosis. Even today, vigorous handwashing practices are repeatedly hammered into the heads of healthcare practitioners because diligence is still an issue. Thankfully, spitting in public is now considered bad manners, attested by the total lack of spittoons in the US, compared to a century ago. Good thing, too, as COVID-19 reminded us that keeping one's spit to oneself is not just good manners, but good for public health.

◆ ◆ ◆

Aside from putting barriers between us and plagues, the next primary approach to defeating them was to attack them directly, thanks to breakthroughs in science that created and discovered antibiotics and antitoxins. Some of these medicines aren't simply employed against microorganisms like bacteria, but act as antifungals, antivirals, and antiparasitics as well. Today, there are more than a hundred types of medicines in this group. The World Health Organization (WHO) maintains a list of medicines deemed essential for a country's healthcare system to best care for its citizenry, and a large chunk of those essential meds battle infectious diseases.

Some might assume that penicillin was the first definitive weapon discovered in our fight against pathogens, but there were several that preceded it and broke significant ground when they were discovered.

The Prussian-born Emil von Behring was a doctor and an assistant to the famed Robert Koch at the Institute for Hygiene in Berlin. In 1888, he developed a way to treat those suffering from diphtheria and tetanus. Not a disease familiar to many these days, diphtheria is prevented by a vaccine that is usually coupled with your routine tetanus shot. In the 1800s, diphtheria was a terrible killer that inflamed a victim's heart, inflicted paralysis, and caused a suffocating membrane to cover the throat. In Spain, the disease was so rampant in 1613 it was nicknamed *El Año de los Garrotillos*, or "The Year of Strangulations."

Much of the disease caused by diphtheria is driven by the toxin created by *Corynebacterium diphtheriae*. Von Behring infected rats, rabbits, and guinea pigs with weakened (attenuated) forms of it, then gathered their serum—the liquid fraction of their blood, minus the red and white blood cells. That light, honey-colored liquid, which contained antibodies to the diphtheria toxin, was then injected into another set of animals that were sickened with fully virulent diphtheria bacteria.

The newly infected animals given the serum didn't die because they gained a passive form of protection against the toxin with the donated serum. In 1891, a child's life was saved using this new method for the first time. The serum was produced in large quantities using animals like sheep and horses. At a time when 50,000 children died annually from diphtheria, it was a miraculous treatment.

Tetanus serum was created soon after, becoming a workable treatment by 1915. Today, antitoxins are used to treat botulism, diphtheria, and anthrax. The same principles of antitoxin treatment are utilized for antivenom therapy to remedy poisonous animal bites, including those from black widow spiders, scorpions, box jellyfish, and cobras. A treatment called passive antibody therapy, in which the serum of patients recovered from an infection is given to other sick patients (also called convalescent plasma therapy), may have been helpful during the COVID-19 pandemic, though data is still forthcoming. Antibodies against infections can not only treat diseases like toxic shock syndrome, but prevent infections during exposures, such as those for hepatitis A and B and botulism. But the antibodies themselves have been employed to treat more than just bites, stings, and infections. Intravenous immunoglobulins from pooled donors treat a variety of disorders, such as ITP (immune thrombocytopenia) and severe immune deficiency diseases.

Another antibody therapy—monoclonal antibodies—has been a game-changer in treatments over the last decade or so, the first one approved by the FDA in 1986. These specially designed antibodies are used to treat several types of cancers (melanoma, breast, and stomach, among many others) and autoimmune diseases (including Crohn's disease, rheumatoid arthritis, and psoriasis).

The antibodies themselves are Y-shaped proteins that bind to a specific protein. In doing so, they can elicit a whole range of effects: switching on or off immune system cascades, destroying cells, blocking or engaging cell activities. The antibodies only bind to a single antigen, hence "mono," and are produced by clones of cells that churn out the antibodies in large amounts. Sometimes they can also be bound to radioactive particles, delivering radioactivity directly to a cancer cell. Others can be bound to a chemotherapy agent. Often, they work alone.

In the realm of cancer therapy, most of us have some understanding of chemotherapy. But the origin of the term chemotherapy itself actually came from

the fight to treat infections, not cancer. At the turn of the twentieth century, antibiotics had yet to establish themselves as a cure for infections. That changed with a physician and scientist named Paul Ehrlich. He was born in 1854 in East Prussia (now Poland) where his father ran a lottery office. During his career, he took advantage of the burgeoning German dye industry to experiment on how cells looked stained with different chemicals. His love of color led to some notable idiosyncrasies, like carrying colored pencil stubs in his pockets. But Ehrlich's work led to what would become the famous Ziehl-Neelsen acid-fast stain for tuberculo-

German physician Paul Ehrlich (1854–1915), who discovered a cure for syphilis

sis. (Unfortunately, he also stained his very own TB bacteria from his sputum, though luckily survived the illness.) Later he collaborated with the aforementioned Emil von Behring, a Nobel Prize–winning physiologist, on serum therapy for tetanus and diphtheria.

But perhaps Ehrlich's most notable discovery happened by accident as he sought a chemical cure to treat a specific disease—a "chemotherapy." Specifically, he hoped to cure sleeping sickness, a disease caused by a microscopic parasite called *Trypanosoma brucei*. He had been working with a chemical called atoxyl (meaning "nontoxic"), ironically an arsenic compound. Ehrlich coined the term "magic bullet" related to his hope of finding that perfect chemical that would hopefully kill a very specific pathogen, the *Trypanosoma* parasite, and not the

Acetaminophen
(aka Tylenol)

For millennia, the only thing that could reduce a fever was an extract of willow tree bark containing a compound related to today's aspirin, or, alternately, an extract of the cinchona tree containing quinine. These febrifugal agents were quite useful, and in the case of the quinine, served to treat malaria as well. The cinchona tree was only available in South America, but later hybrids were cultivated elsewhere due to the high demand.

However, in the 1880s the cinchona tree grew scarce from over-harvesting. The British Empire needed it to quell the disease within their empire, which also cost millions of pounds annually from the detrimental economic effects. Scientists were forced to look to other compounds as fever and pain reducers. The result was acetaminophen, though it was largely ignored in favor of the two compounds from which it was created, acetanilide and phenacetin. The latter ended up as the world's fever reducer of choice, marketed by Bayer. It would go on to compete commercially with aspirin for decades.

It wasn't until the late 1940s that acetaminophen (also called paracetamol and APAP, for acetyl-para-aminophenol) was rediscovered and became a household pain reliever and antipyretic known around the world. But we can thank malaria and its treatment, quinine—or rather, the lack of quinine—for its ultimate discovery.

patient. He ended up testing nine hundred variations of the arsenic compounds on mice. None were particularly effective, but he revisited #606 because it seemed to have an effect on a newly discovered bacterium believed to cause syphilis. In 1910, the medicine called Salvarsan (sometimes simply called "606") was proven to be effective—it killed the syphilis spirochete and left the guinea pigs, rabbits, and mice alive.

In the next few decades, new research would be applied to battle not just the pandemics of old, but daily infections that could upend people's lives. A scratch or bite could kill if those *Staphylococcus* or *Streptococcus* infections spiraled out of control. A German scientist named Gerhard Domagk began working with a group of chemicals called azo dyes that had a characteristic double nitrogen bond. Azo dyes can color textiles, leather, and foods various shades of brilliant orange, red, and yellow. When an azo compound had a sulfonamide group attached (a nitrogen and sulfur link with two oxygen atoms double-bonded to the sulfur, should you need to impress friends at a party), they knew they'd found something special. The sulfonamide group inhibits a bacteria's ability to make folate, a necessary B vitamin. Humans, on the other hand, can obtain folate through their diet. And so another magic bullet was born. The new compound seemed to work in mice infected with *Streptococcus*, otherwise known as strep.

Domagk used the new medicine, called KL 730 and later patented as Prontosil, on his own daughter Hildegard. Suffering from a severe strep infection, she received a shot of Prontosil and recovered, though the drug left a telltale dyed, reddish discoloration at the injection site.

"Sulfa" drugs would go on to be used in a variety of medicines, including antibiotics (trimethoprim and sulfamethoxazole, aka Bactrim), diabetes medicines (glyburide, a sulfonylurea), diuretics (furosemide, or Lasix), pain meds (celecoxib, or Celebrex), and are also used today to treat pneumonia, skin and soft tissue infections, and urinary tract infections, among others.

Domagk's work won him the Nobel Prize in 1935. However, the Nazis, who disapproved of how the Nobel committee tried to help German pacifist Carl von Ossietzky, had their Gestapo arrest Domagk for accepting the prize and forced him to give it back. He was able to receive it later in 1947.

◆ ◆ ◆

By the early twentieth century, the tide was turning against microbes, with humans no longer at the complete mercy of these pathogens. One of the most famous discoveries in this battle was made in 1928 by Alexander Fleming, a Scottish physician-scientist who was also a renowned marksman in the London Scottish Regiment of the Territorial Army. After graduating medical school, he pursued research in order to stay in the rifle club at the medical school. As a captain in World War I, Fleming was frustrated by war wounds that would fester and kill despite the use of antiseptics.

After the war, he continued his research into battling infections. But while Fleming may have been an excellent marksman and army physician, and was seemingly meticulous in his habits, he wasn't, well, tidy. He had a tendency toward a cluttered work area, sometimes forgetting about experiments altogether. One day while working with *Staphylococcus* bacteria growing in petri dishes, he left one near an open window, inadvertently allowing the wind to bring in mold spores that colonized the culture along with the *Staph*. The same

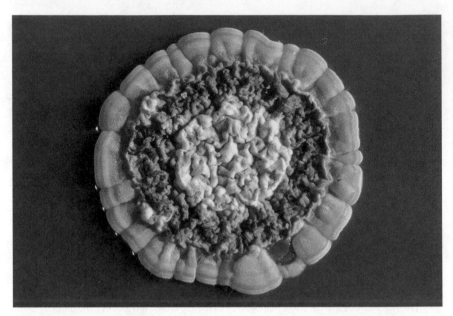

A colony of Streptomyces coelicolor *bacteria, a soil-based bacteria used to make various antibiotics*

thing would happen, for example, if you left an orange on a windowsill to rot. At some point, a powdery white mold would form on the rind. This also happened in Fleming's petri dishes. But strangely the *Staph* colonies near the mold growth were dying.

Fleming then isolated the mold and identified it as one of the *Penicillium* species. He thought it wasn't the mold itself but "mold juice" that was killing off the bacteria. He called the mold juice "penicillin," one of the most monumental discoveries in human existence. Penicillin and its related compounds, including amoxicillin, piperacillin, and methicillin, would prove effective at neutralizing pneumonia, gonorrhea, syphilis, *Listeria, Staph, Strep*, and *Clostridium* species.

Fleming would later famously say, "I did not invent penicillin. Nature did that. I only discovered it by accident." He would receive the Nobel Prize for his lifesaving work.

Another major advancement came in tandem with the discovery of a new antibiotic, streptomycin, leading to a sea change in the way clinical research is done. Although some people might be familiar with the idea of "randomized clinical trials," or RCTs, they likely don't realize how much they've benefited from them.

RCTs are the gold standard by which we test new treatments, typically by dividing patients randomly into two groups—a control group that gets a placebo, and a treatment group that gets a potential cure, vaccine, or treatment. It's a simple and brilliant approach that allows researchers to compare outcomes of the test group to the control group. Although it seems logical and necessary now, it was a thoroughly new concept in the 1940s.

The antibiotic streptomycin was extracted from a soil bacteria by microbiologists Albert Schatz and Selman Waksman at Rutgers University in New Jersey. Unlike penicillin, streptomycin appeared effective against tuberculosis bacteria. However, Schatz and Waksman had difficulty creating enough of the drug for testing when it occurred to them—why not compare streptomycin treatment with the usual treatment for tuberculosis, which was basically bed rest and not much else? Participants were randomized into two groups, one of which received the medicine.

The trial was far from perfect—there was no placebo used, nor was it blinded to patients or caretakers, meaning everyone knew who was in which arm of the trial. Blinding, or shielding people in the trial from knowing who's getting what treatment, is a great way of eliminating bias. Bias is a prejudice in favor for or against something, such as being the research investigator for a new drug that would really boost their career if it was effective, so that person may read a subject's equivocal results as . . . positive, after all. It also wasn't the first comparative clinical trial undertaken (in 1747, James Lind, a Scottish physician in the British Royal Navy, proved that citrus could cure scurvy in sailors, though Lind himself didn't believe the findings). Still, the streptomycin trial was the first published RCT and would set the stage for innumerable RCTs to come, establishing the scientific method as the gold standard of clinical trials.

◆ ◆ ◆

Masks. Distancing. Hygiene. Germ theory. Antibiotics. Vaccines.

All were leaps in the battle against infectious diseases and contagion. But one particular leap—some would say a rocket launch—was understanding how pathogens worked at the level of their basic, biologic war strategy: genetics.

Modern geneticists stand on the shoulders of those who came before to discover the presence of DNA and RNA. Back in 1871, Friedrich Miescher discovered the molecule nuclei (what we know is DNA) with particular phosphorus and nitrogen ratios. By 1910, Nobel laureate Albrecht Kossel discovered the five basic molecule units, or nucleotide bases, that make up DNA and RNA: adenine, cytosine, guanine, thymine, and uracil. DNA was isolated by Oswald Theodore Avery in 1944, and Alfred Hershey and Martha Chase determined that DNA carried our inheritable genetic information using experiments with *E. coli* and bacteriophage viruses. By 1953, Rosalind Franklin, James Watson, and Francis Crick had made the landmark discovery of the double helix structure of DNA.

Discoverers of DNA structure, James Watson (left) and Francis Crick, with their DNA model in 1953

Another huge step came in 1977, when Nobel laureate Frederick Sanger and his team figured out how to sequence DNA using Sanger sequencing. DNA is copied with the help of an enzyme, DNA polymerase, but with modified nucleotides that end the amplification. The result is different sized DNA pieces, which are then separated through a gel filter based on size. The gel can then be read to determine the sequence of nucleotides for a particular section of DNA. (Currently, next-generation sequencing took this labor-intensive method and made it possible to sequence an entire human genome in one day.)

By 1995, the entire sequence of an influenza virus had been sequenced, followed by a yeast in 1996. In 1998, one human chromosome, 22, was fully sequenced as a part of the Human Genome Project. In 2002, the full mouse genome was sequenced, along with the malaria parasite, *Plasmodium falciparum*. And finally, the human genome was fully sequenced in 2003, showing then that we have 20,000–25,000 genes. On average, our coding genes are 85 percent identical to those of mice.

We now have a library of cancer genomes, which enables doctors to choose targeted chemotherapy options for patients. Heritable disease genes are better understood to help with disease management and predict future illnesses. We are able to better clarify how human genetic variation can make us more susceptible or protect us against infection, such as in the case of tuberculosis or Hansen's disease. In hepatitis C infection, the genotype of the specific virus dictates which treatment we use. And we now have the ability to rapidly sequence old pathogens, such as the 1918 influenza virus (see page 237), as well as novel pathogens, like SARS-CoV-2 (see page 116), in order to help develop treatments and vaccines as rapidly as possible. Truly, the understanding of the innate, genetic uniqueness of multiple species in the global biome has revolutionized our world. Genomics

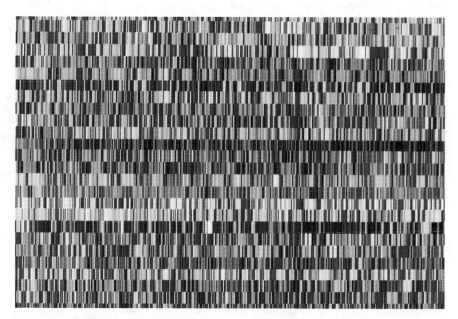

A genetic sequencing of the DNA of a grape

The Stethoscope

The stethoscope is both a tool and a stereotype associated with healthcare practitioners worldwide. There's an old adage that refers to them as "flea collars" when worn by internists, for various unfunny reasons (because internists are the last to leave a dead or dying person, like fleas, or that they tend to wear them around their necks like flea collars). Insider doctor jokes aside, the wearing of the stethoscope is a hallmark of the medical profession.

One might mistake their design for hygienic purposes. After all, distancing from sick patients, even by inches, makes a lot of sense. However, the stethoscope's origins are both charming and slightly scandalous.

René Théophile Hyacinthe Laënnec, a French physician in the nineteenth century, was watching some kids send sound signals to each other through long pieces of wood by scratching the wood with a pin.

That same year, Laënnec was hesitant about examining a young woman's heart via the usual method at the time, called immediate auscultation, which involved pressing his ear against her chest. Remembering the children playing with the wood and pins, he tried something different. He "tightly rolled a sheet of paper, one end of which I placed over the precordium (chest area over the heart) and my ear to the other. I was surprised and elated to be able to hear the beating of her heart with far greater clearness than I ever had with direct application of my ear."

A wooden tube was the first prototype. In the next century, other versions followed that included rubber tubing, the diaphragm, and the binaural form that we recognize today. Not a bad invention for a morally driven doctor who not only wished to auscultate hearts better, but wanted to be more respectful of his female patients!

An original Laënnec
stethoscope, ca. 1820

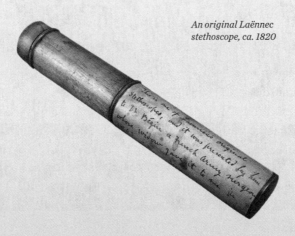

has brought us new ways to define and compare pathogens, among others. What kills or can be killed can often come down to the difference of a single pair of molecules embedded within a strand of DNA.

◆ ◆ ◆

The power dynamic between humans and their pathogenic pursuers has changed enormously in the last hundred years, and not just because of genomic advances. A little more than a century ago, the 1918 influenza pandemic opened the world's eyes to the conditions that could ignite a pandemic. Poor sanitation, the lack of good nutrition, and crowded conditions (the flu was often called "crowd disease") could spell disaster for an entire country. And the plight of the poor put everyone's lives at stake. Programs to improve the quality of living conditions for lower-income groups gained momentum. Several countries enacted free universal healthcare for their citizens. Organizations formed to help track diseases (every US state, as well as the CDC, has a veritable alphabet soup of reportable diseases). An international bureau for fighting epidemics started in Vienna, Austria, in 1919—the origins of the World Health Organization. By 1925, the US was creating a system of disease reporting, and later, national surveys to understand the health of its citizens. Epidemiology—investigations into the causes of disease, why they spread, and their effects on populations—became a mainstay of public health in both small towns and the entire world. Healthcare and pandemic prevention were no longer just a biological imperative; they became a social and moral one as well.

Today, we have preparedness plans for epidemics, and we remain vigilant to new zoonotic diseases ready to spill over and hurt us on a grand scale. We have stockpiles of vaccines and antivirals and are battling to reduce antibiotic resistance. But as COVID-19 has shown, even high-income countries with advanced healthcare, education, and rapid information-spreading public health networks can still succumb disastrously to novel diseases, often as a result of politicization (see page 126). COVID-19 is proof that we still have a long way to go. The battle of medicines and methods versus disease will forever be an unfinished chapter, and so this one will be, too.

SOURCES

For an unabridged list of sources, please visit: workman.com/patientzero

GENERAL

Andiman, Warren A. *Animal Viruses and Humans, A Narrow Divide: How Lethal Zoonotic Viruses Spill Over and Threaten Us.* Philadelphia: Paul Dry Books, 2018.

Centers for Disease Control and Prevention. Accessed February 11, 2021.

Dobson, Mary. *Disease: The Extraordinary Stories Behind History's Deadliest Killers.* London: Quercus, 2013.

Garrett, Laurie. *The Coming Plague: Newly Emerging Diseases in a World Out of Balance.* New York: Farrar, Straus and Giroux, 1994.

Gaynes, Robert P. *Germ Theory: Medical Pioneers in Infectious Diseases.* Washington, D.C.: ASM Press, 2011.

Hempel, Sandra. *The Atlas of Disease: Mapping Deadly Epidemics and Contagion from the Plague to the Zika Virus.* London: White Lion Publishing, 2018.

Kang, Lydia MD and Nate Pedersen. *Quackery: A Brief History of the Worst Ways to Cure Everything.* New York: Workman, 2017.

Khan, Ali. *The Next Pandemic: On the Frontlines Against Humankind's Gravest Dangers.* New York: Public Affairs, 2016.

Kohn, George C. *The Wordsworth Encyclopedia of Plague and Pestilence.* New York: Facts on File, 1995.

Loomis, Joshua S. *Epidemics: The Impact of Germs and Their Power over Humanity.* Nashville: Turner Publishing Company, 2018.

McNeill, William H. *Plagues and Peoples.* Garden City: Anchor Books, 1998.

Nelson, Kenrad E., and Carolyn Masters Williams. *Infectious Diseases Epidemiology, Theory and Practice.* 3rd ed. Burlington, MA: Jones and Bartlett Learning, 2013.

Oldstone, Michael B. A. *Viruses, Plagues, and History: Past, Present, and Future.* New York: Oxford University Press, 2010.

Quammen, David. *Spillover: Animal Infections and the Next Human Pandemic.* New York: W. W. Norton and Company, 2012.

Shah, Sonia. *Pandemic: Tracking Contagions, from Cholera to Coronaviruses and Beyond.* New York: Sarah Crichton Books, 2016.

Skwarecki, Beth. *Outbreak!: 50 Tales of Epidemics that Terrorized the World.* Avon: Adams Media, 2016.

Snowden, Frank M. *Epidemics and Society: From the Black Death to the Present.* New Haven: Yale University Press, 2020.

World Health Organization. "Disease Outbreaks." World Health Organization, May 21, 2020.

Wright, Jennifer. *Get Well Soon: History's Worst Plagues and the Heroes Who Fought Them.* New York: Henry Holt, 2017.

INFECTION: Patient ZERO: Ergotism

Bennett, J. W., and Ronald Bentley. "Pride and Prejudice: The Story of Ergot." *Perspectives in Biology and Medicine* 42, no. 3 (Spring 1999): 333–355.

Fuller, John Grant. *The Day of St. Anthony's Fire.* London: Hutchinson, 1969.

Gabbai, Lisbonne, and Pourquier. "Ergot Poisoning at Pont St. Esprit." *British Medical Journal* 2, no. 4732 (September 15, 1951). doi: 10.1136/bmj.2.4732.650.

Hagan, Ada. "From Poisoning to Pharmacy: A Tale of Two Ergots." *American Society for Microbiology.* November 2, 2018.

Hofmann, Albert. *LSD My Problem Child: Reflections on Sacred Drugs, Mysticism and Science.* 4th edition. Santa Cruz, California: Multidisciplinary Association for Psychedelic Studies, 2017.

Lee, M. R. "The History of Ergot of Rye (*Claviceps purpurea*) I: From Antiquity to 1900." *Journal of the Royal College of Physicians of Edinburgh* 39, no. 2 (June 2009): 179–184.

Nemes, C. N. "The Medical and Surgical Treatment of the Pilgrims of the Jacobean Roads in Medieval Times: Part 1. The Caminos and the Role of St. Anthony's Order in Curing Ergotism." *International Congress Series* 1242 (December 2002): 31–42.

New York Times. "3 Die, Many Stricken by Madness From Poison in Bread in France; Mayor Describes Attacks." August 29, 1951.

Nicholson, Paul. "The Highs and Lows of Ergot." *Microbiology Today* 41, no. 1 (February 2015): 14–17.

Zoonoses: Making the Leap

Burgdorfer, Willy et al. "Lyme disease-a tick-borne spirochetosis?" *Science* 216, no. 4552 (June 1982). doi.org/10.1126/science.7043737.

Headrick, Daniel R. "Sleeping sickness epidemics and colonial responses in East and Central Africa, 1900–1940." *PLoS Neglected Tropical Diseases* 8, no. 4 (2014): e2772. doi.org/10.1371/journal.pntd.0002772.

Hoen, Anne Gatewood et al. "Phylogeography of *Borrelia burgdorferi* in the eastern United States reflects multiple independent Lyme disease emergence events." *Proceedings of the National Academy of Sciences of the United States of America* 106, no. 35 (September 2009). doi.org/10.1073/pnas.0903810106.

Kilpatrick, A. Marm, and Sarah E. Randolph. "Drivers, dynamics, and control of emerging vector-borne zoonotic diseases." *Lancet* 380, no. 9857 (December 2012). doi.org/10.1016/S0140-6736(12)61151-9.

Levi, Taal et al. "Deer, predators, and the emergence of Lyme disease." *Proceedings of the National Academy of Sciences of the United States of America* 109. Vol. 27 (July 2012). doi.org/10.1073/pnas.1204536109.

Luby, Stephen P., Emily S. Gurley, and M. Jahangir Hossain. "Transmission of human infection with Nipah Virus." *Clinical Infectious Diseases* 49, no. 11 (December 2009). doi.org/10.1086/647951.

Morse, Stephen S. et al. "Prediction and prevention of the next pandemic zoonosis." *Lancet* 380, no. 9857 (December 2012). doi.org/10.1016/S0140-6736(12)61684-5.

Moudy, Robin M. et al. "A newly emergent genotype of West Nile virus is transmitted earlier and more efficiently by Culex mosquitoes." *American Journal of Tropical Medicine and Hygiene.* Vol. 77, no. 2 (August 2007).

Parrish, Colin R. et al. "Cross-species virus transmission and the emergence of new epidemic diseases." *Microbiology and Molecular Biology Reviews* 72, no. 3 (September 2008). doi.org/10.1128/MMBR.00004-08.

Plowright, Raina K. et al. "Pathways to zoonotic spillover." *Nature Reviews Microbiology* 15 (August 2017).

University of North Carolina Health Care. "Likelihood of tick bite to cause red meat allergy could be higher than previously thought." *ScienceDaily.* February 24, 2019.

Walter, Katharine S. et al. "Genomic insights into the ancient spread of Lyme disease across North America." *Nature Ecology and Evolution* 1 (August 2017). doi.org/10.1038/s41559-017-0282-8.

Wu, Katherine J. "COVID-19 Reignites a Contentious Debate Over Bats and Disease." *Undark.* May 5, 2020.

Patient ZERO: Ebola

Baize, Sylvain, et al. "Emergence of Zaire Ebola Virus Disease in Guinea." *New England Journal of Medicine* 371 (October 9, 2014). doi: 10.1056/NEJMoa1404505.

Breman et al. "Discovery and Description of Ebola Zaire Virus in 1976 and Relevance to the West African Epidemic During 2013–2016." *The Journal of Infectious Diseases* 214, suppl 3 (October 15, 2016). doi: 10.1093/infdis/jiw207.

Brown, Rob. "The virus detective who discovered Ebola in 1976." *BBC World Service*. July 17, 2014.

Farmer, Paul. *Fevers, Feuds, and Diamonds: Ebola and the Ravages of History*. New York: Farrar, Straus and Giroux, 2020.

Osterholm, Michael, T. et al. "Transmission of Ebola Viruses: What we know and what we do not know." *mBio* 6, no. 2 (March-April 2015): e00137-15. doi: 10.1128/mBio.00137-15.

Quammen, David. *Ebola: The Natural and Human History*. New York: W. W. Norton and Company, 2014.

Rimoin, Anne W. et al. "Ebola Virus Neutralizing Antibodies Detectable in Survivors of the Yambuku, Zaire Outbreak 40 Years after Infection." *The Journal of Infectious Diseases* 217, no. 2 (January 15, 2018). doi.org/10.1093/infdis/jix584.

Stylianou, Nassos. "How World's Worst Ebola Outbreak Began with one Boy's Death." *BBC News* online. November 27, 2014.

Wolfe, Nathan D. et al. "Bushmeat Hunting, Deforestation, and Prediction of Zoonotic Disease." *Emerging Infectious Diseases* 11, no. 12 (December 2005). doi: 10.3201/eid1112.040789.

Yong, Ed. "40 Years Later, Some Survivors of the First Ebola Outbreak Are Still Immune." *Atlantic*. December 14, 2017.

Germ Theory: From Miasma to Microscopes

Aristotle. *The History of Animals*. Bk. 5. Translated by D'Arcy Wentworth Thompson. Oxford: Clarendon Press, 1910. The Internet Classics Archive.

Béchamp, Antoine. *The Blood and its Third Anatomical Element*. Philadelphia: Boericke & Tafel, 1911.

Guralnick, Rob, and David Polly. "Antony van Leeuwenhoek." *A History of Evolutionary Thought*. Berkeley: UC Museum of Paleontology, 1996.

Harvard Library, CURIOSity. "Florence Nightingale." *Contagion: Historical Views of Diseases and Epidemics*. 2020.

Horrox, Rosemary, ed and trans. *The Black Death*. Manchester: Manchester University Press, 1994.

Liu, Ts'ui-jung, and Shi-yung Liu. "Disease and Mortality in the History of Taiwan." In *Asian Population History,* edited by Ts'ui-jung Liu, James Lee, David Sven Reher, Osamu Saito, and Wang Feng, 248–269. Oxford: Oxford University Press, 2004.

McGough, Laura J. "Demons, Nature, or God? Witchcraft Accusations and the French Disease in Early Modern Venice." *Bulletin of the History of Medicine* 80, no. 2 (Summer 2006). doi: 10.1353/bhm.2006.0069.

Partington, J. R. "Joan Baptista van Helmont." *Annals of Science* 1 no. 4, (October 15, 1936). doi.org/10.1080/00033793600200291.

Pasteur, Louis. "On Spontaneous Generation." Address, Sorbonne Scientific Soirée, Paris, April 7, 1864.

Redi, Francesco. *Experiments on the Generation of Insects*. Chicago: The Open Court Publishing Company, 1909.

Snow, John. *On the Mode of Communication of Cholera*. London: Wilson and Ogilvy, 1849.

Varro, Marcus Terentius. *De Re Rustica*. Bk. 1. Translated by W. D. Hooper and H. B. Ash. London: Loeb Classical Library, 1934.

Vitruvius. *Vitruvius: The Ten Books on Architecture*. Translated by Morris Hicky Morgan. Cambridge, MA: Harvard University Press, 1914.

Winternitz, M. "Witchcraft in Ancient India." *Indian Antiquary* 28 (March 1899): 71-83.

Patient ZERO: The Plague

"Bubonic Plague in San Francisco." *Journal of the American Medical Association* 34, no. 20 (November 10, 1900). doi:10.1001/jama.1900.24610200021001g.

Cliff, Andrew, Matthew R. Smallman-Raynor, and Peta M. Stevens. "Controlling the geographical spread of infectious disease: plague in Italy, 1347-1851." *Acta medico-historica adriatica* 7, no. 1 (2009): 197–236.

Hawgood, Barbara J. "Waldemar Mordecai Haffkine, CIE (1860-1930): prophylactic vaccination against cholera and bubonic plague in British India." *Journal of Medical Biography* 15, no. 1 (February 15, 2007). doi: 10.1258/j.jmb.2007.05-59.

Inglesby, Thomas V. et al. "Plague as a Biological Weapon: Medical and Public Health Management." *Journal of the American Medical Association* 283, no. 17 (May 3, 2000). doi:10.1001/jama.283.17.2281.

Kalisch, Philip A. "The Black Death in Chinatown: Plague and Politics in San Francisco 1900-1904." *Arizona and the West* 14, no. 2 (Summer 1972): 113–136. Accessed January 31, 2020.

Kugeler, Kiersten J. et al. "Epidemiology of Human Plague in the United States, 1900–2012." *Emerging Infectious Diseases* 21, no. 1 (January 2015): 16-22. Accessed January 11, 2020. doi:10.3201/eid2101.140564.

Link, Vernon Bennett. *A History of Plague in the United States of America: Public Health Monograph No. 26*. Whitefish, MT: Literary Licensing, 2012.

Mackowiak, Philip, and Paul S. Sehdev. "The Origin of Quarantine." *Clinical Infectious Diseases* 35, no. 9 (November 1, 2002). doi.org/10.1086/344062.

McClain, Charles. "Of Medicine, Race, and American Law: The Bubonic Plague Outbreak of 1900." *Law and Social Inquiry* 13, no. 3 (Summer 1988). Accessed January 15, 2020.

Murimuth, Adam, and Robert of Avesbury. *Adae Murimuth Continuatio Chronicarum; Robertus de Avesbury de Gestis Mirabilibus Regis Edwardi Tertii*. Edited by Edward Maunde Thompson. Cambridge: Cambridge University Press, 2012.

"On the Plague in San Francisco." *Journal of the American Medical Association* 36, no. 15 (April 13, 1906). doi:10.1001/jama.1901.52470150038003.

"Plague Fake Part of Plot to Plunder." *The San Francisco Call* 87, no. 98 (March 8, 1900). Accessed Jan 20, 2020.

Phelan, James D. "The Case Against the Chinaman." *The Saturday Evening Post*. Vol 174. December 21, 1901. Accessed January 15, 2020.

Risse, Guenter B. *Plague, Fear and Politics in San Francisco's Chinatown*. Baltimore: Johns Hopkins University Press, 2012.

Toke, Leslie. "Flagellants." In *Catholic Encyclopedia*. Vol. 6. New York: Robert Appleton Company, 1909.

Wheelis, Mark. "Biological warfare at the 1346 siege of Caffa." *Emerging Infectious Diseases* 8, no. 9 (September 2002). doi:10.3201/eid0809.010536.

Autopsy: From Humoral Theory to Grave Robbing

Allen, Marshall. "Without Autopsies, Hospitals Bury Their Mistakes." ProPublica. December 15, 2011.

Augustine, and Marcus Dods. 1950. *The City of God*. New York: Modern Library.

Bay, Noel Si-Yang and Boon-Huat Bay. "Greek anatomist herophilus: the father of anatomy." *Anatomy and Cell Biology* 43, no. 4 (December 2010). doi: 10.5115/acb.2010.43.4.280.

Eknoyan, Garabed. "Michelangelo: Art, anatomy, and the kidney." *Kidney International* 57, no. 3 (March 2000). doi.org/10.1046/j.1523-1755.2000.00947.x.

Elsevier Health Sciences. "First reported autopsy of patient with MERS coronavirus infection provides critical insights." *ScienceDaily*. February 5, 2016.

Ghosh, Sanjib Kumar. "Giovanni Battista Morgagni (1682–1771): father of pathologic anatomy and pioneer of modern medicine." *Anatomical Science International* 92, no. 3 (June 2017). doi.org/10.1007/s12565-016-0373-7.

Henry, Robert S. *The Armed Forces Institute of Pathology Its First Century 1862–1962*. Washington, DC: Office of the Surgeon General, 1964.

Jacobs, Nathan, Michele Bossy, and Amish Patel. "The life and work of Antonio Maria Valsalva (1666–1723)—Popping ears and tingling tongues." *Journal of the Intensive Care Society* 19, no. 2 (May 2018). doi: 10.1177/1751143717731229.

Lucas, Sebastian. "Investigating Infectious Disease at Autopsy." *Diagnostic Histopathology* 24, no. 9 (September 2018). doi.org/10.1016/j.mpdhp.2018.07.003.

MacGregor, George. *The History of Burke and Hare, and of the Resurrectionist Times: a Fragment from the Criminal Annals of Scotland*. Glasgow: Thomas D. Morison, 1884.

Mitchell, Piers D. *Medicine in the Crusades; Warfare, Wounds, and the Medieval Surgeon*. Cambridge: Cambridge University Press, 2004.

Morgagni, Giambattista. *The seats and causes of diseases: investigated by anatomy, containing a great variety of dissections and accompanied with remarks*. Vols. 1 & 2. Boston: Wells and Lilly, 1824.

Park, Katharine. "The Criminal and the Saintly Body: Autopsy and Dissection in Renaissance Italy." *Renaissance Quarterly* 47, no. 1 (Spring 1994). doi.org/10.2307/2863109.

Ramos-e-Silva, Marcia. "Giovan Cosimo Bonomo (1663–1696): discoverer of the etiology of scabies." *International Journal of Dermatology* 37, no. 8 (August 1998). doi.org/10.1046/j.1365-4362.1998.00400_1.x.

Schwartz, David A., and Chester J. Herman. "The importance of the autopsy in emerging and reemerging infectious diseases." *Clinical Infectious Disease* 23, no. 2 (August 1996). doi: 10.1093/clinids/23.2.248.

Singh, Veena D., and Sarah L. Lathrop. "Role of the Medical Examiner in Zika Virus and Other Emerging Infections." *Archives of Pathology and Laboratory Medicine* 141, no. 1 (January 2017). doi: 10.5858/arpa.2016-0327-SA.

Spencer, Rowena. *Conjoined Twins: Developmental Malformation and Clinical Implications*. Baltimore: Johns Hopkins University Press, 2003.

Taubenberger, Jeffery K., Johan V. Hultin, and David M. Morens. "Discovery and characterization of the 1918 pandemic influenza virus in historical context." *Antiviral Therapy* 12, no. 4 pt. B (May 22, 2008): 581–591.

The University of Edinburgh Anatomical Museum. "William Burke." April 1, 2019.

Vesalius, Andreas. *On the Fabric of the Human Body*. Padua: School of Medicine, 1543.

Patient ZERO: Mad Cow Disease

AP Archive. "UK: Teenager Who Died From Human Equivalent of Mad Cow Disease." YouTube video. 1:45. July 21, 2015.

Arthur, Charles. "Agonising decline that led to first diagnosis of new illness." *Independent*. March 19, 1997.

Bougard, Daisy et al. "Detection of prions in the plasma of presymptomatic and symptomatic patients with variant Creutzfeldt-Jakob disease." *Science Translational Medicine* 8, no. 370 (December 21, 2016): 370ra182. doi:10.1126/scitranslmed.aag1257.

Dietz, Klaus et al. "Blood Transfusion and Spread of Variant Creutzfeldt-Jakob Disease." *Emerging Infectious Diseases* 13, no. 1 (January 2007): 89–96.

Huor, Alvina et al. "The emergence of classical BSE from atypical/Nor98 scrapie." *Proceedings of the National Academy of Sciences of the United States of America* 116, no. 52 (December 26, 2019). doi.org/10.1073/pnas.1915737116.

Liberski, P. P. and P. Brown. "Kuru: its ramifications after fifty years." *Experimental Gerontology* 44, no. 1–2, (January–February 2009). doi:10.1016/j.exger.2008.05.010.

Lindenbaum, Shirley. *Kuru Sorcery: Disease and Danger in the New Guinea Highlands.* London and New York: Routledge, 2013.

_____. "Understanding kuru: the contribution of anthropology and medicine." *Philosophical Transactions of the Royal Society of London. Series B, Biological Sciences* 363, no. 1510 (November 27, 2008). doi:10.1098/rstb.2008.0072.

Maheshwari, Atul et al. "Recent US Case of Variant Creutzfeldt-Jakob Disease—Global Implications." *Emerging Infectious Diseases* 21, no. 5 (May 2015). doi:10.3201/eid2105.142017.

Max, D. T. "Case Study: Fatal Familial Insomnia; Location: Venice, Italy; To Sleep No More." *New York Times Magazine.* May 6, 2001.

McAlister, Vivian. "Sacred Disease of Our Times: Failure of the Infectious Disease Model of Spongiform Encephalopathy." *Clinical and Investigative Medicine* 28, no. 3 (June 2005): 101-104.

NOVA. "The Brain Eater." Directed by Alan Ritsko. Written by Joseph McMaster. BBC TV/WGBH Educational Foundation, February 10, 1998.

Prusiner, Stanley. "Neurodegenerative Diseases and Prions." *New England Journal of Medicine* 344, no. 20: (May 17, 2001). doi:10.1056/NEJM200105173442006.

Anatomy of an Outbreak: Calling in the Public Health Cavalry

Brachman, Philip S., and Stephen B. Thacker. "Evolution of Epidemic Investigations and Field Epidemiology during the MMWR Era at CDC; 1961-2011". *Morbidity and Mortality Weekly Report Supplement* 60, no. 4 (October 7, 2001): 22–26.

Fraser, David W. et al. "Legionnaires' Disease—Description of an Epidemic of Pneumonia." *New England Journal of Medicine* 297 (December 1, 1977). doi: 10.1056/NEJM197712012972201.

Garrett, Laurie. *The Coming Plague: Newly Emerging Diseases in a World Out of Balance.* New York: Farrar, Straus and Giroux, 1994.

Glick, Thomas H. et al. "Pontiac fever. An epidemic of unknown etiology in a health department: I. Clinical and epidemiologic aspects." *American Journal of Epidemiology* 107, no. 2 (February 1978). doi: 10.1093/oxfordjournals.aje.a112517.

Mahoney, Francis J. et al. "Community wide outbreak of Legionnaires' disease associated with a grocery store mist machine." *The Journal of Infectious Diseases* 165, no. 4 (April 1992). doi.org/10.1093/infdis/165.4.xxxx.

Marr, J. S., and C. D. Malloy. "An Epidemiologic Analysis of the Ten Plagues of Egypt." *Caduceus* 12, no. 1 (Spring 1996): 7–24.

McDade, Joseph E. et al. "Legionnaires' disease: isolation of a bacterium and demonstration of its role in other respiratory disease." *New England Journal of Medicine* 297, no. 22 (December 1, 1977). doi: 10.1056/NEJM197712012972202.

Mortlock, Stephen. "The Ten Plagues of Egypt." *The Biomedical Scientist* (January 7, 2019): 18–22.

Sanford, Jay P. "Legionnaires' Disease—The First Thousand Days." *The New England Journal of Medicine* 300 (March 22, 1979). doi: 10.1056/NEJM197903223001205.

Winn, Washington C. Jr. "Legionnaires Disease: Historical Perspective." *Clinical Microbiology Reviews* 1, no. 1 (January 1988): 60–81.

Patient ZERO: Yellow Fever

Baker, Thomas H. "Yellowjack: The Yellow Fever Epidemic of 1878 in Memphis, Tennessee." *Bulletin of the History of Medicine* 42, no. 3 (May–June 1968): 241–264.

Crosby, Molly Caldwell. *The American Plague: The Untold Story of Yellow Fever, the Epidemic that Shaped Our History*. New York: Berkley Publishing, 2006.

Duffy, John. "Yellow Fever in the Continental United States During the Nineteenth Century." *Bulletin of the New York Academy of Medicine* 44, no. 6 (June 1968): 687–701.

Espinosa, Mariola. "The Question of Racial Immunity to Yellow Fever in History and Historiography." *Social Science History* 38, no. 3–4 (Fall/Winter 2014): 437–453.

Foster, Kenneth R., Mary F. Jenkins, and Anna Coxe Toogood. "The Philadelphia Yellow Fever Epidemic of 1793." *Scientific American* 279, no. 2 (August 1998): 88–93.

Humphreys, Margaret. *Yellow Fever and the South*. Baltimore: Johns Hopkins University Press, 1999.

Marr, John S. and John T. Cathey. "The 1802 Saint-Domingue Yellow Fever Epidemic and the Louisiana Purchase." *Journal of Public Health Management and Practice* 19, no. 1 (January–February 2013). doi: 10.1097/PHH.0b013e318252eea8.

Powell, J. M. *Bring Out Your Dead: The Great Plague of Yellow Fever in Philadelphia in 1793*. Philadelphia: University of Pennsylvania Press, 1993.

Sisters of St. Mary at Memphis. *The Sisters of St. Mary at Memphis: with the Acts and Sufferings of the Priests and Others Who Were There with Them During the Yellow Fever Season of 1878*. 1879. Yellow Fever Collection. Edward G. Miner Library, University of Rochester Medical Center, Rochester, New York.

COVID-19: The Origins of SARS-CoV-2 and the Politicization of Plagues

Ackerknecht, Erwin H. "Anticontagionism Between 1821 and 1867." *Bulletin of the History of Medicine,* 22 no. 5 (October 1948): 562–593.

Chen, Laurie. "Left at Home for Six Days: Disabled Chinese Boy Dies After Carer Dad and Brother are Quarantined for Coronavirus Checks." *South China Morning Post*. January 30, 2020.

Correspondence from Thomas Jefferson to Thomas Cooper, October 7, 1814, Series 1: General Correspondence, 1651–1827, microfilm reel 047, Library of Congress, Washington, DC.

Foulkes, Imogen. "Tedros Adhanom Ghebreyesus: The Ethiopian at the Heart of the Coronavirus Fight." *BBC News* online. May 7, 2020.

Gong, Jingqi. "'People'—the person who sends the whistle." Matters News. March 10, 2020.

Hartley, Sarah, and Barbara Ribeiro. "Why Brazil's Zika Virus Requires a Political Treatment." The Conversation. February 16, 2018.

Huang, Chaolin et al. "Clinical Features of Patients Infected with 2019 Novel Coronavirus in Wuhan, China." *Lancet* 395, no. 10223 (February 15, 2020). doi.org/10.1016/S0140-6736(20)30183-5.

Khan, Natasha. "New Virus Discovered by Chinese Scientists Investigating Pneumonia Outbreak." *The Wall Street Journal*. January 8, 2020.

Kapiriri, Lydia, and Alison Ross. "The Politics of Disease Epidemics: a Comparative Analysis of the SARS, Zika, and Ebola Outbreaks." *Global Social Welfare* 7 (September 3, 2020). doi.org/10.1007/s40609-018-0123-y.

Lancet COVID-19 Commissioners, Task Force Chairs, and Commission Secretariat. "Lancet COVID-19 Commission Statement on the occasion of the 75th session of the UN General Assembly." Lancet 396, no. 10257 (October 10, 2020). doi.org/10.1016/S0140-6736(20)31927-9.

Latinne, Alice et al. "Origin and Cross-Species Transmission of Bat Coronaviruses in China." Nature Communications 11, (August 25, 2020). doi.org/10.1038/s41467-020-176.

Milne-Price, Shauna, Kerri L. Miazgowicz, and Vincent J. Munster. "The Emergence of the Middle East Respiratory Syndrome Coronavirus." Pathogens and Disease 71, no. 2 (July 2014). doi: 10.1111/2049-632X.12166.

Press Trust of India. "WHO Expert Team Tasked to Probe Virus Origins to Visit China on January 14." Business Standard. Updated January 12, 2021.

Rush, Benjamin. The Selected Writing of Benjamin Rush. Edited by Dagobert Runes. New York City: Philosophical Library, 1947.

Saag, Michael S. "Misguided Use of Hydroxychloroquine for COVID-19." Journal of the American Medical Association 324, no. 21 (November 9, 2020). doi:10.1001/jama.2020.22389.

Silverman, Ed. "Poll: Most Americans believe the Covid-19 vaccine approval process is driven by politics, not science." Statnews. August 31, 2020.

Smith, Nicola. "'They Wanted to Take Us Sightseeing. I Stayed in the Hotel,' Says First Foreign Official to Enter Wuhan." Telegraph. May 6, 2020.

US Library of Congress. Congressional Research Service. COVID-19 and China: A Chronology of Events (December 2019–January 2020), by Susan V. Lawrence. R46354. 2020.

Van Beusekom, Mary. "Scientists: 'Exactly Zero' Evidence COVID-19 Came from a Lab." University of Minnesota Center for Infectious Disease Research and Policy. May 12, 2020.

World Health Organization. "WHO-convened Global Study of the Origins of SARS-CoV-2." November 5, 2020.

XinhuaNet. "China Detects Large Quantity of Novel Coronavirus at Wuhan Seafood Market." January 27, 2020.

Yao, Yuan, Ma Yujie, Zhou Jialu, and Hou Wenkun. "Xinhua Headlines: Chinese Doctor Recalls First Encounter with Mysterious Virus." XinhuaNet. April 16, 2020.

Yu, Gao, Peng Yanfeng, Yang Rui, Feng Yuding, Ma Danmeng, Flynn Murphy, Han Wei, and Timmy Shen. "In Depth: How Early Signs of a SARS-Like Virus Were Spotted, Spread, and Throttled." Caixin Global. February 29, 2020.

Zhao, Jie, Wei Cui, and Bao-ping Tian. "The Potential Intermediate Hosts for SARS-CoV-2." Frontiers in Microbiology 11 (September 2020). doi: 10.3389/fmicb.2020.580137.

Zhou, Peng et al. "A Pneumonia Outbreak Associated with a New Coronavirus of Probable Bat Origin." Nature 579 (March 2020). doi.org/10.1038/s41586-020-2012-7.

SPREAD: Patient ZERO: HIV

Auerbach, David M. et al. "Cluster of Cases of the Acquired Immune Deficiency Syndrome: Patients Linked by Sexual Contact." The American Journal of Medicine 76, no. 3 (March 1984). doi:10.1016/0002-9343(84)90668-5.

Centers for Disease Control and Prevention. "A Cluster of Kaposi's Sarcoma and Pneumocystis carinii Pneumonia among Homosexual Male Residents of Los Angeles and range Counties, California." Morbidity and Mortality Weekly Report 31, no. 23 (June 18, 1982): 305–307.

_____. "Follow-Up on Kaposi's Sarcoma and Pneumocystis Pneumonia." Morbidity and Mortality Weekly Report, 30, no. 33 (August 28, 1981): 409–420.

———. "Opportunistic Infections and Kaposi's Sarcoma among Haitians in the United States." *Morbidity and Mortality Weekly Report* 31, no. 26 (July 9, 1982): 353–354, 360–361.

———. "*Pneumocystis* Pneumonia-Los Angeles." *Morbidity and Mortality Weekly Report* 30, no. 21 (June 5, 1981): 1–3.

———. "Reports on AIDS." *Morbidity and Mortality Weekly Report*, 30–35 (June 1981–May 1986).

Cohen, Jon. "Early AIDS virus may have ridden Africa's rails." *Science Magazine* 346, no. 6205 (October 3, 2014). doi:10.1126/science.346.6205.2.

Deschamp, M. M. et al. "HIV Infection in Haiti: natural history and diseases progression." *AIDS* 14, no. 16 (November 10, 2000). doi:10.1097/00002030-200011100-00014.

Desvarieux, M., and J. W. Pape. "HIV and AIDS in Haiti: recent developments." *AIDS Care* 3, no. 3 (1991): 271–279. doi:10.1080/09540129108253073.

"The Development of HIV Vaccines." In *The History of Vaccines*. The Historical Medical Library of the College of Physicians of Philadelphia. Updated January 10, 2018.

Faria, Nuno R. et al. "The early spread and epidemic ignition of HIV-1 in human populations." *Science* 346, no. 6205 (October 3, 2014). doi:10.1126/science.1256739.

Farmer, Paul. *AIDS and Accusation: Haiti and the Geography of Blame*. Berkeley: University of California Press, 2006.

Fouron, Georges E. "Race, blood, disease and citizenship: the making of the Haitian-Americans and the Haitian immigrants into 'the others' during the 1980s–1990s AIDS crisis." *Identities: Global Studies in Culture and Power* 20, no. 6 (2013). doi.org/10.1080/1070289X.2013.828624.

Gaillard, E. M. et al. "Understanding the reasons for decline of HIV prevalence in Haiti." *Sexually Transmitted Infections* 82, suppl 1 (April 2006). doi:10.1136/sti.2005.018051.

Gilbert, M. Thomas P. et al. "The emergence of HIV/AIDS in the Americas and beyond." *Proceedings of the National Academy of Sciences of the United States of America* 104, no. 47 (November 20, 2007). doi:10.1073/pnas.0705329104.

Hughes, Sally Smith. *The AIDS Epidemic in San Francisco: The Medical Response, 1981–1984*. Vol. 1. Berkeley: University of California Regional Oral History Office, 1995.

Keele, Brandon F. et al. "Chimpanzee Reservoirs of Pandemic and Nonpandemic HIV-1." *Science Magazine* 313, no. 5786 (July 28, 2006). doi:10.1126/science.112653.

Kuiken, Carla et al. "Genetic analysis reveals epidemiologic patterns in the spread of human immunodeficiency virus." *American Journal of Epidemiology* 152, no. 9 (November 1, 2000). doi.org/10.1093/aje/152.9.814.

Lemey, Philippe et al. "The Molecular Population Genetics of HIV-1 Group O." *Genetics* 167 (July 2004). doi:10.1534/genetics.104.026666.

McKay, Richard A. "'Patient Zero': The Absence of a Patient's View of the Early North American AIDS Epidemic." *Bulletin of the History of Medicine* 88, no. 1 (Spring 2014). doi: 10.1353/bhm.2014.0005.

Pape, Jean W. et al. "Characteristics of the Acquired Immunodeficiency Syndrome (AIDS) in Haiti." *New England Journal of Medicine* 309 (October 20, 1983). doi:10.1056/NEJM198310203091603.

Pape, Jean William et al. "The epidemiology of AIDS in Haiti refutes the claims of Gilbert *et al.*" *Proceedings of the National Academy of Sciences of the United States of America* 105, no. 10 (March 11, 2008): e13. doi:10.1073/pnas.0711141105.

Park, Alice. "The Story Behind the First AIDS Drug." *Time*. March 19, 2017.

Pepin, Jacques. The Origin of AIDS. Cambridge: Cambridge University Press, 2011.

Rerks-Ngarm, Supachai et al. "Vaccination with ALVAC and AIDSVAX to Prevent HIV-1 Infection in Thailand." *New England Journal of Medicine* 361, no. 23 (December 3, 2009). doi:10.1056/NEJMoa0908492.

Schmid, Sonja. "The discovery of HIV-1." *Nature Research* (November 28, 2018): 4.

Sharp, Paul M., and Beatrice H. Hahn. "The evolution of HIV-1 and the origin of AIDS." *Philosophical Transactions of the Royal Society B* 365, no. 1552 (August 27, 2010). doi:10.1098/rstb.2010.0031.

Volkow, P. "The AIDS epidemic and commercial plasmapheresis." Medical Hypotheses 49, no. 6 (December 1997). doi:10.1016/s0306-9877(97)90073-6.

Worobey, Michael et al. "Direct evidence of extensive diversity of HIV-1 in Kinshasa by 1960." *Nature* 455, no. 7213 (October 2, 2008). doi:10.1038/nature07390.

———. "1970s and 'Patient 0' HIV-1 genomes illuminate early HIV/AIDS history in North America." *Nature* 539, no. 7627 (November 3, 2016). doi:10.1038/nature19827.

Indigenous Peoples & The Columbian Exchange: The "Exchange" Was Not Equal

Acuna-Soto, Rodolfo et al. "Drought, Epidemic Disease, and the Fall of Classic Period Cultures in Mesoamerica (AD 750–950). Hemorrhagic Fevers as a Cause of Massive Population Loss." *Medical Hypotheses* 65 (2005). doi:10.1016/j.mehy.2005.02.025.

Brooks, Francis J. "Revising the Conquest of Mexico: Smallpox, Sources, and Populations." *The Journal of Interdisciplinary History* 24, no. 1 (Summer 1993): 1–29.

Callaway, Ewen. "Collapse of Aztec Society Linked to Catastrophic Salmonella Outbreak." *Nature* 542 (February 23, 2017). doi:10.1038/nature.2017.21485.

Chardon, F. A. *Chardon's Journal at Fort Clark, 1834–1839*. Edited by Annie Heloise Abel. Lincoln: University of Nebraska Press, 1997.

Conroy, J. Oliver. "The Life and Death of John Chau, the Man who Tried to Convert his Killers." *The Guardian.* February 3, 2019.

Crosby, Alfred W. "Virgin Soil Epidemics as a Factor in the Aboriginal Depopulation in America." *William and Mary Quarterly* 33, no. 2 (April 1976): 289–299.

Crosby, Alfred W. Jr. *The Columbian Exchange: Biological and Cultural Consequences of 1492.* Santa Barbara: Praeger, 2003.

Denevan, William M. "The Pristine Myth: The Landscape of the Americas in 1492." *Annals of the Association of American Geographers* 82, no. 3 (September 1992): 369–385.

Diamond, Jared. *Guns, Germs, and Steel: The Fates of Human Societies.* New York: W. W. Norton and Company, 2017.

Dollar, Clyde D. "The High Plains Smallpox Epidemic of 1837–38." *Western Historical Quarterly* 8, no. 1 (January 1977): 15–38.

Fenn, Elizabeth Anne. *Pox Americana: The Great Smallpox Epidemic of 1775–82.* New York: Hill and Wang, 2002.

Heagerty, J. J. "The Story of Small-Pox Among the Indians of Canada." *The Public Health Journal* 17, no. 2 (February 1926): 51–61.

Jones, David S. "Virgin Soils Revisited." *William and Mary Quarterly* 60, no. 4 (October 2003): 703–742.

Mann, Charles C. *1491: New Revelations of the Americas Before Columbus.* New York: Vintage, 2006.

Ranlet, Philip. "The British, the Indians, and Smallpox: What Actually Happened at Fort Pitt in 1763?" *Pennsylvania History: A Journal of Mid-Atlantic Studies* 67, no. 3 (Summer 2000): 427–441.

Stearn, Esther Wagner, and Allen Edwin Stearn. *The Effect of Smallpox on the Destiny of the Amerindian.* Boston: B. Humphries, 1945.

Trigger, Bruce G. Natives and Newcomers: Canada's "Heroic Age" Reconsidered. Montreal: McGill-Queen's University Press, 1986.

US Congress. House. *Executive Documents.* Vol. 1. 13th Cong., 2nd sess., 49th Cong., 1st sess., 1814–1855.

368 PATIENT ZERO I SOURCES

Willey, Gordon R., and Demitri B. Shimkin. "The Collapse of Classic Maya Civilization in the Southern Lowlands: A Symposium Summary Statement." Southwestern Journal of Anthropology 27, no. 1 (Spring 1971): 1–18.

Patient ZERO: Typhus

Angelakis, Emmanouil, Yassina Bechah, and Didier Raoult. "The History of Epidemic Typhus." Microbiology Spectrum 4, no. 4 (August 2016). doi: 10.1128/microbiolspec.PoH-0010-2015.

Bourgogne, Adrien. Memoirs of Sergeant Bourgogne: 1812–1813. Independently Published, 2016.

Dworkin, Jonathan, and Siang Yong Tan. "Charles Nicolle (1866–1936): Bacteriologist and Conqueror of Typhus." Singapore Medical Journal 53, no. 11 (2012): 764.

Griffiths, Josie. "'I Dug My Own Sister's Grave.'" The Sun. January 27, 2017.

McKie, Robin. "Return of the Damned After 400 Years." Guardian. January 11, 2004.

Raoult, Didier et al. "Evidence for Louse-Transmitted Diseases in Soldiers of Napoleon's Grand Army in Vilnius." Journal of Infectious Diseases 193, no. 1 (January 1, 2006). doi: 10.1086/498534.

Remarque, Erich Maria. All Quiet on the Western Front. New York: Ballantine Books, 1987.

Stanley, Arthur Penrhyn. Historical Memorials of Canterbury. London: J. M. Dent and Company, 1906.

"Typhus Reemerges as Plague Suspect." Science 283 no. 5405 (February 19, 1999). doi: 10.1126/science.283.5405.1111c.

University of Maryland Medical Center. "Plague of Athens: Medical Mystery May Be Solved." Newswise. January 29, 1999.

Wood, Anthony à. Athenae Oxonienses: An Exact History of All the Writers and Bishops Who Have Had Their Education in the University of Oxford; To Which Are Added the Fasti, or Annals of the Said University. 4 vols. London: Printed for F. C. and J. Rivington; Lackington, Allen, and Company; T. Payne; White, Cochrane, and Company; Longman, Hurst, Rees, Orme, and Brown; Cadell and Davies; J. and A. Arch; J. Mawman; Black, Parry, and Company; R. H. Evans; J. Booth, London: and J. Parker, Oxford, 1813–1820.

Zinsser, Hans. Rats, Lice and History. Milton Park: Routledge, 2007.

Patient ZERO: Measles

Bech, Viggo. "Measles Epidemics in Greenland." American Journal of Diseases in Children 103, no. 3 (March 1962). doi:10.1001/archpedi.1962.02080020264013.

Corney, Bolton G. "The Behaviour of Certain Epidemic Diseases in Natives of Polynesia, with Especial Reference to the Fiji Islands." Transactions of the Epidemiological Society of London 2 (April 9, 1884): 76–95.

Furuse, Yuki, Akira Suzuki, and Hitoshi Oshitani. "Origin of Measles Virus: Divergence from Rinderpest Virus Between the 11th and 12th Centuries." Virology Journal 7, no. 52 (March 4, 2010). doi:10.1186/1743-422X-7-52.

Greenwood, Major. "Epidemics and Crowd-Diseases: Measles." Reviews of Infectious Diseases 10, no. 2 (March–April 1988): 492–499.

Hajar, Rachel. "The Air of History (Part IV): Great Muslim Physicians Al Rhazes." Heart Views 14, no. 2 (April–June 2013). doi:10.4103/1995-705X.115499.

Hirsch, August. Handbook of Geographical and Historical Pathology. Vol. 1. Translated by Charles Creighton. London: The New Sydenham Society, 1883.

Panum, Peter Ludwig. "Observations Made During the Epidemic of Measles on the Faroe Islands in the Year 1846." Translated by A.S. Hatcher. Unpublished manuscript, n.d., typescript. University of Toronto.

Quintero-Gil, Carolina et al. "Origin of Canine Distemper Virus: Consolidating Evidence to Understand Potential Zoonoses." *Frontiers in Microbiology* 10, no. 1982 (August 28, 2019). doi:10.3389/fmicb.2019.01982.

Shulman, Stanford T., Deborah L. Shulman, and Ronald H. Sims. "The Tragic 1824 Journey of the Hawaiian King and Queen to London: History of Measles in Hawaii." *Pediatric Infectious Disease Journal* 28, no. 8 (August 2009). doi:10.1097/INF.0b013e31819c9720.

Squire, William. "On Measles in Fiji." *Transactions of the Epidemiological Society of London* 4 (January 10, 1877): 72–74.

Patient ZERO: Hansen's Disease (Leprosy)

Brody, Saul Nathaniel. *The Disease of the Soul: Leprosy in Medieval Literature.* Ithaca: Cornell University Press, 1974.

Daws, Gavan. *Holy Man: Father Damien of Molokai.* Honolulu: University of Hawaii Press, 1984.

Fessler, Pam. *Carville's Cure: Leprosy, Stigma, and the Fight for Justice.* New York: Liveright Publishing, 2020.

Harris, Gardiner. "Armadillos Can Transmit Leprosy to Humans, Federal Researchers Confirm." *New York Times.* April 27, 2011.

Manchester, Keith. "Tuberculosis and Leprosy in Antiquity: An Interpretation." *Medical History* 28, no. 2 (April 1984). doi:10.1017/s0025727300035705.

Miller, Timothy S., and Rachel Smith-Savage. "Medieval Leprosy Reconsidered." *International Social Science Review* 81, no. 1–2 (2006): 16–28.

Rawcliffe, Carole. *Leprosy in Medieval England.* Woodbridge, England: Boydell Press, 2016.

Stevenson, Robert Louis. *Father Damien: An Open Letter to the Reverend Doctor Hyde of Honolulu.* London: Chatto and Windus, 1890.

Tayman, John. *The Colony: The Harrowing True Story of the Exiles of Molokai.* New York: Scribner, 2007.

Yzendoorn, Reginald. *History of the Catholic Mission in the Hawaiian Islands.* Honolulu: Honolulu Star-Bulletin, limited, 1927.

Patient ZERO: Syphilis

Díaz de Ysla, Ruy. *Tractado Contra el Mal Serpentino.* Seville, 1539.

Dinesen, Isak. *Letters from Africa, 1914–1931.* Chicago: University of Chicago Press, 1984.

Ferris, Kathleen. *James Joyce and the Burden of Disease.* Lexington: University Press of Kentucky, 2010.

Flaubert, Gustave. *The Letters of Gustave Flaubert: 1830–1857.* Edited and translated by Francis Steegmuller. Cambridge, MA: Belknap Press, 1980.

Frith, John. "Syphilis – Its early history and Treatment until Penicillin and the Debate on its Origins." *Journal of Military and Veterans' Health* 20, no. 4 (November 2012): 49–58.

Hayden, Deborah. *Pox: Genius, Madness, and the Mysteries Of Syphilis.* New York: Basic Books, 2003.

Hemarajata, Peera. "Revisiting the Great Imitator: The Origin and History of Syphilis." *American Society for Microbiology.* June 17, 2019.

Hutten, Ulrich von. *De morbo gallico: a treatise of the French disease, publish'd above 200 years past.* London: Printed for John Clarke at the Bible under the Royal Exchange, 1730.

Jones, James H. *Bad Blood: The Tuskegee Syphilis Experiment.* New York: Free Press, 1993.

Majander, Kerttu et al. "Ancient Bacterial Genomes Reveal a High Diversity of *Treponema Pallidum* Strains in Early Modern Europe." *Current Biology* 30, no. 19 (October 5, 2020): 3788–3803.

Quétel, Claude. *History of Syphilis.* Baltimore: Johns Hopkins University Press, 1990.

Schuenemann, Verena J. et al. "Historic *Treponema Pallidum* Genomes From Colonial Mexico Retrieved From Archaeological Remains." *PLOS Neglected Tropical Diseases* 12, no. 6 (June 21, 2018): e0006447. doi.org/10.1371/journal.pntd.0006447.

Tampa, M., I. Sarbu, C. Matei, V. Benea, and S.R. Georgescu. "Brief History of Syphilis." *Journal of Medicine and Life* 7, no. 1 (March 15, 2014): 4–10.

Tarlach, Gemma. "First Ancient Syphilis Genomes Reveal New History of The Disease." *Discover*. June 21, 2018.

Quackery: From Mercury and Bloodletting to Hydroxychloroquine

Baksi, Catherine. "Landmarks in Law: Louisa Carlill and the Fake Flu Cure." *Guardian*. June 25, 2020.

BBC News. "The President Who Made People Take His Bogus HIV Cure." January 22, 2018.

Brody, Saul Nathaniel. *The Disease of the Soul: Leprosy in Medieval Literature*. Ithaca: Cornell University Press, 1974.

Cawthorne, Nigel. *The Curious Cures of Old England: Eccentric Treatments, Outlandish Remedies and Fearsome Surgeries for Ailments from the Plague to the Pox*. London: Robinson, 2018.

Fahim, Kareem, and Mayy El Sheikh. "Disbelief After Egypt Announces Cures for AIDS and Hepatitis C." *New York Times*. February 26, 2018.

Hopkins, Jerry. *Strange Foods: An Epicuran's Guide to the Weird and Wonderful*. North Clarendon, Vermont: Tuttle Publishing, 1999.

Lantos, Paul M., Eugene D. Shapiro, Paul G. Auwaerter, Phillip J. Baker, John J. Halperin, Edward McSweegan, and Gary P. Wormser. "Unorthodox Alternative Therapies Marketed to Treat Lyme Disease." *Clinical Infectious Diseases* 60, no. 12 (June 15, 2015). doi:10.1093/cid/civ186.

Neuman, Scott. "Man Dies, Woman Hospitalized After Taking Form of Chloroquine To Prevent COVID-19." NPR. March 24, 2020.

North, Robert L. "Benjamin Rush, MD: Assassin or Beloved Healer?" Proceedings of Baylor University Medical Center 13, no. 1 (January 2000). doi:10.1080/08998280.2000.11927641.

Pivarnik, Meghan. "History and Folklore of the Four Thieves Vinegar Blend." Herbal Academy. November 27, 2019.

Rafaeli, J. S. "These Guys Flew to Liberia to 'Cure' Ebola Patients with Homeopathy." VICE. November 26, 2014.

Rudge, F. M. "Orders of St. Anthony." *The Catholic Encyclopedia*. Vol. 1. New York: Robert Appleton Company, 1907.

Patient ZERO: Typhoid Mary

Bourdain, Anthony. *Typhoid Mary*. London: Bloomsbury Publishing, 2005.

"Guide a Walking Typhoid Factory; Adirondack Woodsman Found to be Charged with the Germs—Is a Public Menace." *New York Times*. December 2, 1910.

Leavitt, Judith Walzer. *Typhoid Mary: Captive to the Public's Health*. Boston: Beacon Press, 2014.

Marconi Transatlantic Wireless. "A 'Typhoid Mary' Found in Alsace; Young Woman, Brought to Paris, Found to be Saturated with Disease Microbes." *New York Times*. March 30, 1913.

Mallon, Mary. "The Most Dangerous Woman in American, In Her Own Words." NOVA. https://www.pbs.org/wgbh/nova/typhoid/letter.html.

Marineli, Filio, Gregory Tsoucalas, Marianna Karamanou, and George Androutsos. "Mary Mallon (1869–1938) and the History of Typhoid Fever." *Annals of Gastroenterology* 26, no. 2 (2013): 132–134.

Mendelsohn, J. Andrew. "'Typhoid Mary' Strikes Again: The Social and the Scientific in the Making of Modern Public Health. *Isis* 86, no. 2 (June 1995): 268–277.

Soper, George A. "The Curious Career of Typhoid Mary." *Bulletin of the New York Academy of Medicine* 15, no. 10 (October 1939): 698–712.

"'Typhoid Mary' Dies of a Stoke at 68; Carrier of Disease, Blamed for 51 Cases and 3 Deaths, but She Was Held Immune." *New York Times*. November 12, 1938.

CONTAINMENT: Patient ZERO: 1918 Influenza

Abrahams, Adolphe, Norman Hallows, and Herbert French. "A further Investigation into Influenzo-Pneumococcal and Influenzo-Streptococcal Septicæmia: Epidemic Influenzal 'Pneumonia' of Highly Fatal Type and its Relation to 'Purulent Bronchitis'." *Lancet* 193, no. 4975 (July 1919). doi.org/10.1016/S0140-6736(01)22115-1.

Adams, Mikaëla M. "'A Very Serious and Perplexing Epidemic of Grippe': the Influenza of 1918 at the Haskell Institute." *American Indian Quarterly* 44, no. 1 (Winter 2020): 1–35.

Barry, John M. *The Great Influenza*. New York: Viking Press, 2004.

Grant, Peter. "The 1918 Influenza Outbreak at Haskell Institute: An Early Narrative of the Great Pandemic." *Kansas History* 43, no. 2 (Summer 2020). Kansas Historical Society, Topeka, KS.

Grist, N. R. "Pandemic Influenza 1918." *British Medical Journal* 22–29 (December 1979): 1632–33.

Hammond, J. A. B., William Rolland, and T. H. G. Shore. "Purulent Bronchitis: A Study of Cases Occurring Amongst the British Troops at a Base in France." *Lancet* 190, no. 4898 (July 14, 1917). doi.org/10.1016/S0140-6736(01)56229-7.

"The Haskell County Origin Story and the other 'Haskell' in Kansas." Spanish Influenza in Victoria, Canada, 1918–1920. Published September 13, 2018; updated June 17, 2020.

Humphreys, Margaret. "The influenza of 1918: Evolutionary perspectives in a historical context." *Evolution, Medicine, and Public Health* 2018, no. 1 (2018). doi:10.1093/emph/eoy024.

The Indian Leader 21, no. 32 (April 12, 1918). Haskell Institute, Haskell Indian Junior College. Lawrence, Kan.: The College.

Kansas State Board of Health. Division of Vital Statistics. *Summary of Deaths*. Vol. 2. Haskell County. 1918. Kansas Historical Society, Topeka, KS.

———. Division of Vital Statistics. *Summary of Deaths*. Vol. 1. Influenza. 1918. Kansas Historical Society, Topeka, KS.

"'Letter from Camp Funston' Reading." Lesson Plans, Einstein and His Times. NASA's *Cosmic Times*, 1919. Updated March 8, 2018.

Morens, David M., Jeffery K. Taubenberger, and Anthony S. Fauci. "Predominant Role of Bacterial Pneumonia as a Cause of Death in Pandemic Influenza: Implications for Pandemic Influenza Preparedness." *The Journal of Infectious Diseases* 198, no. 7 (October 2008). doi.org/10.1086/591708.

New York Times. "Memories of the 1918 Pandemic from Those Who Survived." April 4, 2020.

Olson, Donald R. et al. "Epidemiological evidence of an early wave of the 1918 influenza pandemic in New York City." *Proceeding of the National Academy of Sciences of the United States of America* 102, no. 31 (August 2, 2005). doi:10.1073/pnas.0408290102.

Opie, Eugene L. et al. "Pneumonia at Camp Funston: Report to the Surgeon-General." *Journal of the American Medical Association* 72, no. 2 (January 11, 1919). doi:10.1001/jama.1919.02610020026010.

Oxford, J. S. et al. "A hypothesis: the conjunction of soldiers, gas, pigs, ducks, geese and horses in Northern France during the Great War provided the conditions for the emergence of the 'Spanish' influenza pandemic of 1918–1919." *Vaccine* 23 (2005). doi:10.1016/j.vaccine.2004.06.035.

Parrish, Colin R. et al. "Cross-species virus transmission and the emergence of new epidemic diseases." *Microbiology and Molecular Biology Review* 72, no. 3 (September 2008). doi:10.1128/MMBR.00004-08.

Reid, Ann H. et al. "Origin and evolution of the 1918 'Spanish' Influenza Virus Hemmagglutinin Gene." *PNAS* 96, no. 4 (February 1999). doi.org/10.1073/pnas.96.4.1651.

Taubenberger, Jeffery K. et al. "Initial Genetic Characterization of the 1918 "Spanish" Influenza Virus." *Science* 275, no. 5307 (March 1997). doi.org/10.1126/science.275.5307.1793.

Trench and Camp 24, no. 6 (March 16, 1918). Fort Riley, KS.

Tumpey, Terrence M. et al. "Characterization of the reconstructed 1918 Spanish influenza pandemic virus." *Science* 7, no. 310 (October 2005). doi.org/10.1126/science.1119392.

US Surgeon General and US Public Health Service. *Public Health Reports*, "Influenza: Kansas–Haskell." Vol. 33, no. 14. Washington, DC: GPO, 1918. Federal Documents Collection. State Library of North Carolina, Raleigh, NC.

Worobey, Michael, Guan-Zhu Han, and Andrew Rambaut. "Genesis and pathogenesis of the 1918 pandemic H1N1 influenza A virus." *Proceeding of the National Academy of Sciences of the United States of America* 111 no. 228 (June 3, 2014). doi/10.1073/pnas.1324197111.

———. "A synchronized global sweep of the internal genes of modern avian influenza virus." *Nature* 508 (February 16, 2014). https://doi.org/10.1038/nature13016.

Worobey, Michael, Jim Cox, and Douglas Gill. "The Origins of the Great Pandemic." *Evolution, Medicine, and Public Health* 2019, no. 1 (January 21, 2019). https://doi.org/10.1093/emph/eoz001.

Vaccines: From Variolation to Messenger RNA

Boylston, Arthur. "The Origins of Inoculation." *Journal of the Royal Society of Medicine* 105, no. 7 (July 28, 2012). doi:10.1258/jrsm.2012.12k044.

Brunton, Deborah. "Pox Britannica: Smallpox Inoculation in Britain, 1721–1830." PhD diss. University of Pennsylvania. 1990.

Bryce, James. *Practical Observations on the Inoculation of Cow Pox.* Edinburgh: William Creech, 1802.

Burckhardt, John Lewis. *Travels in Nubia.* Cambridge: Cambridge University Press, 2010.

Collins, W. J. "M. Pasteur's Experiments with Chicken-Cholera." *Lancet* 116, no. 2988 (December 4, 1880). doi.org/10.1016/S0140-6736(02)33676-6.

Jenner, Edward. *An inquiry into the causes and effects of the variolae vaccinae, a disease discovered in some of the western counties of England, particularly Gloucestershire, and known by the name of the cow pox.* 2nd ed. London: Sampson Low, 1800.

Knight-Ridder. "Pasteur 'Borrowed' From Rival then Lied, Notes Show." *Chicago Tribune.* February 16, 1993.

Ma, Boying. *A History of Medicine In Chinese Culture.* Singapore: World Scientific, 2020.

Mather, Cotton. *Angel of Bethesda, an Essay.* Barre: American Antiquarian Society and Barre Publications, 1972.

McVail, John C. "Cow-Pox and Small-Pox: Jenner, Woodville, And Pearson." *British Medical Journal* 1, no. 1847 (May 23, 1896): 1271–1276.

Skloot, Rebecca. *The Immortal Life of Henrietta Lacks.* New York: Crown Publishers, 2010.

Smith, Kendall A. "Louis Pasteur, the Father of Immunology?" *Frontiers in Immunology* 3, no. 68 (April 10, 2012). doi.org/10.3389/fimmu.2012.00068.

Patient ZERO: Polio

Beaubien, Jason. "The Campaign to Wipe Out Polio Was Going Really Well. . .Until It Wasn't." *NPR Morning Edition.* October 30, 2020.

DeWeerdt, Sarah. "Two Polio Vaccines for Defeating a Paralysing Scourge." *Nature* 587, Suppl. (November 2020): 14.

Ditunno, John F. Jr., Bruce E. Becker, and Gerald J. Herbison. "Franklin Delano Roosevelt: The Diagnosis of Poliomyelitis Revisited." *PM&R* 8, no. 9 (September 2016). doi: 10.1016/j.pmrj .2016.05.003.

Goldman, Armond S. et. "What Was the Cause of Franklin Delano Roosevelt's Paralytic Illness?" *Journal of Medical Biography* 11, no. 4 (November 2003). doi: 10.1177/096777200301100412.

Gould, Tony. *A Summer Plague.* New Haven: Yale University Press, 1997.

Kurlander, Carl. "The Deadly Polio Epidemic and Why It Matters for Coronavirus." *Discover Magazine.* April 2, 2020.

Leake, J. P., Joseph Bolten, and H. F. Smith. "Winter Outbreak of Poliomyelitis: Elkins, W. Va., 1916–17." *Public Health Reports* 32, no. 48 (November 30, 1917): 1995–2015.

Mehndiratta, Man Mohan, Prachi Mehndiratta, and Renuka Pande. "Poliomyelitis: Historical Facts, Epidemiology, and Current Challenges in Eradication." *Neurohospitalist* 4, no. 4 (October 2014). doi: 10.1177/1941874414533352.

Oshinsky, David M. *Polio: An American Story.* Oxford: Oxford University Press, 2006.

Rogers, Naomi. *Dirt and Disease: Polio Before FDR.* New Brunswick, New Jersey: Rutgers University Press, 1992.

Roosevelt, Eleanor. *The Autobiography of Eleanor Roosevelt.* New York: Harper Perennial, 2015.

Roosevelt, Franklin D. "'Another Milestone:' A Letter to Friends At Warm Springs." *The Polio Chronicle.* December 1932. Roosevelt Warm Springs Institute for Rehabilitation Archives.

Tan, Siang Yong and Nate Ponstein. "Jonas Salk (1914–1995): A Vaccine Against Polio." *Singapore Medical Journal* 60, no. 1 (January 2019). doi: 10.11622/smedj.2019002.

Wyatt, H. V. "Before the Vaccines: Medical Treatments of Acute Paralysis in the 1916." New York Epidemic of Poliomyelitis" *Open Microbiology Journal* 8 (2014). doi:10.2174/1874285801408010 144.

Wyatt, Harold V. "The 1916 New York City Epidemic of Poliomyelitis: Where Did the Virus Come From?" *The Open Vaccine Journal* 4, no. 1 (June 2011): 13–17. doi:10.2174/1875035401104010013.

The Infection-Disease Link: From Viruses, Bacteria, and Parasites to Cancer

Borody, Thomas J., and Sanjay Ramrakha. "Fecal microbiota transplantation for treatment of Clostridioides (formerly Clostridium) difficile infection." UpToDate. Updated November 2, 2020.

De Groot, P. F. et al. "Fecal microbiota transplantation in metabolic syndrome: History, present and future." *Gut Microbes* 8, no. 3 (2017). doi:10.1080/19490976.2017.1293224.

Harrison, Ross, and Warner Huh. "Occupational Exposure to Human Papillomavirus and Vaccination for Health Care Workers." *Obstetrics and Gynecology* 136, no. 4 (October 2020). doi: 10.1097/AOG.0000000000004021.

Leroux, Caroline et al. "Jaagsiekte Sheep Retrovirus (JSRV): from virus to lung cancer in sheep." *Veterinary Research* 38, no. 2 (March-April 2007). doi:10.1051/vetres:2006060.

Rous, Peyton. "Transmission of a Malignant New Growth by Means of a Cell-Free Filtrate." *Journal of the American Medical Association* 250, no. 11 (September 16, 1983). doi:10.1001/jama.1983.03340110059037.

Weintraub, Pamela. "The Doctor Who Drank Infectious Broth, Gave Himself an Ulcer, and Solved a Medical Mystery." *Discover.* April 8, 2010.

Patient ZERO: Cholera

Chicago Tribune. "Chicago's Legendary Epidemic." August 22, 2007.

Cholera Inquiry Committee. *Report on the Cholera Outbreak in the Parish of St. James, Westminster, During the Autumn of 1854.* London: John Churchill, 1855.

Engles, Friedrich. *Condition of the Working Class in England in 1844*. Translated by David McLellan. London: Forgotten Books, 2018.

Hempel, Sandra. *The Medical Detective: John Snow, Cholera and The Mystery Of The Broad Street Pump*. London: Granta Books, 2014.

Johnson, Steven. *The Ghost Map: The Story of London's Most Terrifying Epidemic—and How It Changed Science, Cities, and the Modern World*. New York: Riverhead Books, 2006.

Mayhew, Henry. *London Labour and the London Poor*. London: Penguin Classics, 1985.

Rawnsley, H. D. *Henry Whitehead: 1835-1896: a Memorial Sketch*. Glasgow: J. MacLehose, 1898.

Rosenberg, Charles E. *The Cholera Years: The United States in 1832, 1849, and 1866*. Chicago: University of Chicago Press, 1987.

Schwartz, Allan. "Medical Mystery: A President, an Epidemic, and Grand Tour Cut Short." *Philadelphia Inquirer*. December 21, 2018.

Snow, John. "The Cholera Near Golden Square, and at Deptford." *Medical Times and Gazette* 9 (September 23, 1854): 321–322.

———. *On the Mode of Communication of Cholera*. London: John Churchill, 1855.

Whitehead, H. "The Broad Street Pump: An Episode in the Cholera Epidemic of 1854." *Macmillan's Magazine* 13, no. 74 (April 1866): 113–122.

Whitehead, Henry. *The Cholera in Berwick Street*. London: Hope and Company, 1854.

Anthrax & Biological Warfare: The Weaponization of Disease

Carus, W. Seth. *A Short History of Biological Warfare: From Pre-History to the 21st Century*. Scotts Valley, CA: CreateSpace Independent Publishing Platform, 2017.

Frischknecht, Friedrich. "The History of Biological Warfare." *EMBO Reports* 4, suppl. 1 (June 2003). doi:10.1038/sj.embor.embor849.

Guillemin, Jeanne. *Anthrax: The Investigation of a Deadly Outbreak*. Berkeley: University of California Press, 1999.

Keyes, Scott. "A Strange but True Tale of Voter Fraud and Bioterrorism." *Atlantic*. June 10, 2014.

Lardner, George Jr. "Army Report Details Germ War Exercise in N.Y. Subway in '66." *The Washington Post*. April 22, 1980.

Lebeda, F. J. "Deterrence of Biological and Chemical Warfare: A Review of Policy Options." *Military Medicine* 162, no. 3 (March 1997): 156–161.

Meselson, M., J. Guillemin, M. Hugh-Jones, A. Langmuir, I. Popova, A. Shelokov, and O. Yampolskaya. "The Sverdlovsk Anthrax Outbreak of 1979." *Science* 266, no. 5188 (November 18, 1994). doi:10.1126/science.7973702.

Riedel, Stefan. "Biological Warfare and Bioterrorism: a Historical Review." *Baylor University Medical Center Proceedings* 17, no. 4 (October 2004). doi:10.1080/08998280.2004.11928002.

Singer, Jane. "The Fiend in Gray." *Washington Post*. June 1, 2003.

Wampler Robert A., and Thomas S. Blanton, eds. "Anthrax at Sverdlovsk, 1979: US Intelligence on the Deadliest Modern Outbreak." *National Security Archive Electronic Briefing Book* 5, no. 61 (November 15, 2001).

Patient ZERO: Rabies

Adamson, P. B. "The Spread of Rabies into Europe and the Probable Origin of This Disease in Antiquity." *The Journal of the Royal Asiatic Society of Great Britain and Ireland* 2 (1977): 140–144.

Barnitz, Albert, and Jennie Barnitz. *Life in Custer's Cavalry: Diaries and Letters of Albert and Jennie Barnitz, 1867–1868*. Lincoln: University of Nebraska Press, 1987.

Gomez-Alonso, Juan. "Rabies: A Possible Explanation for the Vampire Legend." *Neurology* 51, no. 3 (September 1, 1998): 856–859. doi.org/10.1212/WNL.51.3.856.

Jackson, Alan C. "The Fatal Neurological Illness of the Fourth Duke of Richmond in Canada: Rabies." *Annals of the Royal College of Physicians and Surgeons of Canada* 27, no. 1 (February 1994): 40–41.

King, A. A., A. R. Fooks, M. Aubert, and A. I. Wandeler, eds. *Historical Perspective of Rabies in Europe and the Mediterranean Basin*. Paris: World Organization for Animal Health, 2004.

Lite, Jordan. "Medical Mystery: Only One Person Has Survived Rabies without Vaccine—But How?" *Scientific American* (October 8, 2008).

Sontag, Susan. *Illness as Metaphor and AIDS and its Metaphors*. New York: Farrar, Straus and Giroux, 2013.

Tarantola, Amanda. "Four Thousand Years of Concepts Relating to Rabies in Animals and Humans, Its Prevention and Its Cure." *Tropical Medicine and Infectious Disease* 2, no. 2 (March 24, 2017). doi:10.3390/tropicalmed2020005.

Velasco-Villa, Andres et al. "The History of Rabies in the Western Hemisphere." *Antiviral Research* 146 (October 2017): 221–232. doi:10.1016/j.antiviral.2017.03.013.

Wasik, Bill, and Monica Murphy. *Rabid: A Cultural History of the World's Most Diabolical Virus*. New York: Viking, 2012.

Whitney, Hugh. "What Evil Felled the Duke? A Re-examination of the Death of the 4th Duke of Richmond." *Ontario History* 105, no. 1 (Spring 2013). doi.org/10.7202/1050746ar.

Woodward, Ian. *The Werewolf Delusion*. New York: Paddington Press, 1979.

Tuberculosis: The "All-Consuming" Disease

Cardona, Pere-Joan, Martí Català, and Clara Prats. "Origin of Tuberculosis in the Paleolithic Predicts Unprecedented Population Growth and Female Resistance." *Nature Scientific Reports* 10, no. 42 (January 8, 2020). doi.org/10.1038/s41598-019-56769-1.

Coitinho, Cecilia et al. "Rapidly Progressing Tuberculosis Outbreak in a Very Low Risk Group." *European Respiratory Journal* 43 (2014). doi:10.1183/09031936.001504.

Comas, Iñaki et al. "Out-of-Africa Migration and Neolithic Coexpansion of *Mycobacterium tuberculosis* with Modern Humans." *Nature Genetics* 45, no. 10 (October 2013).

Hayman, John. "Mycobacterium Ulcerans: An Infection From Jurassic Time?" *Lancet* 24, no. 5410 (November 3, 1984). doi.org/10.1016/S0140-6736(84)91110-3.

Hurt, Raymond. "Tuberculosis Sanatorium Regimen in the 1940s: A Patient's Personal diary." *Journal of the Royal Society of Medicine* 97, no. 7 (July 2004). doi:10.1258/jrsm.97.7.350.

Mackowiak, Philip A. et al. "On the Origin of American Tuberculosis." *Clinical Infectious Diseases* 41, no. 4 (August 15, 2005). doi.org/10.1086/432013.

Pomeroy, Ross. "Why Didn't Tuberculosis Wipe Out Humanity in the Paleolithic Era?" RealClear Science. January 9, 2020.

Patient Anti-ZERO: The Last Smallpox Case

Behbehani, Abbasm. "The Smallpox Story: Life and Death of an Old Disease." *Microbiological Reviews* 47, no. 4 (December 1983): 455–509.

Doucleff, Michaeleen. "Last Person to Get Smallpox Dedicated His Life To Ending Polio." National Public Radio Health Shots. July 31, 2013.

Madrigal, Alexis C. "The Last Smallpox Patient on Earth." *Atlantic*. December 9, 2013.

Pallen, Mark. *The Last Days of Smallpox: Tragedy in Birmingham*. Self-published, 2018.

Ricards, Andy. "The Lonely Death of Janet Parker." Narrated by Dermot Carney. 4 chapters. *BirminghamLive*. Podcast audio. August 23, 2018.

Rimmer, Monica. "How Smallpox Claimed its Final Victim." *BBC News*. August 10, 2018.

Sebelius, Kathleen. "Why We Still Need Smallpox." *New York Times*. April 25, 2011.

Wildy, P. and Gordon Wolstenholme. "Henry Samuel Bedson." Inspiring Physicians, Royal College of Physicians. Accessed March 8, 2021.

Williams, Sally. "'It was a Total Invasion': The Virus that Came Back From the Dead." *Guardian*. November 21, 2020.

Medical Advances: From Pandemics to Progress

Barnes, D. S. *The making of a social disease*. University of California Press, 1995.

Bertrand, Jean-Baptiste. *An Historical Account of the Plague at Marseilles in the Year 1720*. Translated by Anne Plumptre. London: R. Taylor and Co., 1805.

Bynum, H. *Spitting blood: the history of tuberculosis*. Oxford University Press, 2012.

Crofton, J. "The MRC randomized trial of streptomycin and its legacy: a view from the clinical front line." *Journal of the Royal Society of Medicine* 99, no. 10 (October 2006). doi:10.1258/jrsm.99.10.531.

Fletcher, Robert. *A Tragedy of the Great Plague of Milan in 1630*. Baltimore, MD: The Lord Baltimore Press, 1898.

Hobday, Richard A., and John W. Cason. "The Open-Air Treatment of *PANDEMIC* INFLUENZA." *American Journal of Public Health* 99, Suppl. 2 (October 2009). doi:10.2105/AJPH.2008.134627.

Matuschek, Christiane et al. "The history and value of face masks." *European Journal of Medical Research* 25, no. 1 (June 2020). doi:10.1186/s40001-020-00423-4.

Sanger, F., G. G. Brownlee, and B. G. Barrell. "A two-dimensional fractionation procedure for radioactive nucleotides." *Journal of Molecular Biology* 13, no. 21 (September 1965). doi:10.1016/S0022-2836(65)80104-8.

Strasser, Bruno J., and Thomas Schlich. "A History of the Medical Mask and the Rise of Throwaway Culture." *Lancet* 396, no. 10243 (July 4, 2020). doi:10.1016/S0140-6736(20)31207-1.

Tan, S. Y., and S. Grimes. "Paul Ehrlich (1854–1915): man with the magic bullet." *Singapore Medical Journal* 51, no. 11 (November 2010): 842–843.

Tan, Siang Yong, and Yvonne Tatsumura. "Alexander Fleming (1881–1955): Discoverer of penicillin." *Singapore Medical Journal* 56, no. 7 (July 2015). doi:10.11622/smedj.2015105.

Wilson, Mark. "The Untold Origin Story of the N95 Mask." Fast Company. March 4, 2020.

ACKNOWLEDGMENTS

A portion of the royalty sales from this book will be donated by the authors to Doctors Without Borders.

If you're interested in doing more to help fight infectious disease, aiding worldwide eradication efforts, and helping communities affected by the COVID-19 pandemic, please consider supporting any of the following organizations:

Doctors Without Borders: doctorswithoutborders.org
National Foundation for Infectious Diseases: nfid.org
Institut Pasteur:pasteur.fr/en
IAVI (International AIDS Vaccine Initiative): iavi.org
Sabin Vaccine Institute: sabin.org
UNICEF: unicefusa.org
MedShare:
impact.medshare.org/campaign/medshare-responds-coronavirus/c272201
Good360: good360.org/coronavirus
Direct Relief: directrelief.org
WHO COVID-19 Response Fund: covid19responsefund.org/en
Center for Disaster Philanthropy: disasterphilanthropy.org
CDC Foundation: cdcfoundation.org

Books like these are never written in isolation, and we are indebted to so many who generously offered their guidance and support during the process. First, a heartfelt thank you to April Genevieve Tucholke, Bernie Su, Thomas Pedersen, Ben, Maia, Phoebe, and to dear friends and family both near and far.

To our wonderful agent, Eric Myers, our fantastic editor, John Meils, and the dedicated team at Workman Publishing, who have done a marvelous job in making this book a reality, including Janet Vicario, Barbara Peragine, Kim Daly, Sophia Reith, Chloe Puton, Moira Kerrigan, and Carol Schneider—a huge thank you for your talent and time. To Kelly Gonzalez for your help with the bibliography, Sarah Simpson-Weiss and Felicity Bronzan for keeping the chaos at bay, Erica O'Rourke, plus Erin Weaver at Deschutes Public Library in Bend, Oregon, and the many librarians at the Leon S. McGoogan Health Sciences Library at the University of Nebraska, for all those endless article requests.

Several people with world-class expertise in the realm of medicine and public health have been instrumental in reading the manuscript at various stages. To Phil Smith, Sara Bares, Bernie Su, Kirk Foster, Elizabeth Schnaubelt, Jasmine Marcelin, Grant Hutchins, Gale Etheron, Fedja Rochling, Renee Young, David Brett-Meyer, and Chris Bruno: We are truly thankful for your wisdom, and even more so for fighting on behalf of your patients and the community around you.

INDEX

Page numbers in bold indicate
an illustration or its caption.

PHOTO CREDITS

Alamy: Albatross p. 19; Album p. 199; Alpha Historica p. 181; Archivart p. 325 (right); Artokoloro p. 305; Avalon.red p. 34; BSIP SA pp. 167, 215, 263, 318; Scott Camazine p. 138; CDC/S.Dupuis p. 21; Sunny Celeste p. 39; Chronicle p. 191; CPA Media Pte Ltd p. 55; Everett Collection Historical p. 230; Everett Collection Inc p. 254; Glasshouse Images p. 221; David Havel p. 33; Grant Heilman Photography p. 48; History and Art Collection p. 6; INTERFOTO p. 187 (left); Juniors Bildarchiv GmbH p.18; Kirn Vintage Stock p. 260; Lamax p. 3; Lebrecht Music & Arts pp. 194, 287; M&N p. 322 (top); John Morris p. 99; Nutu p. 154; Pictorial Press Ltd p. 349; REUTERS p. 13; RGB Ventures/SuperStock p. 235 Science History Images pp. 12, 58, 94, 197 (left), 251; Sueddeutsche Zeitung Photo p. 256; Trinity Mirror/Mirrorpix p. 330; Vintage_Space p. 231; Virginia Museum of History & Culture p. 226 (bottom); Matthijs Wetterauw p. 17 (right); WILDLIFE GmbH p. 14; World History Archive p. 161.

AP Images: Dake Kang p. 113; John Locher p. 205.

CDC: William Archibald p. 53; Alissa Eckert, MSMI; Dan Higgins, MAMS p. 111; Jennifer Oosthuizen p. 280.

Dreamstime: © Mattiaath p. 144.

Getty Images: AFP p. 117; ballyscanlon/DigitalVision p. 269; Bettmann pp. 41, 108, 214, 339; Robert Cianflone p. 114; Fabrice Coffrini/AFP p. 10; FPG/Archive Photos p. 328; Juan Gaertner/Science Photo Library p. 249; George Eastman House/Premium Archive p. 320; image_jungle/iStock p. 129; Jim James–PA Images p. 82; Marka/Universal Images Group p. 322 (bottom); Catherine McGann p. 139 (bottom); Win McNamee p. 125; Mirrorpix p. 271; John Moore pp. 30, 35, 245; Hans Neleman/Stone p. 76; Photo 12 p. 325 (left); Photo 12/Universal Images Group p. 169; Pool p. 247; Popperfoto p. 324 (right); Mark Ralston/AFP p. 118; Anthony Rosenberg/iStock p. 65; San Francisco Chronicle/Hearst Newspapers via Getty Images p. 133; Sciepro/Science Photo Library p. 17 (left); Seyllou/AFP p. 36; swim ink 2 llc/Corbis Historical p. 324 (left); The Washington Post p. 122; Universal History Archive p. 66; Alex Wong p. 301.

Hekotoen International: Courtesy of late D. Carleton Gajdusek. CC BY 3.0 p. 78.

International Leprosy Association: p. 184.

Reuters: Pool New p. 120.

Science Source: Ramón Andrade, 3Dciencia p. 278; William D. Bachman p. 97; Biophoto Associates p. 157; Wesley Bocxe p. 142; A. Barrington Brown p. 354; CDC p. 89; Mauro Fermariello p. 355; Steve Gschmeissner p.178; Gustoimages p. 340; Kateryna Kon p. 50; Russell Kightley p. 329; Ted Kinsman p. 1; Kateryna Kon pp. 25, 307; Dennis Kunkel Microscopy p. 223; Laguna Design p. 224; MOLEKUUL p. 80; National Library of Medicine p. 226 (top); Claude Nuridsany & Marie Perennou p. 352; NYPL p. 257; Science Source pp. 40, 69, 79, 147, 151, 190; Wellcome Images p. 290.

Shutterstock: Giovanni Cancemi p. 282; Kateryna Kon p. 294; Raj Creationzs p. 23; Jojoe Wang p. 46.

State Library and Archives of Florida: Saint George Sound p. 148; Welcome Collective p. 67; Amand p. 313; R. Carswell p. 333; Fernan Federici & Jim Haseloff p. 2; Public Domain Mark p. 217; Saphir, Moritz Gottlieb p. 289; Science Museum, London pp. 9, 311; St Bartholomew's Hospital Archives & Museum p. 193; Tuberous Leprosy p. 179 (left).

Wikimedia: The following images are used under a Creative Commons Attribution CC BY-SA 4.0 License (creativecommons.org/licenses/by-sa/4.0/deed.en) and belong to the following Wikimedia Commons users: Vadim Chuprina p. 128; Intropin (Mark Oniffrey) p. 265, Science Museum Group pp. 42, 74, 102 (left), 102 (right), 103 (left), 103 (right), 209, 244, 284, 296, 313 (inset), 314, 345, 356.

The following images are used under a Creative Commons Attribution CC BY-SA 3.0 (creativecommons.org/licenses/by-sa/3.0/deed.en) and belong to the following Wikimedia Commons users: 500px p. 22; Anagoria p. 51; Author's archive p. 239; Ellicrum p. 188.

The following image is used under a Creative Commons Attribution CC BY-SA 2.5 (creativecommons.org/licenses/by/2.5/deed.en) and belongs to the following Wikimedia Commons user: PLoS Medicine p. 279.

The following image is used under a Creative Commons Attribution CC BY-SA 2.0 (creativecommons.org/licenses/by/2.0/deed.en) and belongs to the following Wikimedia Commons user: FDR Presidential Library & Museum.

Public Domain: 47; © 2003 Samvado Gunnar Kossatz p. 306; Alfvanbeem p. 201; archives.gov p. 343; Pierre Arents p. 179 (right); Bulletin of Unit 731 p. 304; Captmondo p. 203; CDC/PD pp. 16, 26, 27, 52, 131; David Rumsey Historical Map Collection p. 174; Mary J. Dobson p. 309; drawingsofleonardo.org p. 64; Flickr p. 20; Gerald R. Ford Presidential Library p. 126; E. A. Goeldi p. 105; Government Medical College p. 275; Guise p. 197 (right); Jacek Halicki p. 228; Hawaii State Archives p. 60; Ifwest p. 232; Imperial War Museums p. 166; Dominique Jacquin p. 4; James K. Polk Presidential Museum p. 283; Brent Jones p. 299; lacma.org p. 159; Le Petit Journal p. 286; Library of Congress pp. 88, 240; Los Angeles County Museum of Art p. 170; Stéphane Magnenat p. 342; March of Dimes Foundation p. 259; Motacilla p. 156; Munch Museum p. 327; Museu Nacional d'Art de Catalunya p. 236; National Archives at College Park p. 252 (bottom); National Archives Atlanta, GA p. 200; National Human Genome Research Institute p. 272; National Institutes of Health p. 71; NIAID p. 38; Nicke L p. 295; Olybrius p. 204; Iolani Palace p. 177; Powell, J. p. 150; Public Health Image Library pp. 87, 91; Quibik p. 187 (right); Ras67 p. 183; sciencenewsline.com p. 29; Scott and Sons p. 293; Nathan.S.Sethman-1 p. 101; Teri Shors p. 310; Vore Sygdome p. 213; Jacques Joseph Tissot p. 96; U.S. Army Surgeon General's Office p. 72; U.S. National Archives and Records Administration p. 164; Vatican Museum p. 70; Via p. 109; Wegates p. 87 (inset); Li Wenliang p. 115; biolib.de p. 302; whitehouse.gov p. 334

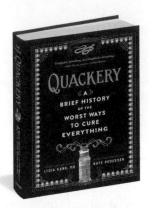

Lydia Kang, MD, and Nate Pedersen are also the authors of
Quackery: A Brief History of the Worst Ways to Cure Everything,
which was named a Best Science Book of 2017 by NPR's *Science Friday.*

Lydia Kang, MD, is an author and a practicing internal medicine physician at Nebraska Medicine in Omaha. Her novels include *Opium and Absinthe: A Novel, The Impossible Girl, A Beautiful Poison, Toxic, The November Girl, Control,* and *Catalyst.* Her short stories appear in *From a Certain Point of View: The Empire Strikes Back* and *Color Outside the Lines: Stories about Love.* For more information, see lydiakang.com.

Nate Pedersen is a writer, librarian, and lecturer living in Portland, Oregon. His nonfiction has appeared in *The Guardian, The Believer, Mental Floss, Fine Books & Collections,* and *Trail Runner.* He edited the short story anthologies *The Starry Wisdom Library, The Dagon Collection,* and *Sisterhood.* For more information, see natepedersen.com.